Springer Series in Physical Environment 16

Springer
Berlin
Heidelberg
New York
Barcelona
Budapest
Hong Kong
London
Milan
Paris
Santa Clara
Singapore
Tokyo

Volumes already published

Vol. 1: Earth Surface Systems
R. Huggett

Vol. 2: Karst Hydrology
O. Bonacci

Vol. 3: Fluvial Processes in Dryland Rivers
W.L. Graf

Vol. 4: Processes in Karst Systems
Physics, Chemistry and Geology
W. Dreybrodt

Vol. 5: Loess in China
T. Liu

Vol. 6: System-Theoretical Modelling in Surface Water Hydrology
A. Lattermann

Vol. 7: River Morphology
J. Mangelsdorf, K. Scheurmann and F.H. Weiß

Vol. 8: Ice Composition and Glacier Dynamics
R.A. Souchez and R.D. Lorrain

Vol. 9: Desertification
Natural Background and Human Mismanagement
M. Mainguet

Vol. 10: Fertility of Soils
A Future for Farming in the West African Savannah
C. Pieri

Vol. 11: Sandstone Landforms
R. Young and A. Young

Vol. 12: Numerical Simulation of Canopy Flows
G. Groß

Vol. 13: Contaminants in Terrestrial Environments
O. Fränzle

Vol. 14: Saturated Flow and Soil Structure
H. Diestel

Vol. 15: Karst in China
Its Geomorphology and Environment
M.M. Sweeting

Vol. 16: Rockglaciers
Indicators for the Present and Former Geoecology
in High Mountain Environments
D. Barsch

Dietrich Barsch

Rockglaciers

Indicators for the Present and Former Geoecology in High Mountain Environments

With 143 Figures and 50 Tables

Springer

Professor Dr. DIETRICH BARSCH
Department of Geography
University of Heidelberg
Im Neuenheimer Feld 348
69120 Heidelberg
Germany

ISSN 0937-3047
ISBN-13:978-3-642-80095-5 e-ISBN-13:978-3-642-80093-1
DOI: 10.1007/978-3-642-80093-1

Library of Congress Cataloging-in-Publication Data. Barsch, Dietrich. Rockglaciers: indicators for the present and former geoecology in high mountain environments/Dietrich Barsch. p. cm. – (Springer series in physical environment, ISSN 0937-3047: 16) Includes bibliographical references and indexes. ISBN-13:978-3-642-80095-5 1. Rock glaciers. I. Title. II. Series. 08641.837 1996 551.3′1–dc20 96-13769

Softcover reprint of the hardcover 1st edition 1996

Cover-design: Springer-Verlag, E. Kirchner

Typesetting: Scientific Publishing Services (P) Ltd, Madras

SPIN: 10527949 32/3136/SPS – 5 4 3 2 1 0 – Printed on acid-free paper

Preface

About 30 years ago in the summer of 1965, I was first confronted with active rockglaciers. In those days, the Department of Geography at the University of Basel had about 20 new students every year. Each second year, Hans Annaheim, the director of the department, and his assistant, Dietrich Barsch, taught, a 2-week field course to introduce the second- or third-year students to practical techniques. In 1965, we went to Lavin/Engadin (Graubünden/Grison, Swiss Alps), where Hans Annaheim worked with a group of students in the valley proper. The smaller group, reinforced by the geodetic engineer (Martin Maurer), lived in a Swiss Army hut at 2600 m. We mapped the periglacial features in the huge cirque of Macun and started detailed measurements on two of the seven rockglaciers detected beforehand on aerial photographs (Barsch 1969a).

Despite the information collected by Jäckli (1957) and despite the classic paper of Wahrhaftig and Cox (1959), single rockglaciers were in those days still being described as peculiar forms and not as regular, typical features of the high mountain (subnival) belt of the central Alps (e.g., Binggeli 1965). Since then, rockglacier literature has multiplied. Before starting this book, I personally had contributed almost 300 printed pages to this "rockglacier paper mountain". Many conflicting views on rockglaciers have been discussed over the years; often influenced by pure semantic approaches towards the "glacier" part of the term "rock glacier". Therefore, it was proposed (Barsch 1988) to use the terminus technicus for this landform written in one word as "rockglacier" to express its singularity as well as its difference from glacial features like (true ice) glaciers.

Today, it seems possible to present a state-of-the art report on rockglacier research. In addition, in a time of increasing use of high mountains for recreational and economic purposes, active rock-glaciers can be used as a valuable indicator of a number of geoecologic factors in the periglacial or geocryogenic belt of high mountains all around the world. They indicate, for instance, the existence of permafrost in their neighborhood, i.e., in the high mountain environment where they are found. At a time when monitoring of global changes becomes more and more important, active rockglaciers are systems with reaction times of decades in response to changes in mean annual air temperature (MAAT) in a delicate environment. In addition to their use in monitoring present-day changes in mountain permafrost, there is valuable information on the late Pleistocene climatic history stored in fossil or relict rockglaciers.

Having worked on a topic over many years, it is difficult to name individuals who have contributed more than others to the knowledge in this field. Asked, I would answer that my respect goes to two groups of mountain men: first, to the cartographers who mapped the high cirques in Switzerland during the second half of the last century; they probably did not have a special name for rockglaciers but mapped them very accurately, even if the scale of their work does not allow any extrapolations of rockglacier velocity from the maps. Second, my respect goes to the early workers on rockglaciers, like Steenstrup, Capps, Taylor, Chamberlin, Siebenthal and both Chaix. Since those days a lot of work has been done. From the published results, only a few can be named here: the classic paper by Clyde Wahrhaftig and Allan Cox, the book by Wilfried Haeberli and the one edited by the John Giardino, John Shroder Jr. and John Vitek.

In the last 30 years, I have made many friends during rockglacier research. Beginning with my students in Basel, I would like to name – quasi as representatives – Peter Fischer and Peter Schneider, who are both living today in Graubünden. Also, there are my first doctorate students like Lorenz King (Gießen) and Wilfried Haeberli (Zürich), who are still working in mountain areas. A number of geodetic engineers such as Martin Maurer, Bruno Baur, Günther Hell and Wolfgang Zick went up with me to many remote places. With others, such as Jussi Baade, Doris, Jörg, Nils, Synje and Peer Barsch, Enrique Buk, Nel Caine, Paul Carrara, Arturo Corte, RichardDikau, Michèle Evin, Wolfgang Flügel, Margrit Grout, Wilfried Haeberli, Hans Happoldt, Helmut Heuberger, Hans Kerschner, Brian Luckman, Lorenz King, Roland Mäusbacher, Gunnar Östrem, Konrad Rögner, Gernot Patzelt, Arnold Schneider, Lothar Schrott, Gerd Schukraft, Achim Schulte, Roland Souchez, Gerhard Stäblein and Dario Trombotto, I spent memorable times on rockglaciers. I would like to thank Frank Ahnert, Peter Birkeland, Denys Brunsden, Francesco Dramis, Klaus Fischer, Otto Fränzle, Gerhard Furrer, John Giardino, Peter Höllermann, Jack Ives, Hartmut Leser, Herbert Liedtke, Bruno Messerli, Dave Moore, John (Jack) Shroder Jr., Olav Slaymaker, Claudio Smiraglia, Clyde Wahrhaftig and Sid White for valuable discussions.

A great challenge and enthusiastic encouragement were given to me by the Interlaken Workshop on Permafrost and Periglacial Environments in Mountain Areas which was organized in September

1991 by Wilfried Haeberli and his great team from Zürich. I want to thank all those who contributed to the great success of this meeting such as Bernard Francou, Pierre Lautridou, Albert Pissart, Johan Ludwig Sollid, Stuart Harris and many others.

To Christian Opp (Leipzig), I owe special thanks for introducing me to the literature written in Russian. Without his help a part of the existing rockglacier literature would have been neglected. Matthias Jakob was an enthusiastic helper during part of the time this book was being written. Eveline Maintz and Steffi Blum helped with the bibliography, Christine Huber typed the tables, and Stephan Scherer was responsible for cartography. Brigitta Oezen finally arranged the technical aspects of the manuscript.

Nel Caine (Boulder) and Wilfried Haeberli (Zürich) read the original manuscript carefully and gave much advice. I am indebted to both friends for their great and encouraging help, which improved this book considerably. Hubert Kroh (Kaiserslautern) helped me by checking the section on rheology. Mel Marcus (Tempe, Arizona) acted as editor of this volume. I am very obliged to him for this work. Bob Allison (Durham) corrected some of my "German" English, but naturally, the responsibility for all faults rests with the author.

Great thanks are due to the foundations and organizations which furnished the necessary funds for my rockglacier research. These are:

- der Schweizerische Nationalfonds (Bern)
- die Deutsche Forschungsgemeinschaft (Bonn)
- die Volkswagenstiftung (Hannover)
- die Daimler-Benz Stiftung (Ladenburg)
- die Arbeitsgemeinschaft für vergleichende Hochgebirgsforschung (München)
- die Universität Basel
- die Ruprecht-Karls-Universität zu Heidelberg.

I would perhaps not have been involved with rockglacier research if my scientific interest in high mountains had not been stimulated by two great men:

Carl Troll
*24. December 1899 in Wasserburg/Inn
†21. July 1975 in Bonn,

who directed my early studies and my doctoral thesis, and

Hans Annaheim
*10. July 1903 in Basel
†13. Januar 1978 in Basel,

who influenced the first part of my scientific career and my *habilitation* thesis in many ways.

Therefore, I dedicate this book to the memory of these two great mountain men and geographers.

Heidelberg, March 1996 DIETRICH BARSCH

Carl Troll (1899–1975)
Archiv: Geographisches Institut Bonn

Hans Annaheim (1903–1978)
Source: Heidi and Hans-Uli Sulser-Annaheim (Basel)

Contents

Introduction 1

1 **Definitions** 4

2 **Historical Development of the Term Rockglacier** 14

3 **Rockglaciers: Description and Morphometry** 18
3.1 General Description 18
3.2 Form Types 19
3.3 Morphometric Parameters 20
3.3.1 Rockglacier Sizes 22
3.3.2 Tongue-Shaped Rockglaciers 22
3.3.3 Lobate Rockglaciers 23
3.3.4 Rockglacier Thickness 24
3.3.5 Surface Relief 26
3.3.6 Rockglacier Surface and Source Area 29

4 **Rockglacier Taxonomy** 31

5 **Rockglacier Distribution** 35
5.1 General Information 35
5.2 Local Rockglacier Distribution 37
5.2.1 The Alps 37
5.2.2 The Mountains of Northern Europe 43
5.2.3 The Pyrenees 46
5.2.4 North American Mountains 48
5.2.5 The Andes of Central and South America . . . 55
5.2.6 The Mountains of Asia 59
5.2.7 The Southern Alps 63
5.2.8 Antarctic Mountains 66
5.2.9 Conclusions 66

6 **Rockglacier Material, Surficial Fabric and Internal Structure** 67
6.1 Rock Type and Grain Size at and below the Surface 67
6.1.1 Rock Type 67

6.1.2 Grain Sizes at the Surface 68
6.1.3 Development of the Bouldery Mantle 68
6.1.4 Surface Fabric 70
6.1.5 Grain Sizes below the Bouldery Mantle 71
6.2 Internal Structure 73
6.2.1 Direct Information 73
6.2.1.1 Excavations, Outcrops, Tunnels 73
6.2.1.2 Smaller Boreholes 78
6.2.1.3 The Deep Borehole through the Rockglacier Murtèl I 79
6.2.2 Indirect Information 82
6.2.2.1 Seismic Information 82
6.2.2.2 Geoelectric Soundings 93
6.2.2.3 Radio-Echo Soundings 100
6.2.2.4 Gravimetry 100
6.2.2.5 Borehole Geophysics and Related Measurements 102
6.2.2.6 BTS Measurements 114
6.2.2.7 Summary: The Inner Core of an Active Rockglacier 118
6.3 The Active Layer on Rockglaciers 119

7 Rockglacier Movement, Velocity, and Rheology 121
7.1 The Horizontal and Vertical Movement of Active Rockglaciers 122
7.1.1 Measurement Methods 122
7.1.2 Annual Horizontal Displacement 123
7.1.3 Long-Term Annual Averages 127
7.1.4 Long-Term Estimates 130
7.1.5 Longer Time Series 131
7.1.6 Monthly and Seasonal Measurements 142
7.1.7 Vertical Displacement 146
7.1.8 Conclusion 156
7.2 Geometry of Movement 158
7.2.1 The General Flow Patterns of Active Rockglaciers 158
7.2.1.1 Gruben Rockglacier 162
7.2.1.2 Macun Rockglacier 163
7.2.1.3 Arapaho Rockglacier 164
7.2.2 Horizontal Velocity on Longitudinal and Cross-Sectional Profiles 166
7.2.3 Surface and Subsurface Velocity 168
7.3 Rheologic Considerations 171
7.3.1 Shear Stress and Strain Rates in Active Rockglaciers 175

7.3.2 The Rheological Description of Active Rockglaciers 176
7.4 Rockglacier Movement and Climate 180
7.5 Discussion of Rockglacier Movement 180

8 Rockglacier Genesis and the Relation to Similar-Looking Landforms 184
8.1 Rockglacier Genesis 184
8.1.1 The Formation of Active Rockglaciers 185
8.1.1.1 Talus Rockglaciers 185
8.1.1.2 Debris Rockglaciers 186
8.1.1.3 Special Rockglaciers 188
8.1.1.4 Problematic Cases 189
8.1.2 Inactive Rockglaciers 189
8.1.3 Relict (Fossil) Rockglaciers 194
8.2 Published Hypotheses of Rockglacier Formation . . . 196
8.2.1 Mass-Movement Hypotheses 198
8.2.1.1 The Bergsturz Hypothesis in General 199
8.2.1.2 Landslide Influences 202
8.2.2 The Glacial Hypothesis 208
8.2.2.1 Debris-Covered Glaciers and Thermokarst 211
8.2.2.2 Transition from True Glaciers to Rockglaciers? 212
8.2.2.3 The Moraine Hypothesis 216
8.2.3 The Periglacial (Blockstream) Hypothesis . . . 219
8.3 True Rockglaciers under Wrong Labels 219
8.3.1 The Östrem Ice-Cored Moraine Concept 220
8.3.2 The Protalus Rampart Concept 220

9 The Age of Rockglaciers 225
9.1 The Age of Active Rockglaciers 225
9.2 The Age of Climatic Inactive Rockglaciers 227
9.3 The Age of Relict (Fossil) Rockglaciers 229

10 Rockglaciers and the High Mountain Environment . . . 231
10.1 Active Rockglaciers and Mountain Permafrost 232
10.2 Rockglaciers in the Coarse Debris Cycle 240
10.2.1 Rockglaciers and Talus Production 240
10.2.2 Rockglacier Size and Source Area 246
10.2.3 Rockglaciers as a Debris Transport System . . 248
10.3 Rockglaciers and Climate 250
10.3.1 Rockglaciers and Present Climate 251

10.3.2 Relict Rockglaciers and Paleoclimate Reconstruction 252
10.3.3 Reactivation of Inactive or Relict Rockglaciers 256
10.3.4 Rockglaciers and Climatic Change 257
10.4 Rockglaciers in the Alpine Hydrological Cycle 259
10.4.1 Rockglaciers as a Water Store 259
10.4.2 Discharge from Rockglacier Permafrost 262
10.4.3 Fluctuations in Rockglacier Permafrost Storage 265
10.5 Rockglaciers as Hazards in Alpine Environments ... 265
10.6 The Environment of Active Rockglaciers 266

11 Summary and Outstanding Problems 269

12 References 273

Index of Place Names 321

Subject Index 327

Introduction

Global climatic change and sustainable development are key words in the discussion of the future of the geosystems on Earth. Both concepts are extremely important for the future development of high mountain systems in which the high mountain environments or cryogenic belts, e.g., the landscapes above the timberline, are especially endangered. Unfortunately, our knowledge of these geosystems is very limited. It is, therefore, essential to define diagnostic landforms which are important as indicators for the geoecology of these systems and for monitoring possible climatic changes in mountain belts. This requires an extensive geomorphic knowledge of the landforms accepted as diagnostic. Otherwise the derived information may be confusing or wrong and of no help for monitoring climate changes or for establishing a more considered and sustainable use of these geosystems.

The predicted global warming endangers the stability of high mountain environments. In most of them, permanently frozen ground (mountain permafrost) is a regular, but not an obvious or easily detected, phenomenon. Many summits, rock walls, and deposits of unconsolidated material are glued together by the ice of mountain permafrost. Melting of this ice would destabilize these areas and cause hazards such as bergsturzes, landslides, and debris flows which may effect even the deep mountain valleys below belts with recent mountain permafrost.

Even without climate warming, high mountain environments worldwide are endangered by uses which are not based on a solid geomorphic and geoecologic knowledge. High mountains have always been a fascination to man. They were a challenge: dangerous and helpful at the same time. Today, the mountain–man relationship has entered a new dimension as mountains experience an increasingly intense use for recreational and other purposes. Often, the nature and stability of delicate and sensible mountain systems are threatened by this development. A warning which should not be overlooked is given by the Mountain Agenda (1992) produced for the Earth Summit in Rio de Janeiro. Unfortunately, even the larger report (Stone 1992) mentions neither mountain permafrost nor rockglaciers, which must be considered for sustainable development in high mountain environments.

Awareness of the delicate nature of high mountain environments has grown during recent decades, but high mountain systems are still often not really understood and even their definition is not known to a wider public.

Following Carl Troll (1941, 1954, 1972, 1973), a high mountain system (*Hochgebirge*) in midlatitudes is defined by the existence of:

1. belts above timberline,
2. glacially sculptured relief (developed during the Pleistocene), and
3. a recent periglacial (or cryogenic) belt.

For high mountain systems in arid and polar zones, where no timberline is developed, Barsch and Caine (1984) added as a diagnostic criterion the widespread existence of rock walls (gradient $> 60°$).

The *Hochgebirgslandschaft*, the high mountain belt or the high mountain environment, in a midlatitude mountain system can, therefore, be defined according to Carl Troll as the landscape above the timberline which bears a glacially sculptured relief and which at present is influenced by cryogenic (periglacial) processes. This is the belt where active rockglaciers exist. They are indicators of discontinuous mountain permafrost (Barsch 1978) and, thus, of high mountain environments. Active rockglaciers give geomorphological evidence about the special geoecological environment where they are found. They are, therefore, a landform of high diagnostic value in mountains. Even if they are at present inactive or relict, they provide information on past environmental and paleoclimatic conditions.

Rockglaciers are not small landforms. In the classification of Barsch and Stäblein (1978), they belong to the lower part of the meso-form class, the class of landforms with an area greater than 0.01 km^2. For their development, under present conditions, a period of several millennia is necessary (Barsch 1983; Haeberli 1990). Thus, the information stored in them is integrated over a long period. On the other hand, active rockglaciers seem to react quickly to environmental changes; this demonstrates their great importance for the evaluation of the predicted global warming which may destabilize the lower parts of belts with mountain permafrost around the world. They are, thus, a sensitive indicator of a possible climate warming and a landform with monitoring importance in high mountain environments.

To summarize, the high mountain environment forms a geosystem governed by delicate balances. Today it is under increasing human pressure, though our present knowledge of high mountain environments or high mountain systems is still very incomplete. In this situation, it is important to evaluate all information which exists about possible indicators of parts or of principal elements of these systems. Rockglaciers form such an indicator. They are the visible expression of mountain permafrost, which is not easily recognized without this indicator. They offer, therefore, important geoecological information on the areas where they are found. In addition, they form a geomorphic system which allows the monitoring of global change in high mountain environments and they are, therefore, an indicator of possible instabilities and natural hazards developing with melting permafrost.

In this book, the available information on rockglaciers as one of the most informative geoecologic indicators of high mountain environments and of their present dynamic behavior has been collected as a basis for better knowledge of these great landscapes and the physical systems which govern them. This may help to ensure correct use of this valuable indicator for the benefit of all high mountains as well as for the people living in them permanently and those who make shorter visits.

1 Definitions

Active, i.e., presently moving rockglaciers (cf. Figs. 1.1–1.4), form an important and fascinating landform in alpine (high mountain) environments. They indicate alpine or mountain permafrost belts, and, thus, possess a high geodynamic and geoecologic information value. Unfortunately, discussions about rockglaciers are often very confusing. Some students, influenced by semantics, believe that rockglaciers are just debris-covered glaciers (Lliboutry 1965, 1986; Klaer 1974; Whalley 1974a); others accept features of quite different origin as rockglaciers. Often, glacial geomorphologists or glaciologists favor a glacial origin; workers in the field of permafrost believe in a periglacial genesis. Those who subscribe to neither of these groups have proposed that there are some types of rockglaciers which belong to the periglacial, and other types that belong to the glacial realm (White 1971b; Potter 1972). The term "continuum" (Johnson 1974, 1983; Corte 1987a; Giardino and Vitek 1988; Chap. 8) is highly favored in this context.

To avoid confusion, it is necessary to define the term rockglacier right at the outset. In geomorphology, this requires that three main concerns be addressed: process, material, and form. The following definition (Barsch 1988, 1992, p.176) accomodates all three:

Active rockglaciers are lobate or tongue-shaped bodies of perennially frozen unconsolidated material (cf. Figs. 1.5, 1.6) supersaturated with interstitial ice and ice lenses that move downslope or downvalley by creep as a consequence of the deformation of ice contained in them and which are, thus, features of cohesive flow.

A similar definition (following Haeberli 1985a) emphasizes even more the process-oriented approach (Barsch 1992, p.176):

Active rockglaciers are the visible expression of steady-state creep of ice-supersaturated mountain permafrost bodies in unconsolidated materials. They display the whole spectrum of forms created by cohesive flows.

Both definitions are regarded as equivalent; they will be used throughout this volume.

Both of these definitions include the important three concerns of geomorphology:

1. The process of rockglacier formation is the creep of mountain permafrost that is supersaturated with ice, i.e., it contains more ice by volume (vol%)

Fig. 1.1. Tongue-shaped debris rockglacier Weissmies (Laggintal, Simplon, Wallis, Swiss Alps). The rockglacier develops out of the right lateral moraine of the Weissmies glacier below an arete which is intensely shattered and produces great amounts of talus. Aerial photograph of the Schweizerische Landestopographie Bern (No. 3773, 25 September 1967)

than the normal pore space of the material in an unfrozen state. Deformation of the ice then creates the landform of a "rockglacier".

2. The material in active rockglaciers is unconsolidated but frozen (mostly ice-supersaturated and only partly ice-saturated) rock fragments, which are derived from such sources as talus slopes, moraine deposits (till/debris), and occasionally from elsewhere (e.g., mining dumps).

Fig. 1.2. Lobate talus rockglacier in the Front Range of Colorado, photographed in August 1980

Fig. 1.3. Active talus rockglaciers below talus slopes on Prins Karls Forland (western Svalbard Archipelago). Photographed by Johan Ludvig Sollid, Oslo

Fig. 1.4. Well-developed surface relief on the rockglacier Murtèl I (Upper Engadin, Graubünden, Swiss Alps). The relief is probably due to the highly compressive flow found here

3. The landform "rockglacier" is normally a tongue- or lobate-shaped feature whose shape is controlled by the surrounding topography and by the input of rock fragments, as talus or till/debris (cf. Figs. 1.5, 1.6). Its shape is not genetically controlled in the sense that the tongue-shaped form occurs only below glaciers (cf. Sect. 3.3 and Chap. 8). Its surface relief (cf. Fig. 1.4) is characterized by furrows and ridges which are indicative of the flow processes. In addition, the surface relief of active rockglaciers may be interpreted as expressing a compressive stress (i.e., decelerating movement; cf. Chap. 3). This hypothesis has already been proposed by Wahrhaftig and Cox (1959). Extending flow occurs at all places where the underlying topography allows, and there the surface relief seems to be more subdued.

It is encouraging that this approach has a classic root: Clyde Wahrhaftig, who is the senior author of the benchmark paper published at the beginning of the present period of rockglacier research (Wahrhaftig and Cox 1959) wrote in his foreword to a collection of papers on rockglaciers edited by Giardino, Shroder and Vitek (1987):

"I recommend that *rock glacier* be restricted to those accumulations of blocky detritus, extending outward and downslope from talus cones or from glaciers or the terminal moraines of glaciers, that have extended thus in large

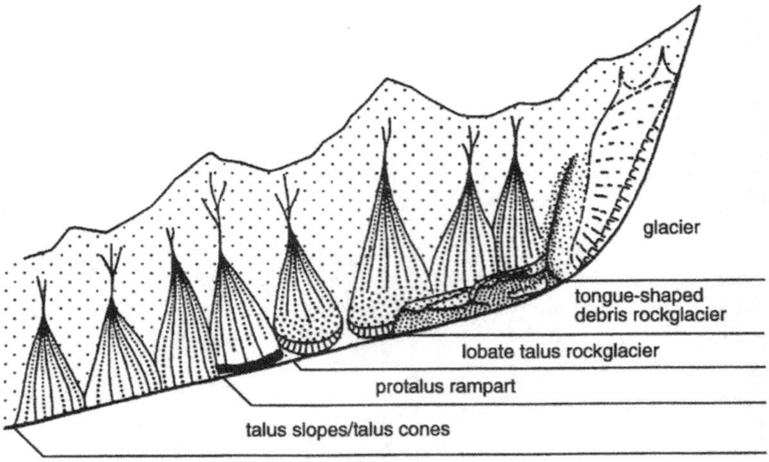

Fig. 1.5. Areal distribution of glaciers, tongue-shaped, and lobate rockglaciers, protalus ramparts and talus slopes in a high mountain environment with discontinuous mountain permafrost. (Modified from Höllermann 1983a)

part th(r)ough[1] the deformation of interstitial ice or clear ice bodies in them. Such accumulations have to be thick enough, and therefore stand high enough above the adjacent terrain in front and alongside, for the shear stress to give reasonable velocities (0.1–2 m/year) for the rock glacier to have grown to its present length in the colder episodes of the Holocene."

This is fully congruent with the definitions given above, which includes information on the processes (movement, type of movement), the material (internal composition), and the form of these great features in alpine environments.

In an active state, rockglaciers move at a typical velocity of 0.1–1 m a^{-1} (cf. Chap. 7). Inactive rockglaciers display no movement today, but they still contain frozen material. Two possibilities may account for this inactivity:

1. Rockglaciers may be inactive due to the melting of ice contained in them. This type of rockglacier inactivity will be called *climatic inactive* (cf. Fig. 1.7). In inactive rockglaciers of this sort, the unfrozen mantle at the top of the front slope is more than 10 m think (Barsch 1973).
2. Alternatively, rockglaciers may become inactive, even in areas of continuous permafrost (Ellis 1982; Calkin et al. 1987; Sollid and Sörbel 1992). In these cases, where extensive melting of the internal ice is unlikely, rockglaciers become inactive if they extend so far from their source area that the tangential stress due to the slope, the thickness of the deposit, its bulk density etc. decreases below the limit required for movement. This may be caused,

[1](r) was added by D.B.

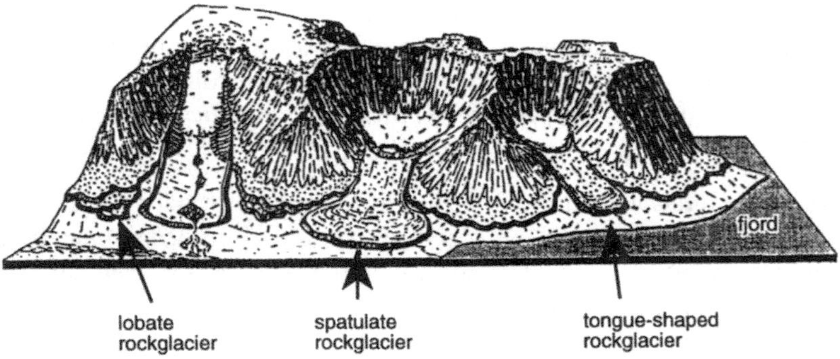

Fig. 1.6. Different rockglacier types in a polar mountain environment. (Modified from Humlum 1982)

Fig. 1.7. Climatic inactive talus rockglacier Albana (Upper Engadin, Graubünden, Swiss Alps). The rockglacier became inactive around 1970 following a decrease in velocity since 1950. Front height is about 70 m

for instance, by a reduction in talus production in the headwalls above the rockglacier and/or by decreasing incorporation of ice into the system. Inactivity is in these cases not directly climatically controlled. These rockglaciers are, as Calkin et al. (1987, p. 73) observe, "climatically insensitive". They will be called *dynamic inactive* (cf. Fig. 1.8).

Fossil or *relict rockgalciers* (Figs. 1.9, 1.10) are defined as former rockglaciers, in which all ice (the total amount of interstitial ice and all ice lenses) has been melted; they are, therefore, characterized by collapse structures at their surface, and their surface relief is much more subdued than on active and inactive rockglaciers.

The terms "relict" and "fossil" are used synonymously depending on the background of the author: In a geologic context, "fossil" means "buried", in geomorphic use, it expresses "not recently formed". In this book, the term "relict" is slightly preferred.

There is a decisive difference between active and inactive rockglaciers, on the one hand, and relict/fossil ones, on the other. Active and inactive rockglaciers have well-developed forms and so are difficult to differentiate on aerial photographs. Thus, some authors have considered active and inactive rock-

Fig. 1.8. Dynamic inactive talus rockglaciers on Stuphallet (Bröggerhalvöya, western Svalbard Archipelago). The lobate rockglaciers are found in a continuous polar permafrost environment. They creep out of the talus slopes. Their inactivity has been proven by measurements (Sollid and Sörbel 1992). (Photographed by Johan Ludvig Sollid, Oslo)

Fig. 1.9. Fossil Blockgletscher Alp Placer (Scarltal, Lower Engadin, Graubünden, Swiss Alps). The rockglacier is situated at the present-day timberline. It is a late Pleistocene feature (probably Younger Dryas)

glaciers as a single class of "intact" rockglaciers. This is acceptable, even if it involves a loss of information, because climatic inactive rockglaciers seem to display recent to subrecent changes in the altitude of the belt of discontinuous mountain permafrost (cf. Sect. 9.2). In contrast, relict rockglaciers display collapsed, subdued, and older (Pleistocene) forms due to the melting of the interstitial ice (cf. Sect. 9.2).

Most rockglaciers occur either below talus slopes or below the end moraines of small glaciers (cf. Figs. 1.5, 1.6). Postponing a discussion of these to Chapter 4, we define according to Barsch (1987a, 1992) the following types:

1. The *talus rockglacier* (first type) is situated below talus slopes (cf. Fig. 1.3) and transports mainly frost-shattered rock fragments.
2. The *debris rockglacier* occurs below the end moraines of (mostly) small glaciers (cf. Fig. 1.6) and transports mainly morainic or glacial debris (till).

A discussion of the geometry, rheology, and genesis of rockglaciers and their relationship to similar forms will take place in the later chapters. Here, only a few terminological points will be added.

Active rockglaciers are Holocene features in the periglacial realm. Their ages will be discussed later (Chap. 9). They are partly related to postglacial

Fig. 1.10. Air photograph with some of the fossil rockglaciers in the Val di Campo (Bernina, Graubünden, Swiss Alps). Photograph of the Schweizerische Landestopographie Bern (No. 2280, 25 August 1964)

(Holocene) climatic periods. Unfortunately, the terminology on the Holocene climate history is not always used in the same sense. Thus, "Neoglaciation" means in North America the time between 5000 and 3000 B.P., in Scandinavia 1750–1890 A.D. and in the Alps ca. 1600–1850 A.D. (Grove 1988). It is partly synonymous with the "Little Ice Age" (North America, ca. 1250–1890 A.D; Scandinavia, 1750–1890 A.D; Alps, ca. 1600–1850 A.D.), which comprises the period called in German "neuzeitlich" (Kinzl 1932). In this book, the exact dates are always given and related to radiocarbon years (if not otherwise stated).

Regarding the altitudinal differentiation of landforms, climate, or permafrost, the term "belt" is used. For the equivalent horizontal distribution, the term "zone" (e.g., periglacial zone) is applied. Geomorphic belts (especially above the timberline) have been discussed for a long time in the Alps.

Thus, the term "alpine" is in common use to describe these belts. It comprises a general, worldwide meaning. In this book, therefore, the terms alpine permafrost or mountain permafrost are used synonymously.

2 Historical Development of the Term Rockglacier

The landform which is today called a rockglacier has been known to scientists for more than a century. With changing models of explanation, the term describing it has been altered several times. In addition, when terms from our everyday language are used, misunderstandings are prone to occur, because – if the correct definition is not known – these terms may lead to the introduction of models which are inappropriate to the phenomena under discussion.

Probably the first description of an active rockglacier was given by Steenstrup (1883) more than 100 years ago (Humlum 1982; Vitek and Giardino 1987). Steenstrup (1883, p.81) described in northern Greenland: *"Paa Umanaksklippen findes endvidere paa Sydsiden en Steenskraaning..., der har en saa fuldstaendig Lighed med en Moraenemasse daekket Brae..., iggennem hvor lange Tider en saadan 'doed Brae' kan bevares"*. Although Steenstrup apparently believes that these features are "dead glaciers", he had observed true rockglaciers as had been stated in a letter by Ole Humlum, who mentioned the features in one of his own papers (1988b).

The next descriptions of a rockglacier were those of Spencer (1900), as a "peculiar form of talus", and Rohn (1900) as "movement of talus slopes". Even though these descriptions have not been taken up directly in the literature, the expressions "runaway talus" (Slaymaker and McPherson 1977, p.173) or "creep of alpine permafrost" (Haeberli 1985a) may be partly related to the old terms used by Spencer and Rohn.

The first explicit comparison of a "rockglacier" with a true glacier is given by Chamberlin and Salisbury (1906, p.474) as "talus glacier", a term also used by Siebenthal (1907, p.19). Apparently, it was already a very common expression as shown by: "a small valley has been cut out, and in this has formed the very pretty example of the talus glacier shown (Fig. 3)". The photo indeed displays a well-developed and apparently active rockglacier from a small side valley in the Huerfano Basin (Sanger de Christo Range, Colorado). In a similar form, the term was revived by H.T.U. Smith (1973, p.69ff) as "piedmont talus glacier".

The modern and now generally accepted term "rockglacier" was first used (Giardino and Shroder 1978, p.446) by Cross and Howe (1905) in their geologic description of the Silverton quadrangle (San Juan Mountains, Colorado). However, in later work (Cross and Howe 1906; Howe 1909), they refer to the same features as *rock streams*.

The first written definition is that in the classic paper of Capps (1910), who *expressis verbis* says (p. 359, 360): "The special agents of degradation with which I wish to deal at present, however, I have called *rock glacier.*" Capps provides a good discussion, which includes many of the concerns that remain controversial today. His use of expressions such as "rock glaciers resemble true glaciers" (p. 362) and "movement of the masses in a glacier-like way" (p. 364), indicates a belief that rockglaciers are quite different from true "ice" glaciers, even if they appear superficially similar. He was probably well aware of the influence of debris on the flow of glaciers, which had been discussed by Russel (1895) in the same journal. At its time, Capps' work was clearly the most comprehensive treatment of rockglaciers and remained so for decades. In his paper " 'Rock glaciers' or chrystocrenes", Tyrrell (1910) responded to Capps. He accepted rockglaciers as true periglacial phenomena, explaining their interstitial ice as frozen water from springs which flows into the talus. His proposition of the term "chrystocrene", following a suggestion by T.C. Chamberlin, has not been accepted.

With translations into German (*Blockgletscher*: Högbom 1914, p. 356; Krebs 1925; Salomon 1929), French (*glacier rocheux*: de Martonne 1920) and Spanish (*litoglaciar*: Catalano 1923), the use of the term rockglacier has been strengthened, despite the fact that it is confusing, leading students to believe that these features are true glaciers. Thus, Mercanton (1923, p. 254) writes about the term rockglacier: "*dénomination impropre au fonds mais parlante*". The international use of this term has been further enforced by the Russian term *kamennye gletcery* (Gorbunov and Titov 1986), literally "stone glacier", which is restricted to true rockglaciers and is distinct from (ice) glaciers (lednik). Other propositions such as *talus glacier* (Chamberlin and Salisbury 1906), *chrystocrene* (Tyrell 1910) or the equivalent of *Blockgletscher* (blockglacier; Smith 1973 p.70) have been turned down.

This applies also to another suggestion concerning terms derived from "*Blockstrom*" such as block stream, boulder stream, rock stream (Patton 1910; Högbom 1914; Salomon 1916; Meyer-Harrassowitz 1918); *coulée de blocs* (Chaix 1919; Cailleux 1947); *colate di pietra* (Nangeroni 1929). The term "*Blockstrom*" and especially "blockmeer/blockfield" (Caine and Jennings (1968) is used for periglacial deposits formed during the last glaciation in the mountains of Central Europe. "*Blockmeere*", "*Blockfelder*"or "*Blockströme*" have been known in German geomorphologic and geologic literature from as early as 1820 (Goethe, cf. Penck 1894, p. 239). Also, Darwin in Voyage of the Beagle describes "streams of stones" in the Falkland Islands (due to earthquakes). Discussions about these features are given by von Cotta (1843, p. 174) or Naumann (1854 p. 232, 233), who believe in a catastrophic (perhaps rather cataclysmic) genesis. Partsch (1882) discusses all block deposits in relation to glaciers, later Blankenhorn (1896; von Lozinski 1910) describes them as relicts from colder climates in the past. Chelius (1896) discussed the "*Felsenmeere*" (or "*Felsenströme*", p. 645) in the Odenwald (near Heidelberg and Darmstadt).

A comprehensive study of these features was done by Schott (1931) who favored a recent, i.e., Holocene, formation. Rother (1965), Braun (1969) or Wilhelmy (1974) later corrected Schott's view. With few exceptions, blockstreams are not rockglacier deposits. They are relicts of periglacial solifluction sheets, from which fine-grained material has been lost by piping or slope wash.

Unfortunately, the term *blockstrom* has also been occasionally used up to the present day to denote real rockglaciers in the Alps (Boesch 1951; Domaradzki 1951; Höllermann 1964; Binggeli 1965; Schweizer 1968a; Louis 1969; Jäckli 1978). In English, the terms, rockglacier and blockstream, have been used synonymously: Ives (1940) uses the former, Kesseli (1941) the latter for the same feature. This leads to confusion. Therefore, it should be accepted that the terms rockglacier and blockstream describe different phenomena [cf. Högbom (1914 p. 356), even if Krebs (1925 p. 103) again mixes both terms]. This is why the proposition of Wilhelmy (1974), following partly Grötzbach (1965a) and Outcalt and Benedict (1965), to reactivate the term blockstream (*Blockstrom*) instead of rockglacier is an unfortunate one that would lead to misinterpretation. This is also true for the unnecessary and artificial genetic differentiation between rockglaciers (*Blockgletscher*) beneath talus slopes (in this book, the talus rockglacier) and boulder tongue (*Blockzunge*) beneath glaciers (in this book, the debris rockglacier).

Over the years, besides many difficult concepts (cf. Chap. 8), a number of very difficult terms have been proposed, e.g., "special form of moraine" (Gerhold 1964, 1965), "ablation complexes" (Johnson 1974), "glacier ice-cored debris system rockglacier" (Johnson 1992 and earlier), "glacier-cored, tongue-shaped rockglacier" (Calkin et al. 1987, p. 73) and "ablation complex rockglacier type" (Evin 1987b). These are not only terrible constructions, they also have genetic implications. According to the photograph presented by Evin (1987b), the "ablation complex rockglacier type" is nothing but a debris rockglacier (Barsch 1987a) or – less elegantly – a "rockglacier beneath glaciers" (Barsch 1969a, 1977a). The description "near glacial end member of the periglacial facies" (Madole 1972, p. 123) is also clumsy, but, at least, not genetically wrong (Chap. 8).

In the terminology of Humlum (1982), the talus rockglacier is a talus-derived rockglacier whereas, the debris rockglacier a glacier-derived one. The terms proposed in this book are shorter, which is always an advantage. Also, the terms used by Humlum again include and indicate a wrong model: the debris rockglacier is not "glacier-derived", it only transports debris deposited by a glacier. More recently, Gerber (1994) proposed naming the debris rockglacier "creeping moraine" (*Kriechmoräne*), with rockglacier reserved for talus rockglacier only. After a century of research and use, this is an exaggerated proposition based on a limited sample (less than ten features). It is neither correct nor helpful, since it would require the renaming of the many large famous landforms which have been called rockglaciers for decades.

Today, the term rockglacier (German: *Blockgletscher,* French: *glacier rocheux*; Italian: *ghiacciai rocciosi* or *ghiacciai di pietre*; Spanish: *glaciar de escombros* or *litoglaciar*; Russian: *kamennye gletcery*) is widely used and well accepted despite its semantic connection to "normal" (ice) glaciers. But this connection (cf. Sect. 7.2) is more an areal than a genetic one. Today, a change of term would meet great resistance. Therefore, and to emphasize the fact that we are dealing with an independent phenomenon, Barsch (1988) proposed writing the term rockglacier as one word.

Thus, the term rockglacier will be used for the landforms that represent the visible expression of the creep of ice-supersaturated mountain permafrost according to the definition in Chapter 1. Features with a similar phenotype, but of quite a different origin, will be discussed later (Chap. 8).

3 Rockglaciers: Description and Morphometry

Rockglaciers have been described in all major high mountain systems of the world (cf. Chap. 5). According to the topographic, climatic and other influences on the creep of mountain permafrost, a wide variety of forms exist. The sizes and forms of active rockglaciers may be quite different, not only in different mountain systems but also locally. Nevertheless, a basic description can be given; a genetic approach, however, which has often been connected with a discussion of size and form in the past, is not intended.

3.1 General Description

Active rockglaciers stand high (at least 10 to 20 m) above the adjacent terrain, which allows no conclusion to be drawn regarding the rockglacier thickness. Their front and side slopes are steep, mostly steeper than the angle of repose (> 35°). On aerial photographs, the front and side slopes appear "lighter" than their upper surface reflecting the exposure of "fresh" (unweathered) material and the lack of lichens, weathering rinds, and iron staining. In addition, the upper surface is normally bouldery with a microrelief of furrows and ridges ("porridge-like" appearance), which is the surface expression of a decelerating viscous or plastic flow (cf. Chap. 7). Fine material is only visible on the front and side slopes; it is missing from the upper surface.

The following seem to be characteristic of an active rockglacier:

1. Steep and relatively light-coloured front and side slopes.
2. A darker, bouldery surface with furrows and ridges.
3. The mean gradient of the upper surface is much less than that of the front and side slopes.
4. The upper end may either grade into a steep talus slope or be a depression between the talus slope and the rockglacier (which itself may now or in the past have been occupied by a glacier or an ice-field).

There is no rule without exceptions and these characteristics are not universally true of all rockglaciers. Some notable exceptions include:

1. The side slopes may not be well defined where talus slopes grade to the rockglacier side.

2. The height of the side slopes may be very low in places where the rockglacier crosses a threshold of bedrock.
3. The uppersurface may consist of comparatively fine-grained material (< pebble size) in areas especially favourable for rockglacier development. This is the case in the Andes west of Mendoza where bedrock disintegrates into fine fragments and where the climate allows the existence of mountain permafrost without an insulating boulder layer ("Balch" Ventilation, cf. Sect. 10.1).
4. The surface relief of debris rockglaciers may be subdued. This is sometimes caused (e.g., in the Alps or in the Canadian Rocky Mountains) by the glacier advance of 1850, which moved in special cases onto active rockglaciers e.g., the rockglacier La Tscheppa in the upper Engadin, (Barsch 1977a; Sect. 8.1.1). It may also be the expression of special developments, e.g., by steady-state talus input as assumed for the talus rockglacier Grosses Gufer near the Eggishorn at the Aletsch glacier, Wallis, Swiss Alps (Messerli and Zurbuchen 1968) or by extending flow.

3.2 Form Types

In rockglacier research two main form types have often been distinguished (cf. Chap. 1; Figs. 1.5, 1.6) and here they are referred to as:

1. Tongue-shaped rockglaciers
2. Lobate rockglaciers

This distinction, derived from Domaradzki (1951) and Wahrahaftig and Cox (1959), differentiates the main forms well. They may be defined by the length width ratio (l:w), where the length is measured in the direction of flow. If $l:w > 1$, the rockglacier is tongue-shaped; if $l:w < 1$, it is lobate.

This difference in form has been interpreted in a variety of ways (Wahrhaftig and Cox 1959). The only widely accepted one is that lobate rockglaciers are normally found below small cliffs and rock walls, where talus production is low. On the other hand, they may also be found beneath small glaciers, where the accumulation of morainic material is equivalently low and where the frozen material deposited as Holocene moraines moves downvalley as rockglaciers (Barsch 1969a). All other interpretations regarding age, (Wahrhaftig and Cox 1959), genesis (Outcalt and Benedict 1965), or internal ice content (White 1976) are not generally valid (cf. Chap. 8). Too often, the distinction between rockglacier types has been taken as a sequence of development, such that the tongue-shaped rockglacier is regarded as a more developed lobate one. Parson (1987) points out that in the Blanca Massif (Sangre de Christo Range, Colorado) a discriminant analysis reveals that the distinction is real. This suggests that site characteristics are important in explaining the different form-types e.g., lobate or tongue-shaped (cf. P.G. White 1979). This seems congruent with other experiences.

Wahrhaftig and Cox (1959) introduced a third form which they called the spatulate rockglacier and which was described as tongue-shaped with a broader lower part. Since this form resembles a piedmont glacier, like the Malaspina glacier, Humlum (1982) named it a piedmont rockglacier. Often, this form is not well developed and can best be considered as a tongue-shaped form (Parson 1987).

A more useful additional class is the complex type, which describes those rockglaciers which consist of a complex of aggregated rockglaciers that would be difficult to describe in other terms. There are several possibilities of aggregation. A rockglacier may be composed of several different rockglaciers (multipart), it may split in to several lobes (multilobe, cf. Humlum 1982), it may form a sequence of several units of slightly (?) different age (multiunit, like the El Salto rockglacier cf. Barsch and King 1989; or the Galena Creek rockglacier, cf. Barsch 1987b) or their material can be derived from different soures (multiroot, e.g., from different talus slopes). Thus, in pure geometric terms, the shape of a rockglacier may be classified according to three categories:

1. tongue-shaped
2. lobate, and
3. complex.

In addition, the third class comprises four subtypes:

3.1. multipart rockglacier,
3.2. multilobe rockglacier,
3.3. multiunit rockglacier,
3.4. multiroot rockglacier.

3.3 Morphometric Parameters

The morphometric features of a rockglacier are summarized in Fig. 3.1 Length (measured in the direction of flow), width (measured perpendicular to length), and thickness at the front slope are the most commonly measured characteristics of rockglaciers and also define in the part the form types. Very important information is given by the gradient of the front slope of a rockglacier especially in relation to the angle of repose of the material involved. Other parameters, like maximum thickness or planimetric area of a rockglacier as well as the planimetric area of its source zone, are very helpful, but they require more sophisticated geophysical and geodetic surveys. The altitude of the foot of the

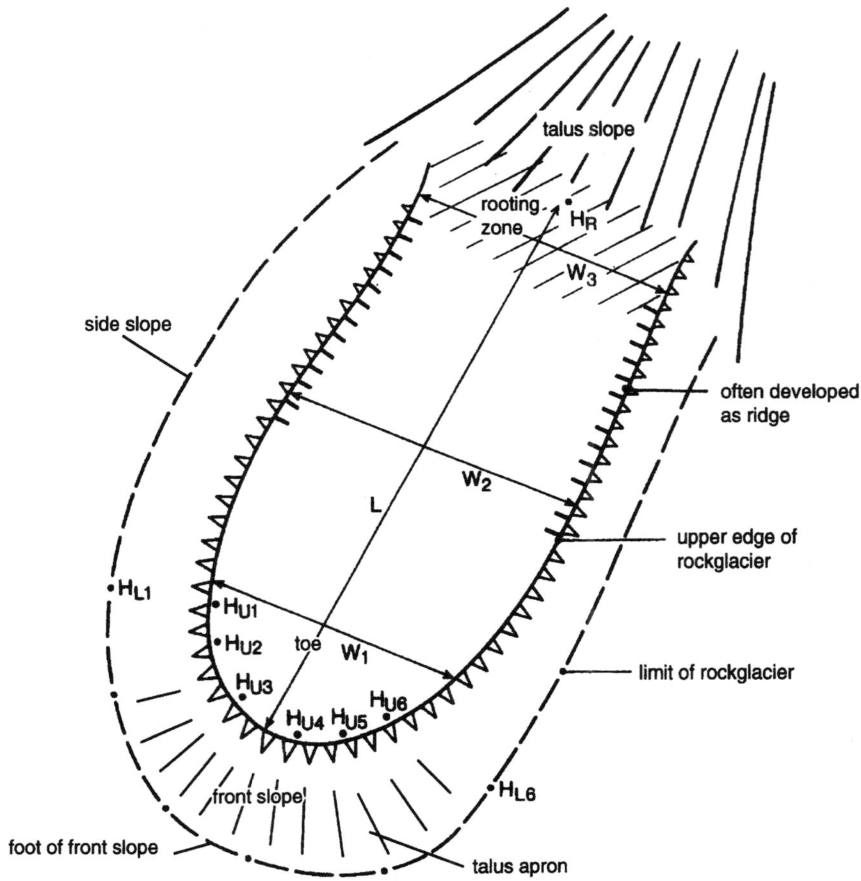

L : length of rockglacier surface

W : width of rockglacier surface $= \dfrac{W_1 + W_2 + W_3}{3}$

H_U: upper height of front slope $= \dfrac{H_{U1} + \ldots + H_{U6}}{6}$

H_L : lower height of front slope $= \dfrac{H_{L1} + \ldots + H_{L6}}{6}$

H_R: height of rooting zone

Fig. 3.1. Schematic diagram of geometric parameters and descriptive terms of a normal (active) talus rockglacier. In the case of a debris rockglacier, a moraine or (today) the relict of a glacier is situated at the base of the talus slope. The length is measured along the stromstrich (center flow line)

front slope, the altitude of the upper part of the rock wall above the rockglacier demand at least good topographic maps, which are still not available for all high mountains around the world.

Additional geometric information may be supplied in relation to the aspect of the rockglacier, the aspect of rock walls, mean slope of rockglacier surface, mean gradient of talus slope mean slope of rock wall, mean slope below the rockglacier, and its altitude relative to the local snow-, tree- or forestline.

The surface relief of a rockglacier is often more easily described in qualitative rather than quantitative terms. Nevertheless, it is necessary to try to evaluate the surface relief quantitatively, because such data are very important characteristics of rockglaciers.

3.3.1 Rockglacier Sizes

Rockglaciers are normally several tens or some hundreds of meters long. The same values are applicable for their width. They cover areas of several hectares (1 ha = 0.01 km^2). Their front slopes are at least 10 to 20 m high (sometimes more than 100 m), and in the active state are steeper than 35° which is more than the angle of repose of the incorporated material in an unfrozen state. Following the classification of Section 3.2, there are two main form classes of rockglaciers according to their geometry; tongue-shaped and lobate. The other class, the complex rockglacier, has not been used hitherto, but if very large rockglaciers are described, it is often relatively certain, that they are complex ones which are, for instance, built up by several units.

A very intense study of rockglacier geometry has been done by G.P. White (1979). In the San Juan Mountains (Colorado, latitude ca. 36° N), he mapped 613 active and inactive rockglaciers from aerial photographs (Table 3.1). One of the main results of this study is the observation that lobate and tongue-shaped rockglaciers are not differentiated by the altitude of their snouts, but the tongue-shaped rockglacier is situated in a more favorable environment. The average ratio of the lengths of tongue-shaped and lobate forms in this area is 2.72:1.

3.3.2 Tongue-Shaped Rockglaciers

Tongue-shaped rockglaciers are very spectacular phenomena and are, therefore, often described as typical rockglaciers. They can be more than 1 km long, especially if they are situated in relatively dry, continental mountains. In most cases, they are debris rockglaciers following the definition in Chapter 1. According to Table 3.1, in the San Juan mountains they have an average length of 642 m (P.G. White 1979). Calkin et al. (1987) report that – measured from their rooting zone – tongue-shaped debris rockglaciers are 100–1200 m long (Brooks Range, Alaska). The assumption put forward by the authors that these rockglaciers might have been even longer "before disturbances by Holocene glacier advances" is very questionable (cf. Sect 8.2). Their thickness is estimated to vary between 50 and 100 m. The bouldery layer is 0.8–2 m thick, the total relative surface relief comprises 5–10 m.

Table 3.1. Geomorphometric data of 613 rockglaciers in the the San Juan Mountains (Colorado) from P.G. White (1979)

	Orientation[a] (°)	Head alt. (m)	Snout alt. (m)	Length (m)	Slope (°)	Source wall height (m)	Source wall rim alt. (m)
Mean	358	3726	3626	350	17.2	229	3995
SD	75	175	186	291	5.2	105	169
Range	–	2896	2804	49	5	30	3139
		–4206	–4145	–2347	–34	–622	–4359
Lobate	354	3723	3652	235	17.8	219	3983
Tongue-shaped	9	3733	3563	642	15.4	257	4026
Active	1	3795	3697	340	17.2	228	4054
Inactive	354	3617	3516	366	17.2	232	3903

Active and inactive rockglaciers reach down to nearly the same altitude (difference: 181 m).Lobate and tongue-shaped rockglaciers do not differ significantly in the altitude of snout, in source wall rim altitude, nor in slope angle.[a] Most rockglaciers display a northern exposure, therefore, orientation in general is around N

Humlum (1982) mentions that tongue-shaped rockglaciers on Disco Island (Greenland) are 500 to 6000 m long. Their thickness is estimated as 20–75 m and the front slopes are 35° to 50°. The furrow and ridge topography has a relative relief of 2–5 m.

From the Zailijskiy Alatau (Central Asia), Gorbunov (1983) mapped 179 tongue-shaped rockglaciers, about 40% of the total number of rockglaciers. Their average length is 630 m, their average width 280 m, and their average surface comprises 0.190 km^2 (= 190 000 km^2). The longest one is the northwards exposed Prijamosy shcheli (Issyk river basin); it starts at 3500 m a.s.l. and ends at 2500 m. Its length is reported to be 4400 m and its average width to be 400 m (max. 1200 m, min. 220 m). The area covered by this rockglacier is around $1.6 \times 10^6 m^2$. Its volume is calculated to $83 \times 10^6 m^3$ (average thickness is apparently assumed to be 50 m). Thus, assuming 60 vol% of ice, it contains about 30 to $40 \times 10^6 m^3$ of talus and debris.

3.3.3 Lobate Rockglaciers

Lobate rockglaciers are in most cases (ca. 90%) developed below talus slopes. They are the typical forms of "runaway talus" and, thus, they belong mainly to the group of talus rockglaciers (Chap. 1). Table 3.1 indicates that lobate

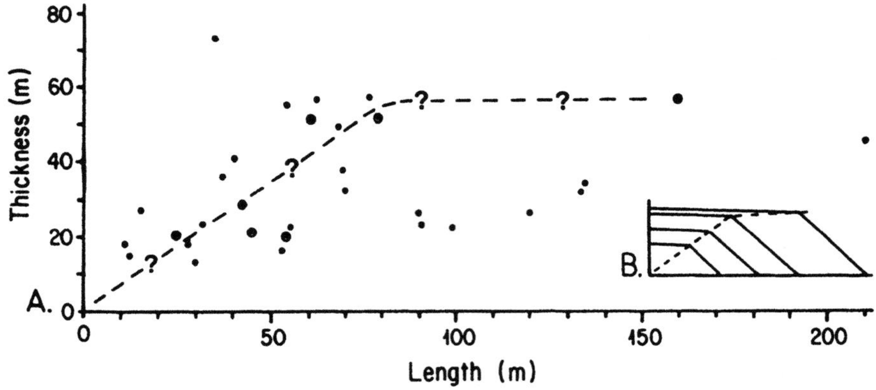

Fig. 3.2. Assumed length/thickness ratio of lobate rockglaciers in the Brooks Range (Alaska) according to Calkin et al. (1987 p. 69). Thickness is taken as the relative height at the front slope. As discussed in the text without geophysical soundings this can only be a rough estimate of the real thickness

rockglaciers in the San Juan Mountains have an average length of 236 m. According to Calkin et al. (1987), lobate rockglaciers in the central Brooks Range (Alaska) can extend 15–200 m, and are 15–17 m thick. Their length/thickness ratio was calculated to be 3:2 (cf. Fig. 3.2). Similar values are given by Humlum (1982) for lobate rockglaciers from Disco:

- Length: 30–300 m
- Thickness: 10–30 m
- Front slope: 35–50°
- Surface relief: 2–5 m (only on the larger ones)
- Surface slope: 5–25°

Gorbunov (1983) reported that the lobate rockglaciers in Zailinjskiy Alatau are on the average 380 m long and 180 m wide. They cover on average an area of about 70 000 m^2. They form 60% of the total number of rockglaciers. The same relationship has been reported by Barsch (1977d) from the Swiss Alps. However, in Banff National Park (Canada), lobate rockglaciers are less than half of the total (45%, cf. Luckmann and Crockett 1978).

3.3.4 *Rockglacier Thickness*

The determination of the thickness of an active rockglacier involves several problems. The most obvious is related to the fact, that direct observations of the rockglacier thickness are normally not possible (cf. Chap. 6). Therefore, values presented are estimates (Table 3.2) or are from indirect evidence. Even

Table 3.2. Assumed average thickness of active rockglaciers in the Swiss Alps as a function of the size of their surface according to Barsch (1977a,d)

Area (km^2)	Av.thickness (m)
< 0.01	20–30
0.1–0.01	40–50
> 0.1	60–80

these values are scientifically sound if drilling or geophysical investigations have been done. A more severe problem has been presented by the results of the great drilling project on the rockglacier Murtél I (Graubünden/Swiss Alps) by Wilfried Haeberli (cf. Chap. 6). According to these results, is necessary to differentiate between the thickness of the active rockglacier and the thickness of mountain permafrost. At the drilling site, the active rockglacier is about 30–32 m thick; mountain permafrost reaches at least 52 m (Wagner 1992). Up until now, this differentiation was not known. Thus, besides all other uncertainties, even geophysically based estimates of rockglacier thickness (Barsch and Hell 1975; Haeberli 1985a) can be looked upon only as a first approximation. It is more or less an estimate of the total mountain permafrost thickness at the place, i.e., it cannot detect whether or not the mountain permafrost below a zone of weakness is incorporated in the creep process. With normal geophysical measurements, this part of the mountain permafrost is included in the estimate of the rockglacier thickness.

Normally, thickness is estimated at the front of a rockglacier. Often, the front slope is strongly reduced in relation to the average thickness. This is especially the case if the rockglacier crosses a rock threshold (cf. the longitudinal profiles of the rockglaciers Macun I or Murtèl I in Chap. 6). In other cases, enormous front slope heights have been reported, e.g., 150 m at the El Salto rockglacier (Barsch and King 1989). Thus, the values of rockglacier thickness between 10 and 100 m based purely on the height of the rockglacier front slope are no more than a first and rough approximation. Calkin et al. (1987) tried to use a length/thickness ratio to determine the thickness of lobate rockglaciers in their study area (Fig. 3.2). In principle, this is a logical concept, because the thickness of a rockglacier (in this case of a lobate rockglacier) should increase with size up to a special threshold (cf. Chap. 7). In the Calkin et al. diagram, an upper threshold at a length of ca. 80 m is proposed. This leads to the question as to whether or not a lower threshold also exists i.e., a minimum thickness, which is slope-dependent for a lobate (or tongue-shaped) rockglacier. In the case of the dynamically inactive lobate rockglacier (cf. Chap. 1), a minimum thickness seems to be necessary to cause further creep. In any case, this diagram has to be applied as cautiously as the relationship between area and thickness used by Barsch (1977a,d) in the Swiss Alps for some calculations (Table 3.2).

Therefore, correct information on the thickness of rockglacier permafrost is only possible if geophysical methods are used (cf. Chapt. 6). The data collected so far are summarized in Table 3.2. In the debris rockglacier Macun I, mountain permafrost thickness is about 100 m; in the talus rockglacier Murtèl I it is about 50 m. The latter has been confirmed by the 1987 drilling investigation (Haeberli et al. 1988), but the thickness of the rockglacier itself, that is the active moving part above the shear planes in the " weak zone", is only 28 to 32 m (Wagner in Haeberli 1990; Wagner 1992; cf. Chaps. 6, 7). Summarizing, it may be said that rockglacier permafrost can easily be more than 50 to 100 m thick, but that the rockglacier thickness (i.e., the thickness of the creep phenomenon) may be not more than 30 to 50 m, i.e., less than 50–60% of the thickness of possible rockglacier permafrost.

3.3.5 Surface Relief

Surface relief of active rockglaciers is characterized by transverse ridges and furrows, which are bent downvalley or downslope as shown in Fig. 3.3. Frequently, lateral longitudinal ridges are observed, which are often found in

Fig. 3.3. Surface relief on the well-developed rockglacier Cerro Negro, Las Bronces, Chilenean Andes near Santiago

different flow deposits (cf. Sect 8.2.2.3). Active rockglaciers without marked surface relief are rare. One example is the rockglacier Großes Gufer near the Aletsch Glacier (Wallis, Swiss Alps; Messerli and Zurbuchen 1968).

The relative surface relief varies between 1 and 10 m. It is more prominently developed on larger rockglaciers than on smaller ones. A difference between lobate and tongue-shaped (or between talus and debris) rockglaciers of the same size has not been observed. Apparently the topographic influences on creep (gradient, breaks of slope, rock thresholds etc.) are more important than the location of a rockglacier (below a talus slope or below a moraine deposit). The transverse ridges on active rockglaciers in the Alaska Range are 2–10 m high and 2–50 m apart (Wahrhaftig and Cox 1959). A contour map of the rockglacier Murtèl I, which presents a well-developed surface relief, is presented in Fig. 3.4.

The furrow and ridge topography displays, as a rule, a regular picture. Ridges are mainly perpendicular to the direction of flow. Only at the sides do longitudinal ridges occur which are often (falsely) interpreted as lateral moraines of a former glacier. These levèes have been described from active, inactive, and relict rockglaciers. Like similar, but much smaller, levèes along the channels of debris flows, they seem to indicate that in the outer parts of the flow the ratio of ice to mineral material is lower than in the inner parts. Thus, the central part moves faster and leaves these outer ridges behind. On relict rockglaciers, the levèes stand higher than the interior parts. This difference can be used to calculate the minimum ice content of the rockglacier during its active state.

Often the surface of a lobate talus rockglacier slopes back towards the talus slope (Calkin et al. 1987). This seems to be due to the fact that the creep is not necessarily directly related to the input (of snow and talus): it may be faster than the input of talus and snow or ice.

In general, the surface relief of active rockglaciers is an expression of a compressive (decelerating) flow. Its formation on active rockglaciers has led to quite a lot of speculation, even if it is generally agreed that it is connected with the movement of the rockglacier. Thus, three main hypotheses have been proposed for the formation of the ridges (Loewenherz et al. 1989):

1. Differential movement of discrete layers of debris (Ives 1940).
2. Compressive flow (Wahrhaftig and Cox 1959; Potter 1972; Haeberli 1985a).
3. Debris input variations (Barsch 1977a,d).

According to rheological considerations, Loewenherz et al. (1989) concluded that the first two assumptions can produce ridges even under a constant talus input. In both cases, perturbations (as a decrease in the slope of the basal plane, etc.) are necessary to create instabilities. The observed surface relief is the visible expression of these instabilities (cf. Chap. 7).

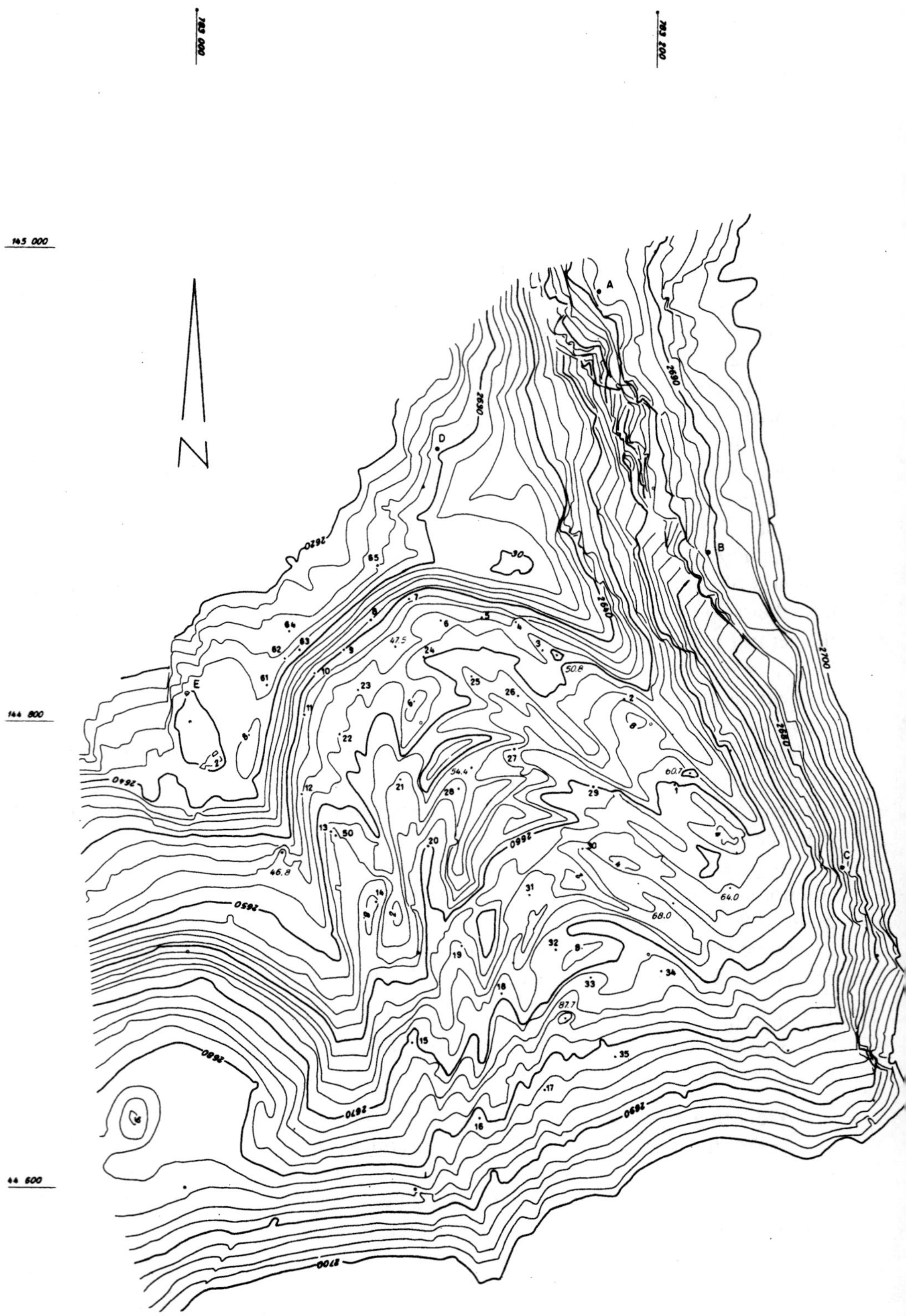

783 000
783 200
145 000
144 800
44 600
N
A
B
C
D
E
2620
2630
2640
2650
2660
2670
2680
2690
2700
47.5
50.8
54.4
60.7
64.0
68.0
46.8
87.7

3.3.6 Rockglacier Surface and Source Area

Rockglaciers are an alpine debris-transport system. In most cases, they are still in areal contact with the source of the material they transport. There has to be, therefore, a close relationship between the size of a rockglacier, the size of the source area for its material, and the intensity of talus production in the source area. This is already indicated by the fact that above tongue-shaped rockglaciers in the Brooks Range (Alaska) the cirque walls are 250 m ($\pm$140 m) high (Ellis and Calkin 1983), whereas cirques without tongue-shaped rockglaciers have walls with a height of only 110 m ($\pm$60 m). Unfortunately, the relationship between rockglacier surface (A_R) and source areas (A_S) has been discussed in only a few papers. Wahrhaftig and Cox (1959) present a diagram for the Wood River area (Alaska Range). Their equation (adapted to the metric system) is as follows:

$$A_S = 1.36A_R + 22.320,$$

where A_S = area of rock walls above the talus slopes (m^2); A_R = area of active rockglaciers (m^2)

Thus, the source area is about 1.4 times larger than the rockgalcier surface in the Alaska Range. This relationship probably does not hold for all mountain systems. It is controlled not only by bedrock resistance, but also by relief and climate, which influence not only the weathering intensity but also – together with slope – the velocity of creep.

In the Swiss Alps, Barsch (1977d) estimates the ratio of source area to rockglacier surface, for active talus rockglaciers, 1 to 3 as and for active debris rockglaciers as 1 to 4, ie.,

$$\text{debris rockglacier}, A_S = (1 \text{ to } 3) \times A_R;$$

$$\text{talus rockglacier}, A_S = (1 \text{ to } 4) \times A_R.$$

Gorbunov (1983) mentions that the most important rockglacier in the Zailijskiy Alatau, the Prijammoy shcheli rockglacier, has a source area of about 7 km^2 of rocky slopes. In case of a maximum glacier melting, this source area would increase to 11–12 km^2. The rockglacier itself has a surface area of $1.6 \times 10^6 m^2$. The relationship between source area and rockglacier surface area can be calculated as:

$$A_S = 4.4 \times A_r \quad (A_S = 7 \times 10^6 m^2)$$

$$A_S = 7.2 \times A_r \quad (A_S = 11.5 \times 10^6 m^2)$$

◄

Fig. 3.4 Countour map of the rockglacier Murtel I (Upper Engadin, Swiss Alps) from Barsch and Hell (1975 p. 114). The plan is derived photogrametrically from air photographs taken in 1971

It is certainly problematic to calculate relationships for such a large rockglacier because its development covers long periods and the real relationships may be masked by other process combinations. Thus, it is not known if Pleistocene debris or other materials have been incorporated in the rockglacier. Nevertheless, at least the relationship for the present situation (7 km^2) is not too far from the other assumptions.

It would be extremely interesting to calculate these relationships for mountains in different climatic zones as well as for different rock types. In addition, the study should incorporate not only the active but also the inactive and the relict rockglaciers. It would present valuable information on the dynamics of rock break up as well as on the reconstruction of paleoclimatic mountain environments (cf. Sect. 10.3.2).

4 Rockglacier Taxonomy

According to the definitions in the foregoing chapters, active rockglaciers are the visible expression of the creep of supersaturated mountain permafrost. The process combination which leads to the formation of an active rockglacier is controlled by thermal boundaries that allow the formation of ice-rich permafrost in unconsolidated materials in mountain terrain. Material properties (ice type, etc.) are clearly of minor importance. A taxonomy or a classification of rockglacier types has, therefore, to follow certain rules: it should avoid unnecessary genetic and material complications, it should not contain special hypotheses, and it should be descriptive and easily applicable.

Unfortunately, most of the proposals for a rockglacier classification system published hitherto do not conform to these rules, but describe the phenotypes in relation to genetic hypotheses. The classification by Corte (1987 a, Fig. 2.1; cf. also Fig. 4.1) is such an example. It mixes different features and processes, which do not belong together. Thus, out of seven rockglacier types (summarized in four groups), five types contain landforms which are not rockglaciers. For instance, the so-called glaciogenic rockglaciers are just debris covered glaciers (cf. Chap. 8). They are generally not a form of permafrost creep, but types of heavy glacier downmelting and thermokarst development connected with the accumulation of a surficial morainic cover by ablation. Groups 2 and 3 in this classification (cryogenic snow-debris avalanche rockglaciers and cryogenic gelifluction debris mantle rockglaciers) are difficult to understand, because gelifluction is not necessarily connected with perennial frozen ground. Also, rock streams are not rockglaciers (cf. Chaps. 2,8)

Often in science, if conflicting hypotheses exist and if clear basic definitions are missing, a "continuum" is proposed. A discussion of a continuum of forms or processes is only appropriate if boundary conditions are clearly stated and if they allow the existence of transitions from one end of the continuum to the other. In rockglacier research, such a continuum has been proposed by Johnson (1974). It comprises all sorts of forms created by glacial, periglacial, or mass-wasting transport processes. This disguises the lack of clear basic definitions, which describe the process of rockglacier formation (cf. Chaps. 1,8). It also hides the fact that individual parts of the proposed continuum belong to quite different geomorphic systems.

Another approach has been used by Evin (1987b). She defines the following three types of rockglaciers based on the grain-size characteristics of the surface material:

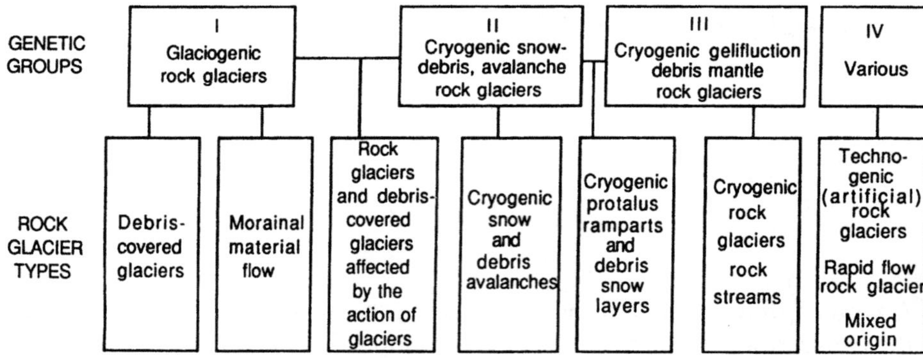

Fig. 4.1. Rockglacier taxonomy proposed by Corte (1987a). See text for discussion

- *Type 1* consists of blocky talus, usually of granite, gneiss, or very resistant sandstones or limestones. Large boulders form the rockglaciers ridges and, in the active state, ice fills the voids between the blocks. Relict rockglaciers of this type are characterized by huge collapse structures and by flat smooth ridges inside a high external ridge.
- *Type 2* is the earthy or silty rockglacier. It is characterized by a small pore volume, and contains little ice. If it is relict, the collapsing is not very important. Usually, this type develops in schists, partly in limestones.
- *Type 3* is an intermediate form composed of different rocks.

This terminology seems exaggerated because it tries to combine too much. If the differentiation between types 1 and 2 regarding the ice content is correct, a velocity difference between them might be expected, but this has not yet been shown empirically. Therefore, the classification regarding the grain sizes at the rockglacier surface can only be used as a descriptive qualifier, e.g., as "blocky" or "earthy". The first term can be applied if the surface of a rockglacier consists mainly of boulders (fragments >200 or >256 mm, depending on the classification used), the latter one applies if the surface is formed mainly of smaller fragments. If boulders of more than 3 m diameter are dominant at the surface, it is possible to speak of a coarse blocky rockglacier.

The following system (cf. Fig. 4.2) is based on purely descriptive parameters that can easily be controlled by field evidence. Starting with the already proposed two main types (talus rockglacier and debris rockglacier; cf. Chap. 1), a third class has to be introduced because rockglaciers can transport materials which have been deposited by other processes. The materials could be an accumulation by man, e.g., mining dumps according to Gorbunov (1983). Another possibility is materials deposited by debris flows. This has been described by Ono and Watanabe (1986) from the Japanese Alps. The feature considered is called a protalus rampart (Sect. 8.3). It exists at an altitude at which mountain permafrost cannot be precluded (Sone 1990). Thus, we have three main types of rockglaciers:

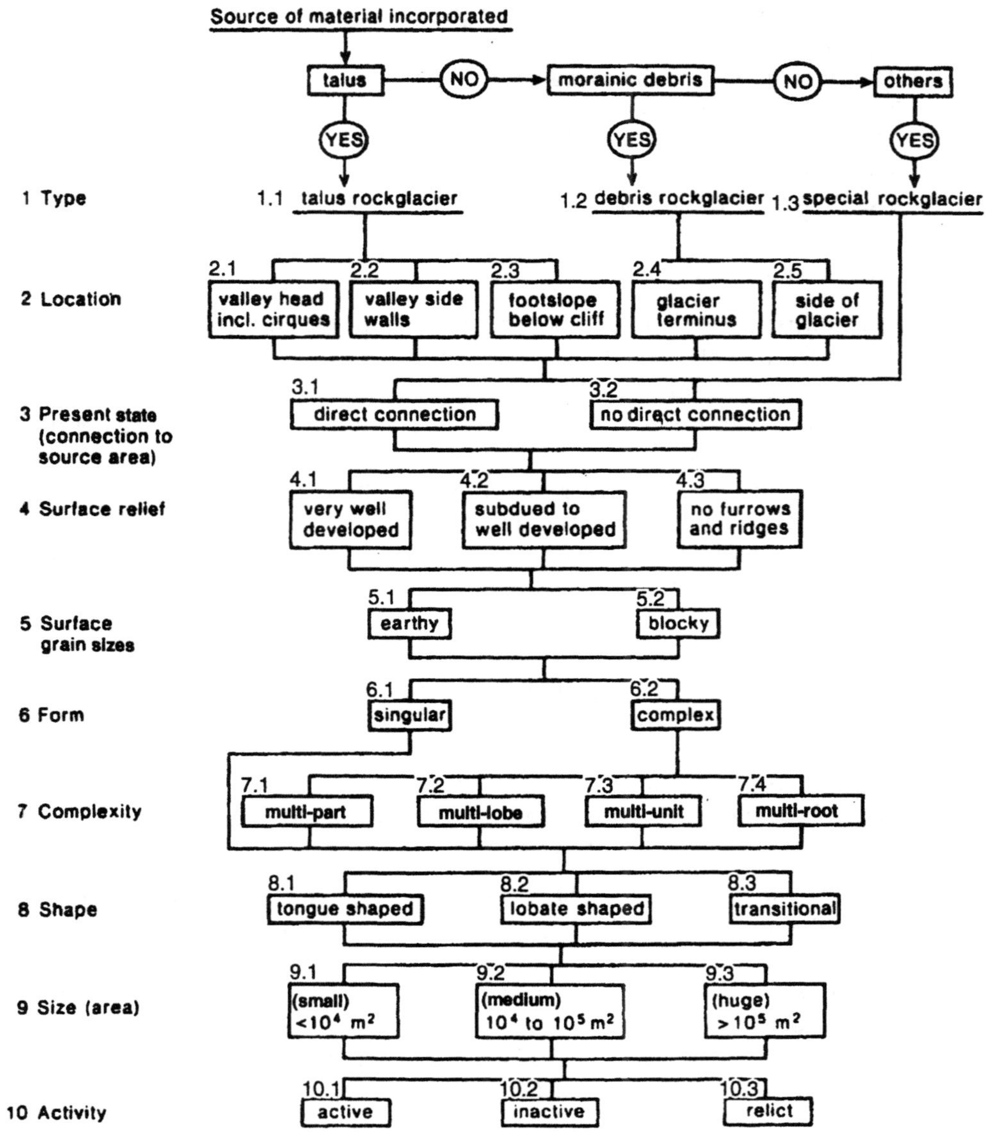

Fig. 4.2. Rockglacier information system. It displays the relevant descriptive information on a rockglacier which can easily be collected in the field or from air photo interpretation. (Barsch 1987a)

- Talus rockglacier,
- Debris rockglacier,
- Special rockglacier.

In addition to classification by type, it is possible (and necessary) to further classify rockglaciers according to the following factors:

- The geomorphic location of the rockglacier,
- The present state of the rockglacier,
- The surface relief of the rockglacier,
- The surface grain sizes of the rockglacier,
- The form of the rockglacier,
- The complexity of the rockglacier,
- The shape of the rockglacier,
- The size of the rockglacier,
- The activity of the rockglacier

The geomorphic location describes the situation of the rockglacier in relation to the source area of the material, to which the rockglacier is or is not directly connected (present state). The latter is normally the case if the rockglacier is moving faster than material (talus) is produced or – in the case of the debris rockglacier – if the glacier has melted.

Surface relief and surface grain sizes have been already discussed. The form of a rockglacier refers to its shape (cf. Chap. 3). Complexity (cf. Chap. 3) is given if the rockglacier consists of different parts (multipart) of different lobes (multilobe), of different units (multiunit), or if it derives its material from different source areas (multiroot). Size and activity give additional information.

This classification is summarized in Fig. 4.2, which is based on a discussion between Arturo Corte and the author (Corte 1987a; Fig. 2.2). It presents a convention for the (field) description of a rockglacier and contains ten levels of information. All this information is easily determined in the field (or from air photographs); together they give an important set of data.

Two examples may demonstrate this:

1. The well-known rockglacier Murtèl I (Barsch and Hell 1975; Haeberli et al. 1988, cf. Fig. 1.4) is a talus rockglacier originating in a cirque with direct connection to the talus slope. It displays a very well developed surface relief and a blocky surface; it is a singular feature, tongue-shaped, of medium size and it is active.
2. The rockglacier Weissmies (Messerli and Zurbuchen 1968; Barsch 1977a; cf. Fig. 1.1) is a debris rockglacier flowing out from the lateral moraine of the Weissmies glacier (Wallis) with no direct connection to the present glacier (the glacier has been melting since 1856!). It has a well-developed surface relief and a blocky surface; it is a singular feature, tongue-shaped, of medium size and it is active.

5 Rockglacier Distribution

5.1 General Information

Active rockglaciers have been reported in all major high mountain systems of the world, but still today a systematic summary of their distribution is missing. An extremely valuable study on the distribution of rockglaciers in Europe and North America has been published by Höllermann (1983a).

Rockglaciers are found as far north as Greenland (Steenstrup 1883; Humlum 1982) and in the North American Arctic (Wahrhaftig and Cox 1959; Calkin et al. 1987); in the south they exist down to the mountains of Antarctica (Mayewski and Hassinger 1980; Mjagkov 1980; Barsch et al. 1985). Thus, an especially long transect is developed in the American Cordilleras stretching from Alaska to Tierra del Fuego. Rockglacier concentrations (cf. Fig. 5.8) are reported from Canada (Luckmann 1976; Luckmann and Crockett 1978) and from the United States (Ives 1939; S.E. White 1971b). Heine (1976) supplied data about some Mexican mountains. Information on true rockglaciers is rare for the tropical mountains south of Mexico. But south of latitude 25°, many rockglaciers have been described in the Andes of Argentina and Chile (Rivelli, pers. comm.; Corte 1976a; Marangunic 1976; Corte and Espizua 1981).

In the Old World, active rockglaciers are known to exist in Scandinavia (Barsch and Treter 1976; Sollid 1989; Svensson 1989), and in the Alps (Höllerman 1964; Barsch 1969a; Guiter 1972; Kerschner 1976; Evin 1983a; Jorda 1983; Haeberli 1985a; Smiraglia 1985). Rockglaciers found in the Middle East mountain ranges have been mentioned by Klaer (1962b) and Messerli (1967), from the Hindukush by Grötzbach (1965a), from the Khumbu Himalaya in Nepal (Jakob 1992a,b), from the southern Alps of New Zealand by Jeanneret (1975).

The African continent seems to be almost devoid of rockglaciers; some have been reported in the Rif near Bab Taza (Mensching 1955) and in the middle Atlas Mountains (Raynal 1960). In the Rif, the lower limit of these relict forms is about 1000 m; in the middle Atlas, they are asumed to be older (penultimate glaciation) and are not found below 2200 m. Relict rockglaciers are also mentioned from the Hoggar Mountains, from the Tibesti (Rognon 1967; Messerli 1972), and from Mt. Kenya (Mahaney 1980). The forms from the Hoggar and the Tibesti seem to be very old (at least penultimate glaciation) according to Messerli (1972). In recent years, Lewis and Hanvey (1993) and

Lewis (1994) have reported relict rockglaciers and protalus ramparts (i.e., embryonal rockglaciers) from the Drakensberg (South Africa). It is not clear yet if the protalus ramparts are correctly identified (Grab, pers. comm.). On the other hand, the descriptions of the Rose Hill rockglacier and the Chesney Wold rockglacier at 31° S (Bottleneck, Drakensberg) seem convincing. They occur between 2060 and 1930 m a.s.l. and between 2135 and 1840 m a.s.l., respectively. Radiocarbon dating puts both rockglaciers in the last glaciation (21 000 ^{14}C years B.P). The fossil rockglaciers at ca. 4000 m at Mt. Kenya (latitude 0 °) probably also belong to the last glaciation (Würm–Weichsel–Wisconsin). As a first approach to the paleoclimatic situation, the gradient for the lower limit of active rockglaciers from Mt. Kenya to the Drakensberg can be estimated to be 65 m per degree latitude. This gradient is less steep than the present gradient for the lower limit of discontinuous mountain permafrost and active rockglaciers in Asia (Cheng 1983; see below)

In summary, it can be said that maritime mountains with large amounts of snowfall throughout the year are unfavourable for the development of rockglaciers (cf. Sect. 10.7). Firstly, all localities that would be favorable for the existence of rockglaciers in the cryogenic belt (situated between the equilibrium line on glaciers and the altitude of MAAT –2 °C) are occupied by glaciers. Secondly, the cryogenic belt is small regarding its vertical extension in these areas. Therefore, continental, comparatively dry mountains with cold winters are more favorable environments for the development and the existence of rockglaciers in the periglacial-cryogenic belt. The number of rockglaciers tends towards zero in maritime high mountain ranges and increases towards continental ones. This rule is also true within single mountain ranges: the number of rockglaciers increases from more maritime environments to the more continental ones (e.g., the Alps, cf. Sect. 5.2.1).

Only in the high mountains of the humid tropics, do rockglaciers not seem to be too well developed, or have not been described yet.

The local relief plays a very important part in terms of the formation of rockglaciers. Talus rockglaciers can only form at places where flat valley floors exist below talus slopes at an altitude permitting mountain permafrost. The same is true for debris rockglaciers, which require that glaciers accumulate morainic debris in the belt of discontinuous mountain permafrost (Barsch 1977a, 1983a; Evin 1987a). Local topographic controls are of great importance in areas where alpine permafrost belts are small, which is usually the case in maritime climates.

The climatic, especially the microclimatic, aspect is given by the exposure of rockglaciers. In most areas, rockglaciers prefer locations with reduced solar radiation input. Thus, they are found mainly in midlatitude mountains of the northern hemisphere in northern (NW–N–NE) exposures (Fig. 5.1).

The distribution of active and inactive rockglaciers provides much information on this geosystem itself. This can be illustrated by considering the distribution in some major mountains. Rockglacier distribution is discussed here using data from previously published studies with the objective of pre-

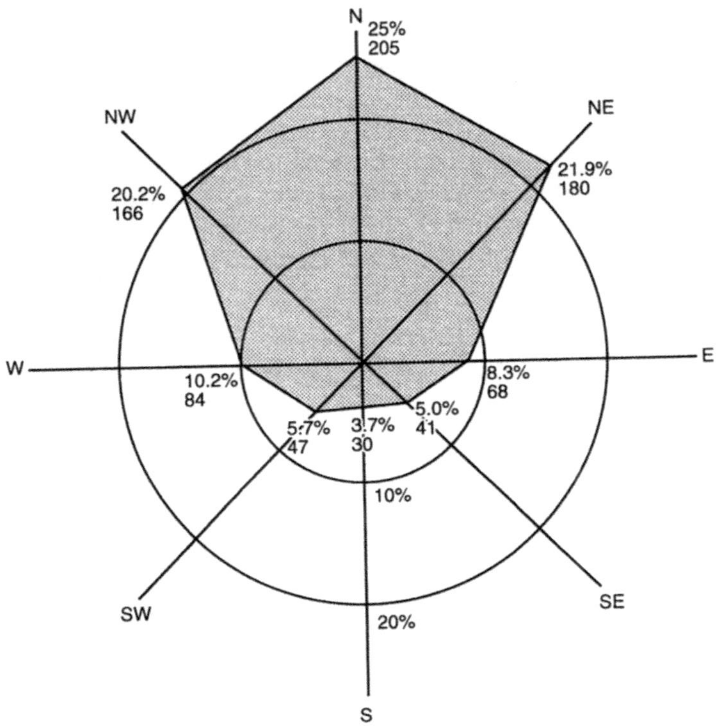

Fig. 5.1. Exposure of active rockglaciers in the Swiss Alps (partly unpubl. data from Barsch; cf. Dieling 1973; cf. also Fig. 5.2) as an example from midlatitude mountain systems of the northern hemisphere

senting a general picture of the regional rockglacier distribution, not a catalogue of every individual form. Unfortunately, in every mountain system large areas exist in which rockglaciers have not been studied. Thus, part of the general picture is influenced by the regional "state of the art" of rockglacier research.

5.2 Local Rockglacier Distribution

5.2.1 The Alps

Chaix (1919) was the first to publish on rockglaciers in the Alps. He studied the rockglaciers Val Sassa and Val dell'Acqua (Engadin, Graubünden). Rockglaciers are a typical and well-developed landform in this part of the central Alps. Two more papers by Chaix (1923, 1943) were published in the following years. Publications on rockglaciers increased after 1950 (Cailleux 1947;

Michaud and Cailleux 1950; Boesch 1951; Domaradzki 1951; Capello 1947; Jäckli 1957; Pillewizer 1957; Höllermann 1964; Gerhold 1967b; Barsch 1969a).

These papers already describe large parts of the Alps where rockglaciers are frequently found, normally in northern exposures (Figs. 5.1, 5.2). These are the central western Alps, especially the area around Wallis and Graubünden (Switzerland), the central eastern Alps in Austria, and the area around Mont Blanc (Brionçonais, Chambeyron) in France. Two maps (Figs. 5.3, 5.4) are presented as different examples of rockglacier distribution in high mountain systems. The first one (Fig. 5.3) displays the number of rockglaciers in the

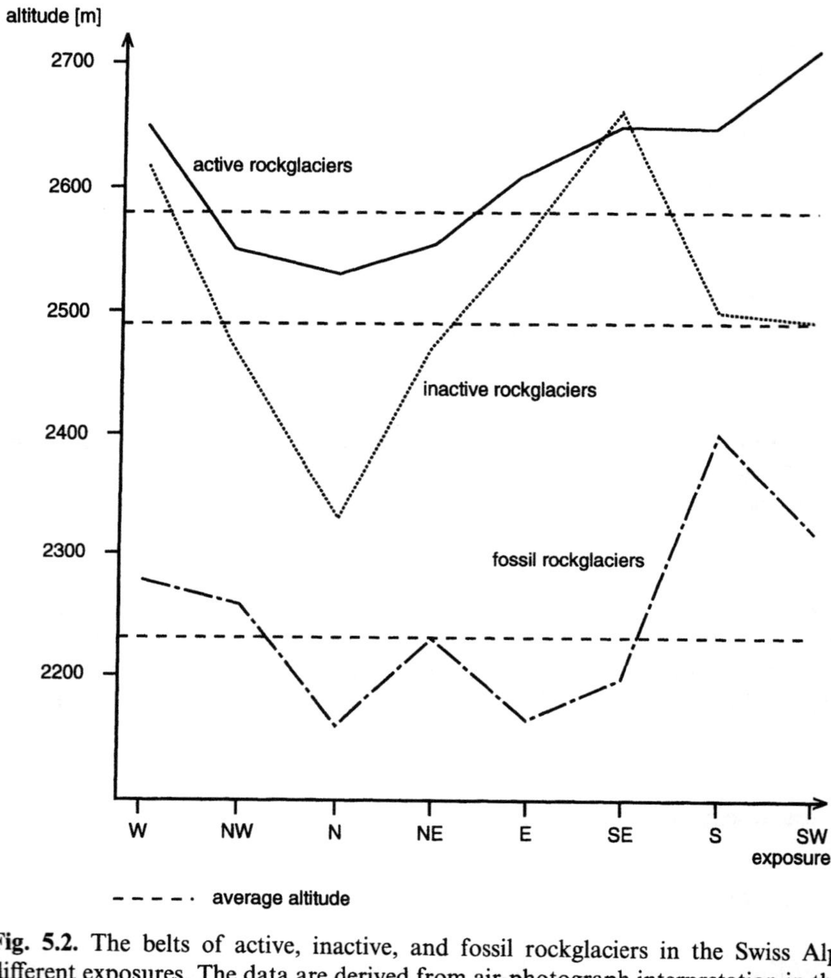

Fig. 5.2. The belts of active, inactive, and fossil rockglaciers in the Swiss Alps on different exposures. The data are derived from air-photograph interpretation in the late 1960s. Both inactive and active rockglaciers are thought to be Holocene features. Thus, inactive rockglaciers indicate the lower limit of the Holocene mountain permafrost fluctuations. Unfortunately, the number of examples in the inactive rockglacier class is small in relation to the other classes due to methodological problems

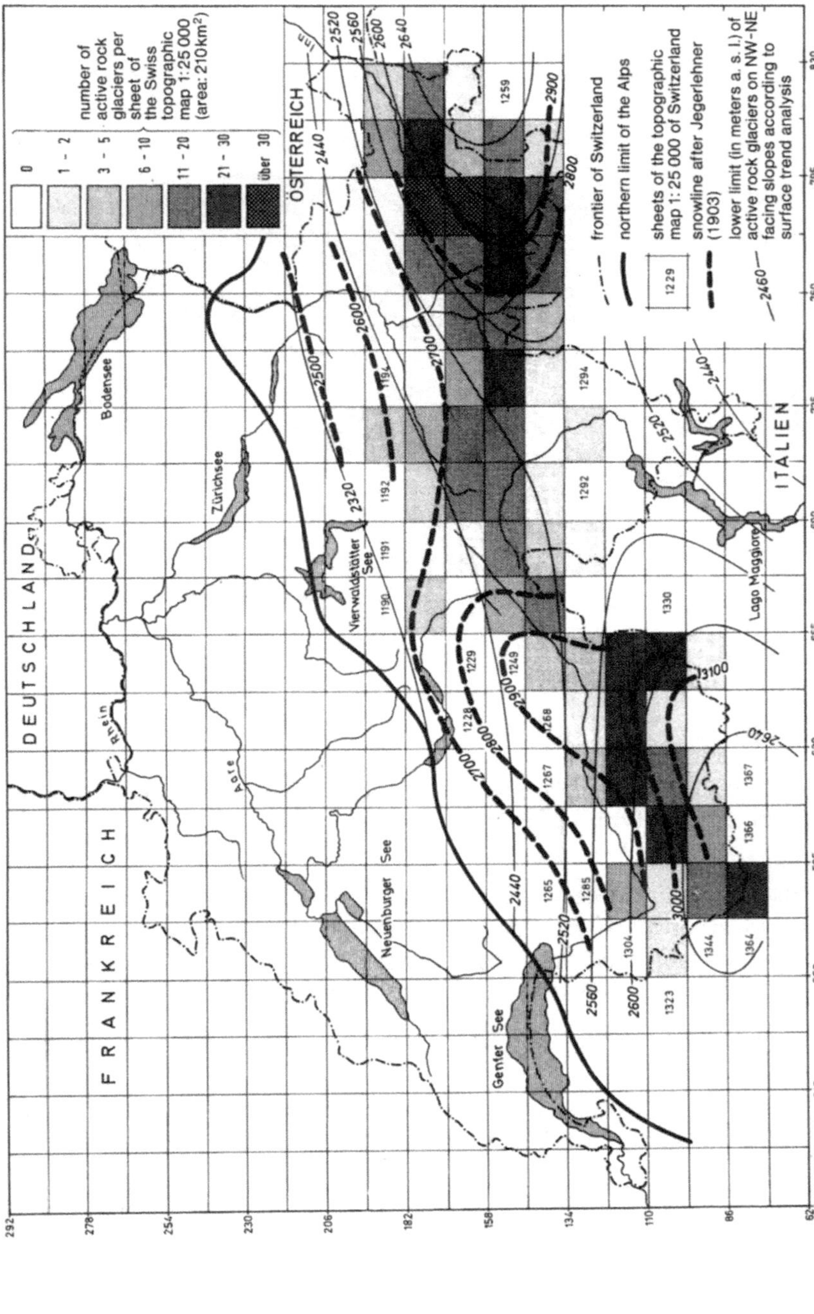

Fig. 5.3. Rockglacier distribution in the Swiss Alps (Barsch 1980). The number of active rockglaciers per sheet of the topographic map of Switzerland (Landerskarte der Schweiz) are presented. Rockglaciers are concentrated in the central parts of the Swiss Alps (Graubünden and Wallis). Their lower limit as presented is the result of a surface trend analysis in comparison to the ELA (equilibrium line altitude on glaciers representing snowline) of 1850/1856 following data given by Jegerlehner (1903)

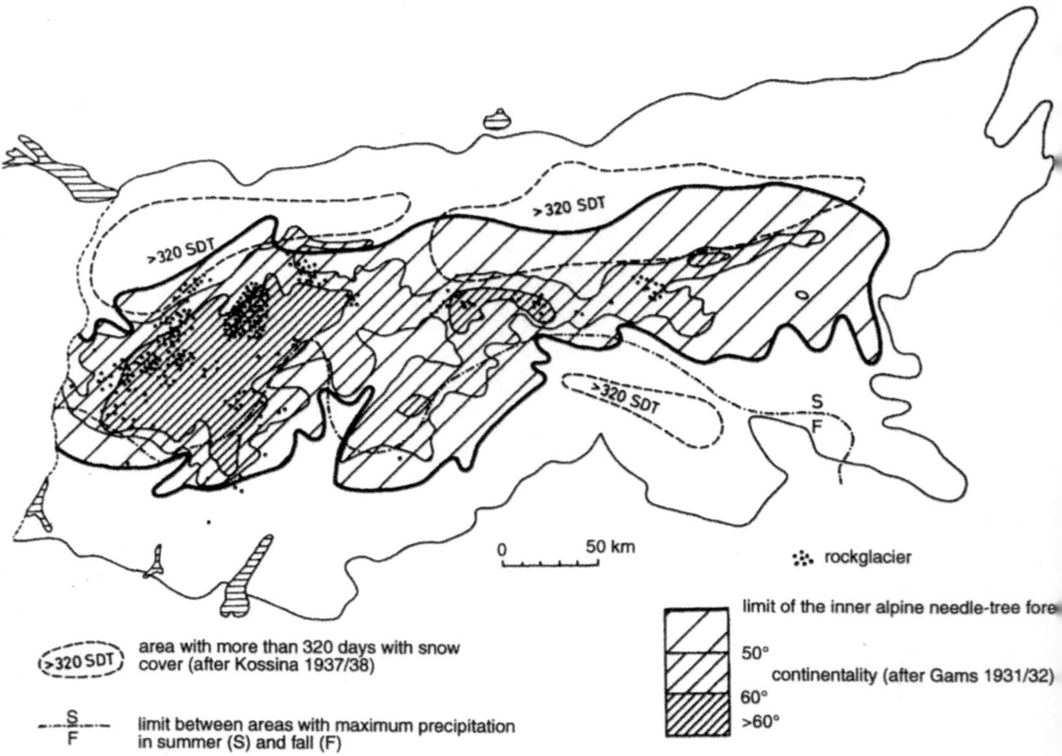

Fig. 5.4. Rockglacier distribution in the Austrian Alps in relation to climatic and biogeographic parameters. (Höllermann 1983a)

Swiss Alps (Barsch 1980); the second (Fig. 5.4) describes the distribution in part of the Austrian Alps according to Höllermann (1983a). All these maps are based on aerial-photographic interpretation followed by a ground check. Thus, it is possible that the quality of the data is not always the same. The reasons for this are:

1. The status (activity or inactivity) of a rockglacier is often difficult to detect from air photographs.
2. Apparently similar features may have been accepted as rockglaciers.
3. Rockglaciers in the shadows of high walls have frequently not been detected.

Despite these obvious problems, the two maps again display the fact that rockglaciers are concentrated in continental areas, which are comparatively dry and where the snowline is lifted in relation to neighboring areas, and glacierization is less pronounced due to reduced precipitation and high solar radiation. In the case of the Swiss Alps (Fig. 5.3), it can be seen that the lower limit of active rockglaciers increases towards the central Alps, but with a lower gradient than the snowline. The data are summarized in Table 5.1. The data for the altitude of the snowline are taken from Jegerlehner (1903), who calculated

Table 5.1. Altitudinal difference between the snowline on northerly exposed glaciers (equilibrium line altitude = ELA) around 1850/1856 according to Jegerlehner (1903) and the lower limit of active rockglaciers with northern (NW–N–NE) exposure along two N–S-profiles through the Swiss Alps according to Barsch (1980)

Y-coord.	Western profile					Eastern Profile				
	Snowline (m)	n	Lower limit of rockglacier (m)	n	Diff.	Snowline (m)	n	Lower limit of rockglacier (m)	n	Diff.
N 182–170						2740	1	2547	86	193
170–158						2836	2	2559	58	279
158–146						2853	2	2632	59	221
146–134	2712	3	2443	4	269	2902	5	2575	25	327
134–122	2675	1	2434	7	241	2734	4	2515	4	224
122–110	2987	3	2576	63	411					
110–98	3122	10	2557	31	565					
S 98–86	2898	9	2707	3	191					

The Y-coordinates are given according to the Swiss national grid; 182 corresponds to ca. 46°42′N. For further information, see text.

more or less the ELA (equilibrium line altitude) on glaciers on topographic maps. (1:50 000) established between 1870 and 1885. All values for glaciers with northern exposure (NW–N–NE) and areas of more than 100 ha (= 1 km^2) have been used. The data represent (*cum grano salis*) relatively conservative estimates for the ELA of 1850/1856, i.e., the situation during one of the important Holocene glacier maxima in the Alps. Even if the number of cases (regarding the ELA) is not sufficient, it shows that towards the central Alps the difference between the ELA and the lower limit of rockglaciers is increasing to maximum values of 565 m and 327 m (cf. Sect. 10.7).

In Fig. 5.5, rockglacier distribution has been summarized for the eastern Schober Mountains in Austria according to Lieb (1986). The map shows that nearly every cirque is occupied by a rockglacier, including inactive and relict ones (Table 5.2). In nearly the same area, Nagl (1976) discusses the distribution of rockglaciers in space and time. He differentiates between active, inactive, and fossil rockglaciers and develops an interesting diagram (1976, p. 117). In Fig. 5.6, the information is presented according to our present-day knowledge. The diagram displays the increasing altitude of rockglacier activity during late glacial times, i.e., during Older and Younger Dryas (Daun, Egesen, cf. Sect. 10.3.2) and the jump to Holocene rockglacier altitudes.

Relict (fossil) and inactive rockglaciers are described from many places at lower altitudes than the active ones (Barsch and King 1975; Kerschner 1976, 1985; Tatenhove and Dikau 1990). For instance, relict rockglaciers are found between 1500 and 1800 m (Würm) and between 2000 and 2300 m (Late Würm). The inactive rockglaciers between 2600 and 2800 m are assumed to be Holocene features like the active ones (cf. Chap. 9).

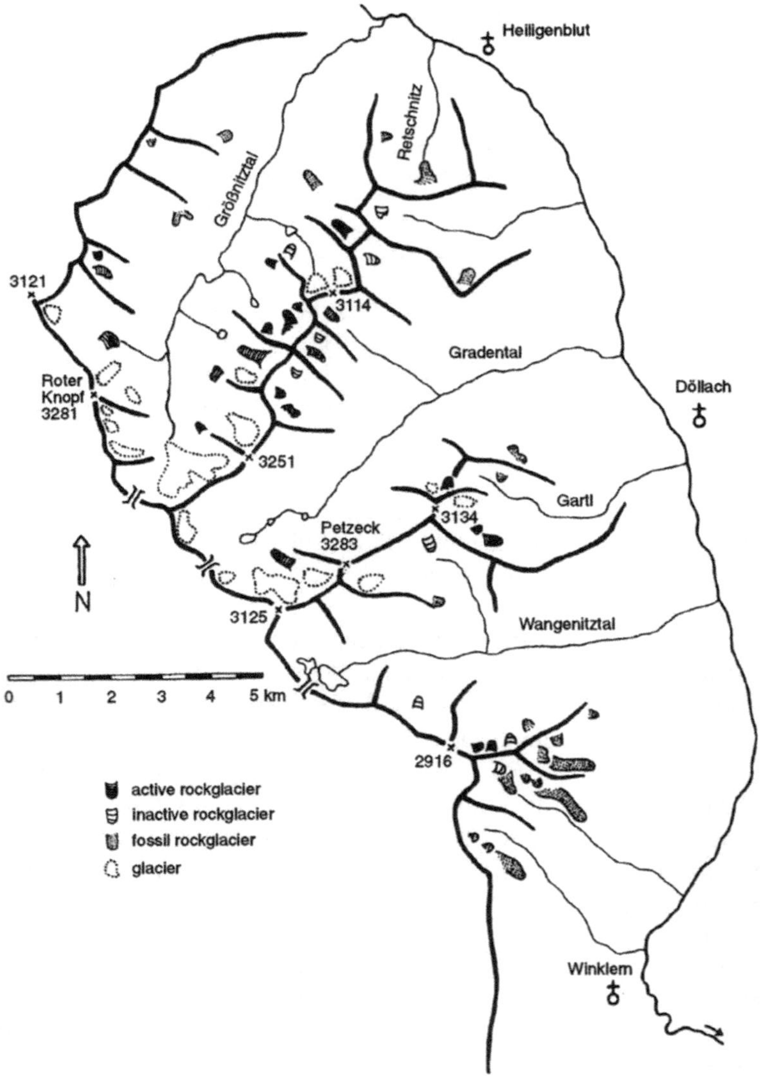

Fig. 5.5. Rockglacier distribution in the eastern Schober Mountains (Austrian Alps) according to Lieb (1986). Practically every cirque that is not occupied by a glacier bears a rockglacier. Active, inactive, and fossil rockglaciers are differentiated; they display different belts of rockglacier activity

Table 5.2. The lower limit of active and inactive rockglaciers in the Schober Mountains (eastern Tirol and Kärnten, Austria) from Buchenauer (1990, Table 42)

Exposure	Active rockglacier Ø (m a.s.l.)	n	Inactive rockglacier Ø (m a.s.l.)	n	Difference (m)
N	2470	7	2440	12	30
NW	2535	9	2456	7	79
W	2585	4	2483	9	102
SW	2733	7	2666	5	67
S	2730	4	2645	2	85
SE	2750	7	2597	7	146
E	2575	3	2495	2	80
NE	2578	5	2422	3	145
	Ø 2620	Σ 46	Ø 2526	Σ 47	Ø 92

The data for the eastern Schober Mountains are taken from Lieb (1987) by Buchenauer

5.2.2 The Mountains of Northern Europe

In northern Europe, rockglaciers have only recently been studied in detail, although the first description of a rockglacier was given by a Danish scientist more than 100 years ago (Steenstrup 1883; Chap. 2). The observation of a

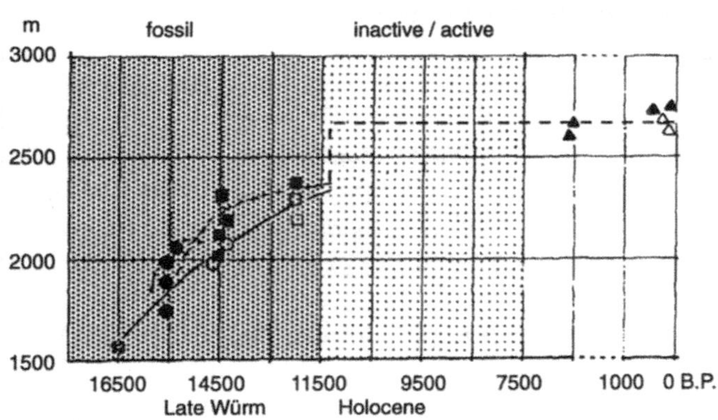

Fig. 5.6. Altitudes of fossil, inactive, and active rockglaciers in the Tauern (Austria) after Nagl (1976, p. 117). The diagram has been modified to include present knowledge. The beginning of the Holocene is accepted as 11 500 calender years B.P. Egesen represents Younger Dryas, Daun the Older Dryas. Inactive and active rockglaciers are both Holocene features. At the beginning of the Holocene an abrupt rise of the lower limit of discontinuous mountain permafrost is assumed to be due to the abrupt warming at the end of late glacial times. (Patzelt 1972, 1977a)

"peculiar phenomenon" from the Hornsund area in Spitsbergen (Svalbard) "resembling something between a talus and a moraine" (Werenskiold 1953, p. 309) did not stimulate any research, even though these "talus terraces" (Chandler 1973) are now recognized as talus rockglaciers (Swett et al. 1980; Birkenmajer 1982; Lindner and Marks 1985; Tolgensbakk and Sollid 1988; Svensson 1989).

In the mountains of the Scandinavian mainland, rockglacier-related research was retarded for a long period of time, because debris rockglaciers were considered as ice-cored moraines. These features have been studied very intensely by Östrem (1964a,b, 1965, 1966; cf. the discussion in Barsch 1971b, Östrem 1971; Sects. 8.2.2.3 or 8.3). However, they have not been assumed to be associated with either permafrost or debris rockglaciers. Therefore, the valuable data and information collected by Östrem need to be thoroughly discussed in a new context.

The paper by Barsch and Treter (1976) revealed that even talus rockglaciers exist in Scandinavia; they are widespread in Rondane (central Norway). It can be concluded that rockglaciers will be reported more frequently in the future (Sollid 1989; Svensson 1989) especially if active "protalus ramparts" are accepted as embryonal rockglaciers (Sect. 8.3) and are, thus, included in the discussion. Active alpine permafrost is today accepted as a regular phenomenon in the Scandinavian Mountains (King 1976, 1983). Better information on rockglaciers will help to establish a better knowledge of the geoecology of the high alpine belt in these mountains.

In some publications, the question of a glacial origin of rockglaciers (cf. Sect. 8.2) has been discussed in relation to Scandinavia too (Whalley 1976a; Griffey and Whalley 1979). In the case of the Stupbreen (glacier) in Troms, Norway, neither the description nor the illustrations clearly demonstrate that the author is dealing with a rockglacier at all. On the illustrations, only a mass of debris can be identified. A critical ground check seems to be necessary, especially since this feature is probably not active (Whalley 1976a, p. 54). The assumption that it contains ice has not been proved either.

On the Lyngen Peninsula, Norway, the features discussed are debris rockglaciers according to the description by the authors (Griffey and Whalley 1979). The questionable genetic model offered will be discussed in Section 8.2.2.

Summarizing, active rockglaciers occur at altitudes above 1450–1500 m in winter-cold continental mountains like Rondane (Barsch and Treter 1976). Relict forms are found in maritime parts such as the coast of the Lofoten (northern Norway) according to Peulvast (1981). They probably developed at the end of the last glacial period (Weichsel/Wisconsin). Rockglacier distribution in Norway is summarized by Sollid and Sörbel (1992). According to their map (cf Fig. 5.7), most of the estimated 150 known rockglaciers occur north of latitude 68°. Most of them are inactive or fossil (relict). It would be very helpful to have more information on the lower limit of active rockglaciers in Norway's maritime northwest. These data should be discussed in relation to the lower

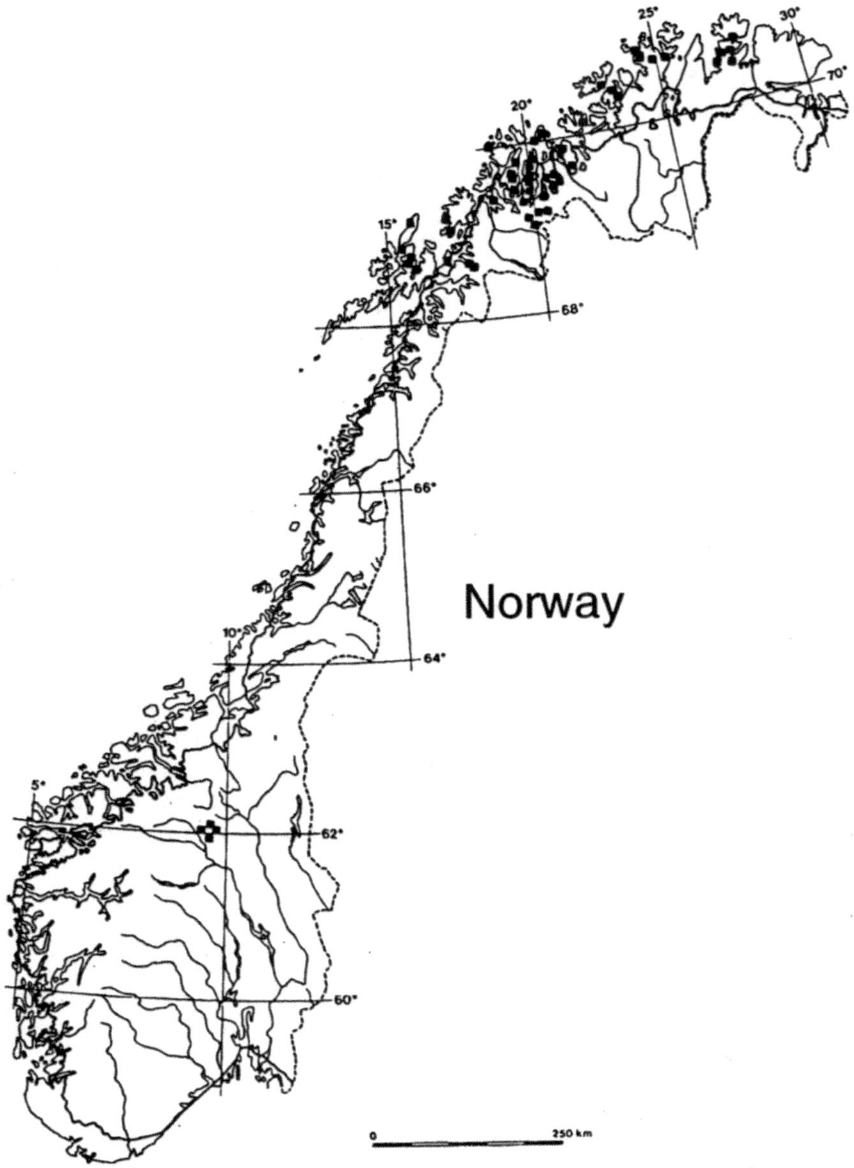

Fig. 5.7. Rockglacier distribution in Norway from Sollid and Sörbel (1992, p. 219). Rockglaciers outside the Younger Dryas end moraines are relict features. Every *dot* marks more than one rockglacier

limit of active rockglaciers in the Rondane mountains and the active debris rockglaciers mapped by Östrem as ice-cored moraines (Barsch 1971b; Östrem 1971).

Eyles (1978) reports rockglaciers from Esjufjöll (southeast Iceland). Svensson (1989) doubts this interpretation, because he assumes that Iceland lacks permafrost. Indeed, the description by Eyles is not completely clear. In particular, the question of inactivity is not discussed. Using the published photographs, the described forms seem to be rockglaciers, but are probably inactive talus rockglaciers. Facing northeast, their terminus is at about 875 m a.s.l. and ca. 200 m below the present regional snowline. Present MAAT is calculated to be + 1 °C using data of a station about 30 km to the southwest. Thus, it cannot be excluded that at least inactive rockglaciers exist in this area, and not only in the more continental areas of northern Iceland where rockglaciers seem to be frequent (Martin and Whalley 1987b). Therefore, it would be appropriate to check the mountains of Iceland for discontinuous alpine permafrost, which is according to Schunke (1975) widespread in central Iceland.

In the continuous permafrost environment of Spitzbergen and the Svalbard archipelago (Liestöl 1962), there are probably more than 500 rockglaciers (Kristiansen and Sollid 1986). They have often been discussed but partly under different names (Werenskiöld 1953; Liestöl 1962; Chandler 1973; Swett et al. 1980; Birkenmajer 1982; Lindner and Marks 1985; Kristiansen and Sollid 1985; Tolgensbakk and Sollid 1988; Svensson 1989; André 1992). For the most part, they are talus rockglaciers (Fig. 5.8). Their distribution is summarized by Sollid and Sörbel (1992). As Fig. 5.9 demonstrates, they are mainly found in the less glacierized, western parts of Spitzbergen. At the foot of the main cliffs, which rise above the strand flat, they exist near present-day sea level. According to Sollid and Sörbel (1992), active and inactive forms exist side by side. Being in a continuous permafrost environment, the inactive forms are thought to be "dynamically inactive" (cf Chap. 1). Inactivity is probably caused by the recent inadequate talus production from the rock walls above these rockglaciers. It has not yet been finally determined whether some of the forms in the extreme west of Spitzbergen (e.g., on Prins Karls Forland) really have a Weichselian age as Sollid and Sörbel (1992) propose.

5.2.3 The Pyrenees

According to Höllermann (1983a), rockglaciers in the Pyrenees are more widespread than could be expected from published reports (Barrere 1952; Viers 1961, 1962; Taillefer 1963, 1968a,b; Angely 1967; Soutadé 1973/74, 1981). Their number increases from west to east, i.e., from the more humid areas to the more continental and drier parts. Höllermann (1983a, p. 19) presents a distribution map for the central part of the eastern Pyrenees (Fig. 5.10).

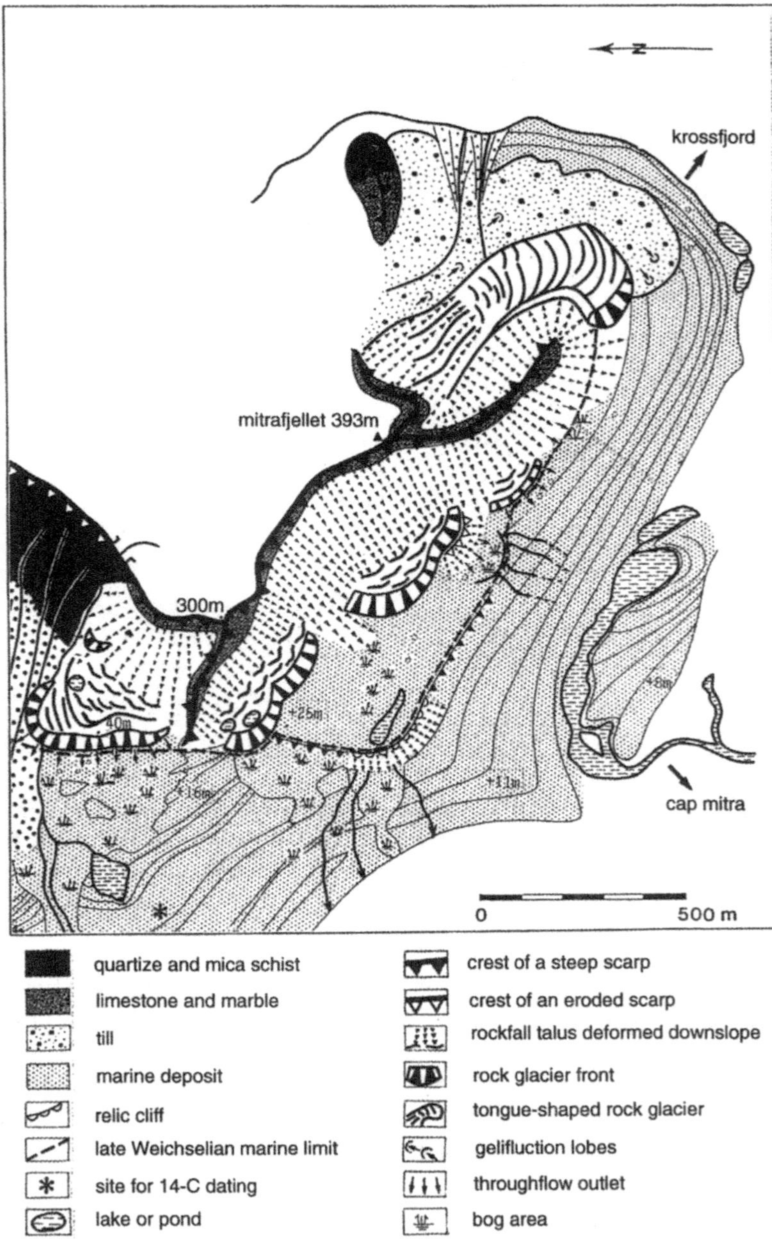

Fig. 5.8. Detailed geomorphic map of Mitrafjellet at Krosfjord in W Spitzbergen from André (1992, p. 52). The map displays well the development of rockglaciers below the talus slopes under the main cliffs. Only if there is enough space are tongue-shaped rockglaciers formed

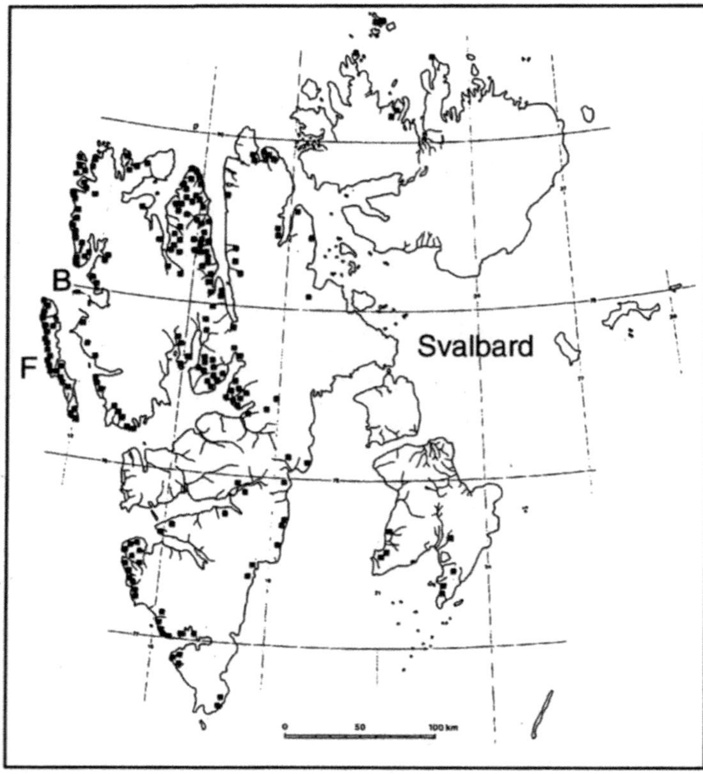

Fig. 5.9. Rockglacier distribution on the Svalbard archipelago from Sollid and Sörbel (1992). Most rockglaciers occur on Spitzbergen, particularly below the cliffs along the coasts. The Bröggerhalvöa (*B*) and Prins Karl Forland (*F*) are famous for their rockglaciers

Most of these rockglaciers are relict or inactive. Höllermann (1983a) doubts whether active ones still exist (Barrere 1952). In recent years, some interesting studies on these rockglaciers have been done by Spanish colleagues (Chueca 1992). In his study on the Spanish central Pyrenees, Chueca (1992) demonstrates that rockglacier distribution is not random, but mainly controlled by rock type. Of the relict rockglaciers here, 56% are found in areas with crystalline rocks (Sect. 6.1.1).

5.2.4 North American Mountains

Many rockglaciers have been reported from very different locations in the high mountain systems of North America including Mexico. Figure 5.11 displays a map published by Höllermann (1983a) that presents localities at which rock-

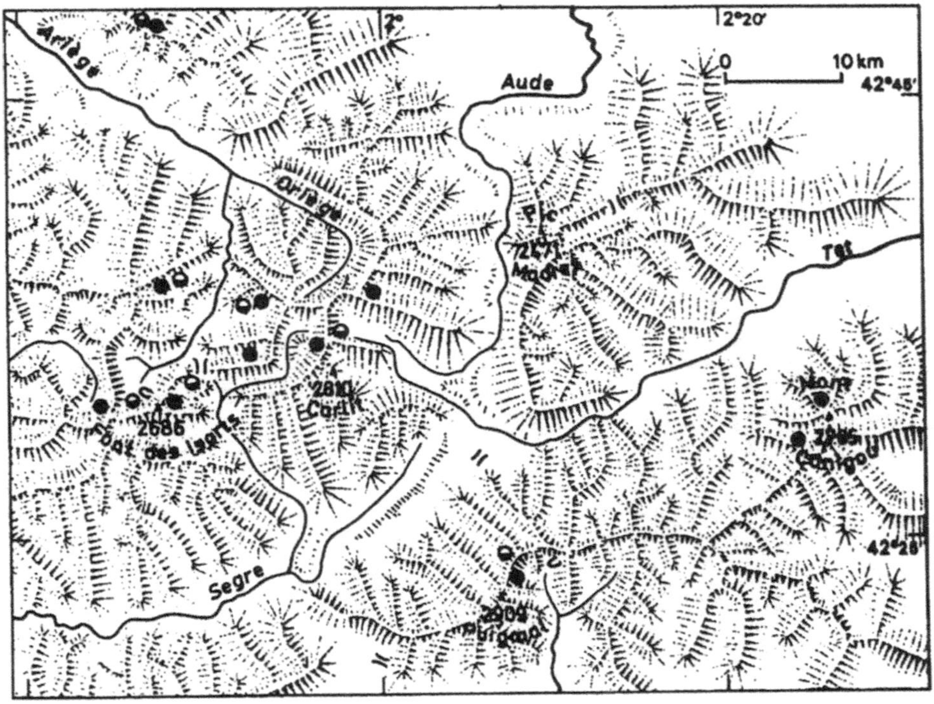

● tongue-shaped rockglacier ◒ lobate rockglacier ⲩ earthy (fine-grained) talus stream, ("Coulées boueuses")

Fig. 5.10. Rockglacier distribution of relict rockglaciers in the eastern Pyrennees. (Höllermann 1983a)

glaciers have been studied. This map cannot be considered complete because a systematic rockglacier inventory has not yet been done. It is more indicative of areas for specialized research than of the real rockglacier distribution. The same can be said of Greenland and the Queen Elizabeth Islands where probably many more rockglaciers will be found in the future. This is indicated by Humlum (1988b), who reports on 1700 rockglaciers (active, inactive, and relict) from Disco Island on the west coast of Greenland at ca. 70° N. He believes the rockglacier density to be very high with 20 per 100 km^2. However, this is of the same order of magnitude as the rockglacier density in the central parts of the Swiss Alps (Barsch 1977c) or other mountains. The distribution of rockglaciers on a small peninsula north of Disco fiord is presented in Fig. 5.12. According to Humlum (1988b), there are 167 rockglaciers; the majority (143 or 85%) are talus rockglaciers, the rest (24 or 15%) are debris rockglaciers. About one-third (52) of all the rockglaciers are relicts (fossils); the others are active or inactive but their present status is not always unambiguously detectable from air photographs. This example demonstrates possible rockglacier densities and the importance of rockglaciers in some periglacial environments.

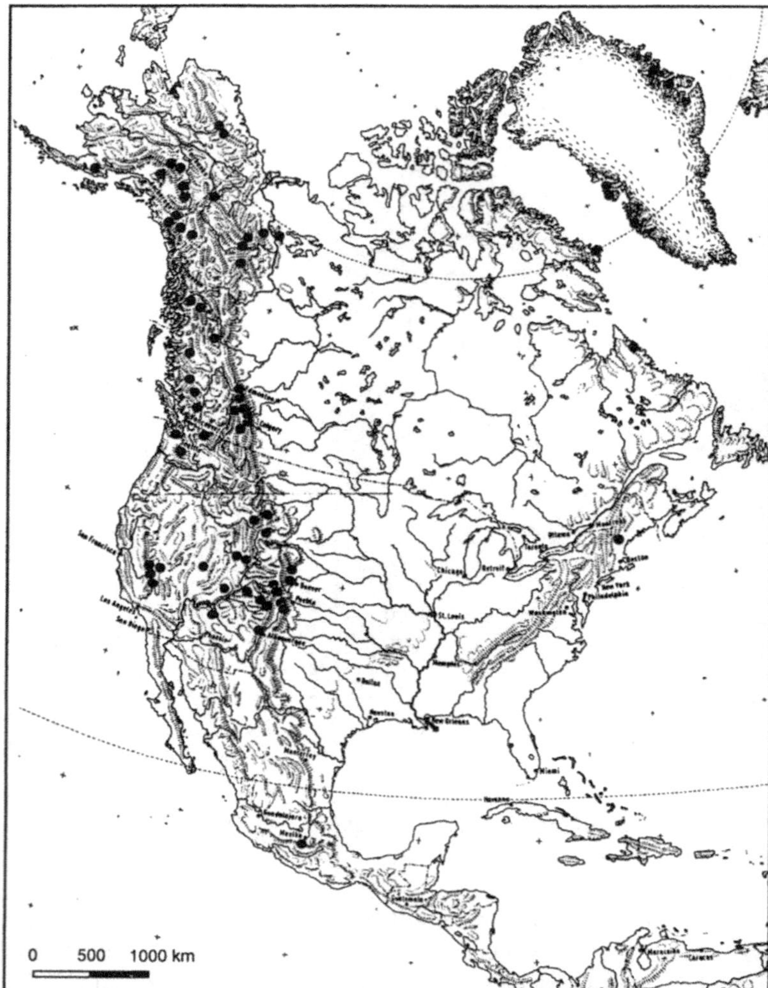

Fig. 5.11. Rockglacier distribution in North America from Höllermann (1983a). The rockglaciers in Greenland have been added mainly according to Humlum (1982). The map is not complete, it represents mainly the areas of special investigations. More rockglaciers will be reported in years to come

In North Alaska, Ellis and Parker (1979; cf. Ellis 1982) have mapped the central part of the Brooks range. The classical study by Wahrhafting and Cox (1959) was carried out further south, in the Alaska Range, where ca. 200 rockglaciers were detected on the Healy quadrangle (1:250 000) alone. Höllermann (1983a) tried to evaluate the data of both papers. He presents a dia-

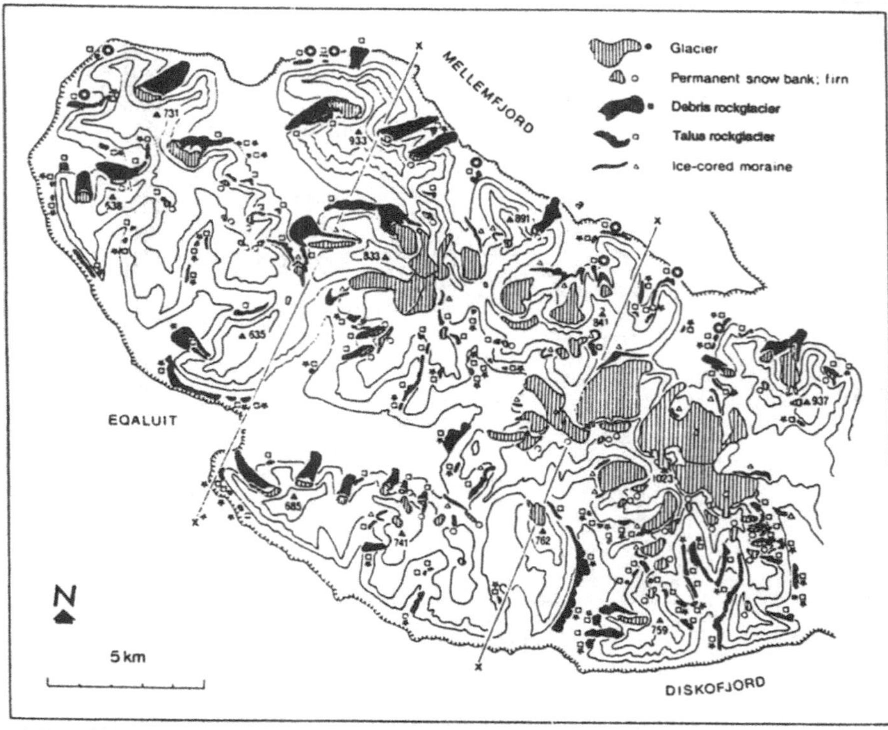

Fig. 5.12. Rockglacier distribution on a small peninsula of Disco Island (Greenland). (After Humlum 1988b). Because the genetic interpretation of the original paper cannot be accepted, the legend has been changed to the terminology used in this book; also, the so-called ice-cored moraines, if they are from the Östrem type (cf. Chap. 8), have to be accepted as small debris rockglaciers. The talus rockglaciers mark exactly the lower end of most talus slopes. The map displays possible rockglacier densities in special areas

gram of the vertical distribution of rockglaciers in this area (Fig. 5.13). The cross section of the Alaska Range is characterized by a sudden rise of all altitudinal limits, which is caused by the change in climate from a more maritime on the southern side to a continental one on the northern side. The fact that the lower limit of active rockglaciers does not change over about 5° latitude is very astonishing. Although the Alaska Range is located in the zone of discontinuous permafrost and the Brooks Range in the continuous permafrost zone, active rockglaciers are only found above the timberline. This may be due to climate, but it is more probable that it is caused by topographic controls (lack of rock walls with flat foot slopes in front of the lower ends of the talus slopes) than by a low rate of rock weathering (talus production) on the walls.

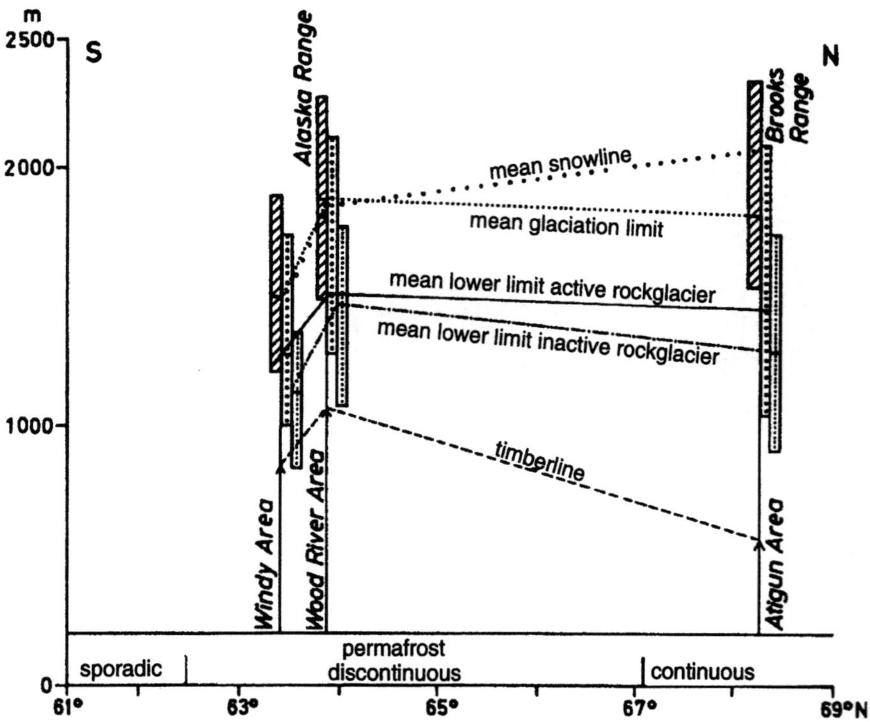

Fig. 5.13. Diagram of lower limits of inactive and active rockglaciers in Alaska. (Höllermann 1983a)

Calkin et al. (1987) report that 75% of all lobate and 90% ("almost all") of the tongue-shaped rockglaciers are located north of the central Brooks Range. This is probably due to the fact that precipitation mainly comes from the south. Thus, it causes more intense glacier development on the southern side where glaciers can occupy localities which would otherwise have been used for rockglacier development. This is in accordance with the fact that the mean altitude of active rockglaciers is almost the same south and north of the divide, although the lower limit of the tongue-shaped rockglaciers rises (from 1300 to 1450 m) as does the mean altitude of cirque glaciers.

All rockglaciers in the Brooks Range are found in the belt of continuous mountain permafrost where MAAT is –14 °C at 1500 m a.s.l. Thus, the low altitudinal differences between south- and north-facing active rockglaciers are explained (lobate ones, 150 m; tongue-shaped forms, 50 m). The same altitudinal differences are given for active/inactive (lobate or tongue-shaped) rockglaciers. The occurrence of rockglaciers seems to be mainly controlled by topography (existence of flat shelves below high cliffs without glaciers) and by talus production (in the case of talus rockglaciers). Inactivity in areas like this can only be dynamic inactivity. Unfortunately, the authors did not discuss this aspect. The authors give the following values: for tongue-shaped rockglaciers, the altitudinal range is 1300 to 1800 m a.s.l.; for lobate forms it is 900 and 2000 m a.s.l.

The Pacific northwest (35 to 60° N) is – due to the high precipitation, especially in winter – heavily glacierized. This is an area in which the periglacial belt is not too well developed at the windward side. Thus, the few possible rockglacier-like forms are only found at the lee side (Alley and Young 1978). This is in accordance with the occurence of "ice-cored moraines" (Fig. 5.14) mapped using air photographs by Östrem and Arnold (1970); the majority of these ice-cored moraines are interpreted as debris rockglaciers (Barsch 1971b; cf. also Sect. 8.3).

Further east, in the Cirque Lake are (63 N, 130 W; Selwyn and Hess Mountains, Yukon, Canada), Kershaw (1978) reports 41 rockglacier (including 34 lobate ones) in an area of about 120 km^2. That is about 3 per km^2. As usual, most of them on northern exposures: the average altitude of their lower ends is 1730 m. The four southerly exposed rockglaciers reach down to 1805 m. The small altitudinal difference of 75 m can probably be explained by the low number of south-facing rockglaciers as well as by the still relatively high latitude, where the differences between north and south facing slopes are not so marked as in midlatitudes. This seems to be partly true also for the relict rockglaciers which are found at Olympic Mountains on the northeast slope and also at Mt. Rainier on southwest facing slopes of lower elevation.

From a theoretical point of view, the Rocky Mountains are very favourable for the development of rockglaciers. They receive less precipitation than the mountains to the west, i.e., they are drier, and are thus less glaciered. They possess a rugged relief above the present timberline, which is favorable for rockglacier development. Probably only a small proportion of the rockglaciers in this area have been reported so far.

In the northern Rocky Moutains active rockglaciers are widespread (Broscoe and Clissold 1967; Johnson 1969, 1972; Gardner 1972, 1978; Gardner and Bajewski 1987; Dingwall 1973; Papertzian 1973; Osborn 1975; Luckman 1976; Ommanney 1976a,b; Luckman and Crocket 1978; also Table 5.3). In the Banff and Jaspers National Park, Canada, about 119 rockglaciers (Table 5.4) have been studied by Luckman and Crocket (1978). In the very continental mountains of Alberta, apparently active rockglaciers exist down to an altitude of 2100 m a.s.l. Some of them are even below the timberline.

Rockglaciers are also found in the Absoraka Mountains, in the Wasatch Range, and especially in the Colorado Front Range (Fig. 5.15). Together with the San Juan Mountains, the last-mentioned area is one of the classical places of rockglacier research in the United States of America.

In the Colorado Front Range, the rockglacier studies by S. White summarize a large amount of knowledge on these features. A still longer tradition in rockglacier research is found in the San Juan Mountains. The famous paper by Howe was published as long ago as 1909. Recent studies have been undertaken by P.G. White (1976, 1979). They reveal the large number of rockglaciers in that area (Fig. 5.16, Table 3.1).

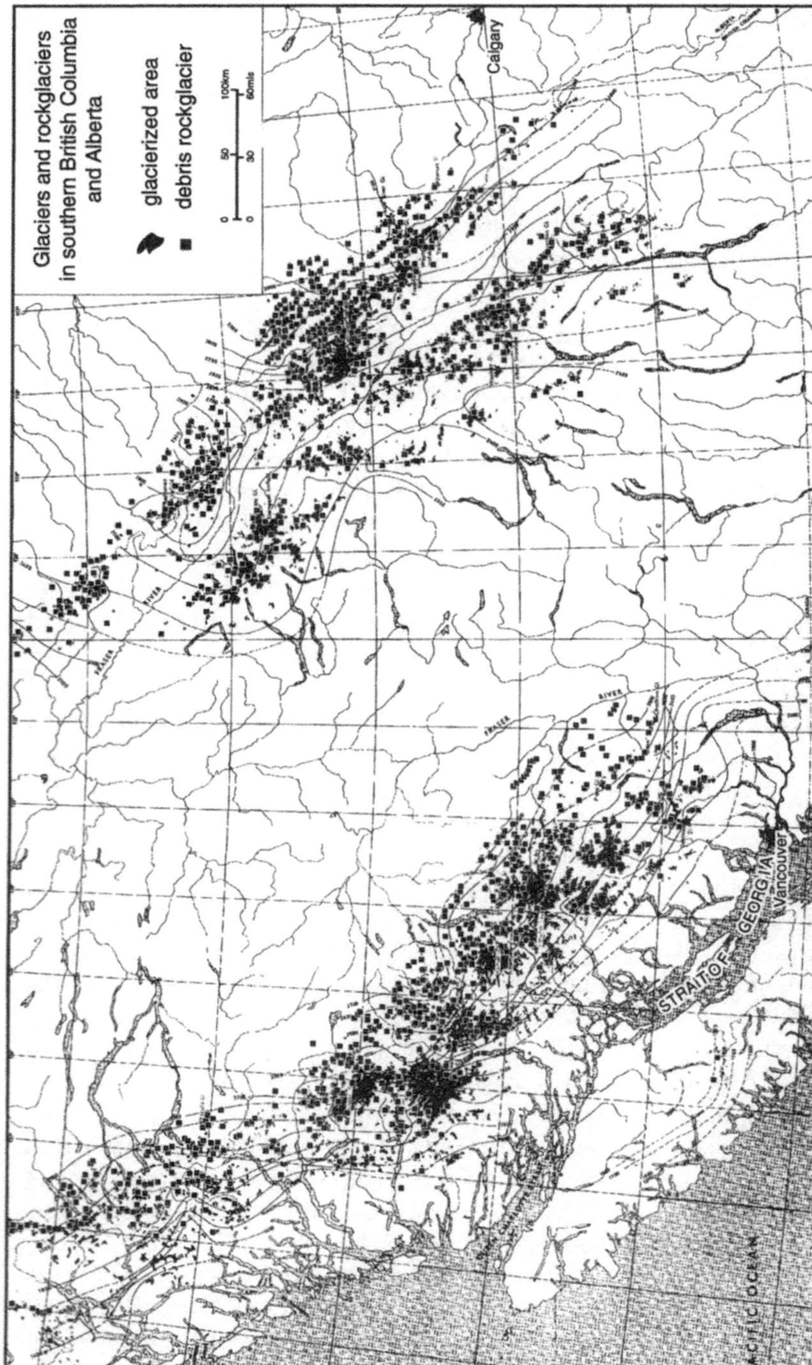

Fig. 5.14. The distribution of debris rockglaciers in British Columbia and Alberta after Östrem and Arnold (1970, map in annexe). The map displays the glacier distribution and the so-called ice-cored moraines which are interpreted as debris rockglaciers. (cf. Sect. 8.3; Barsch 1971b: östrem 1971)

Table 5.3. Some data on the lower limit of rockglaciers in North America

Locality	Latitude	Altitude (m a.s.l.)	Exposure	Source
Brooks Range	ca. 68°	ca. 990[a]	N	Calkin et al. (1987)
Alaska Range	ca. 63°	1070[b]	N	Wahrhaftig and Cox (1959)
Jasper Natl. Park	52° 50′	Ø 2125	–	Luckman and Crocket (1978)
Absoraka Mts.	44° 38′	2700	N	Potter (1972)
Colorado Front Range	ca. 40°	3330	NE	S.E. White (1971b)
San Juan Mts.	37° 30′	Ø 3697	–	P.G. White (1979)
Sierra Neovulcanico	ca. 19°	ca. 4600[c]	–	Heine (1994)

[a]Continuous permafrost zone.
[b]Discontinuous permafrost zone.
[c]Inactive debris and talus rockglaciers.

Table 5.4. Rockglaciers in Banff and Jasper National Parks (Canada). (Luckman and Crocket 1978)

Type	N	Mean elevation (m)			Length (m)		Width (m)		L/W (mean)
		Head	Mid	Toe	Max	Min	Max	Min	
Lobate	33	2245	2181	2120	850	100	1350	150	0.52
Tongue-shaped	76	2304	2219	2135	2100	300	850	100	1.97
Spatulate	10	2329	2201	2075	1300	500	750	250	2.04
All rockglaciers	119	2290	2207	2125	2100	100	1350	100	1.44

The mountains of California are also famous for their great rockglaciers. The state of knowledge on their distribution is summarized by Höllermann (1983a). Two examples of his maps are reproduced here. Figure 5.17 describes the rockglacier distribution in the Sierra Nevada and Fig. 5.18 displays a detailed geomorphological map of Mt. Whitney.

Only a few reports about rockglaciers are available from the high mountains of Mexico. Heine (1976) describes relict (fossil) and probably inactive rockglaciers from the Nevada de Toluca (4558 m, WSW from Mexico City). These inactive rockglaciers are found at ca. 4200 to 4400 m altitude, i.e., 600 to 800 m below the estimated present-day snowline at ca. 5000 m.

5.2.5 The Andes of Central and South America

In the mountains of the South American continent, information has long been available on mountain permafrost (Knoche 1932). Despite this the rockglacier discussion is fairly recent and connected with Arturo Corte. He worked on the

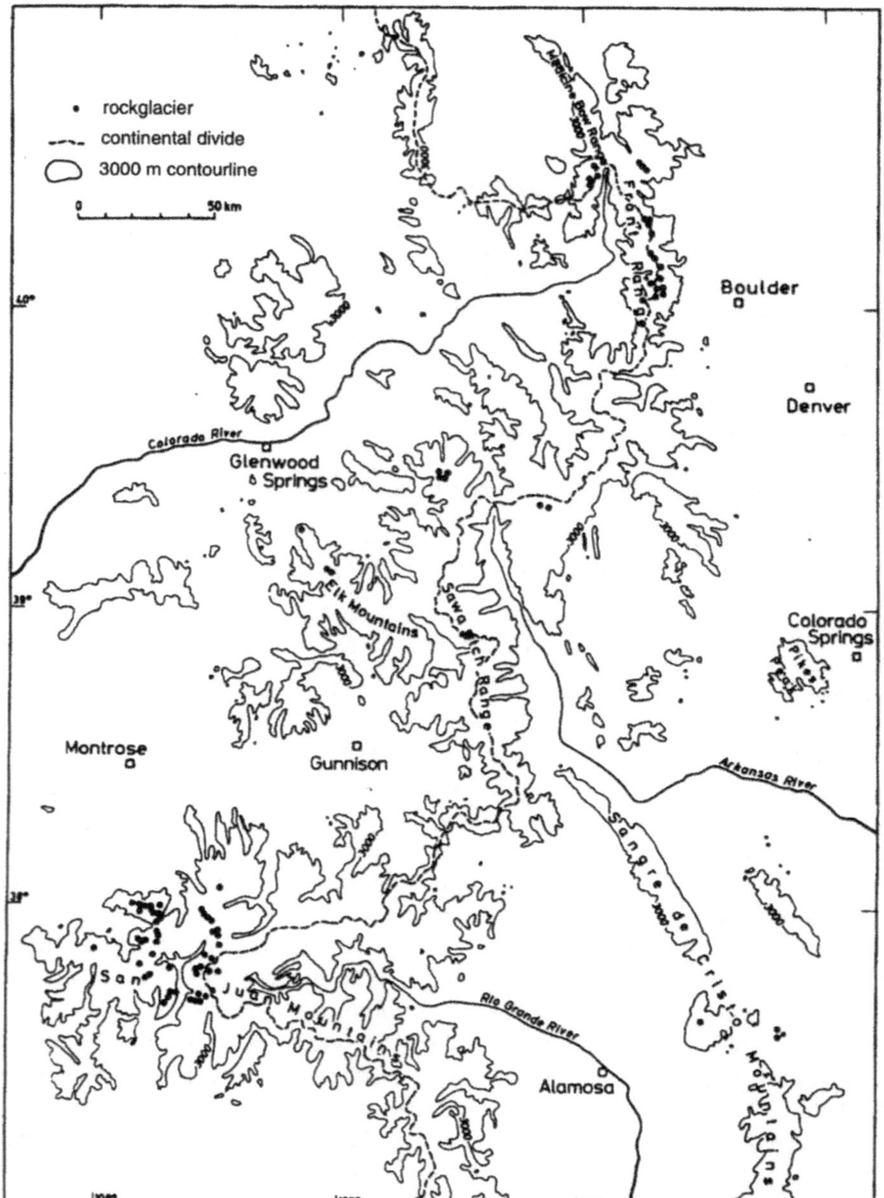

Fig. 5.15. Rockglacier distribution in Colorado from Höllermann (1983a). The continental, summertime-dry mountains of Colorado are rich in rockglaciers and in rockglacier permafrost. Due to mining activities, access to high alpine environments is good; much very valuable information has been collected in these areas

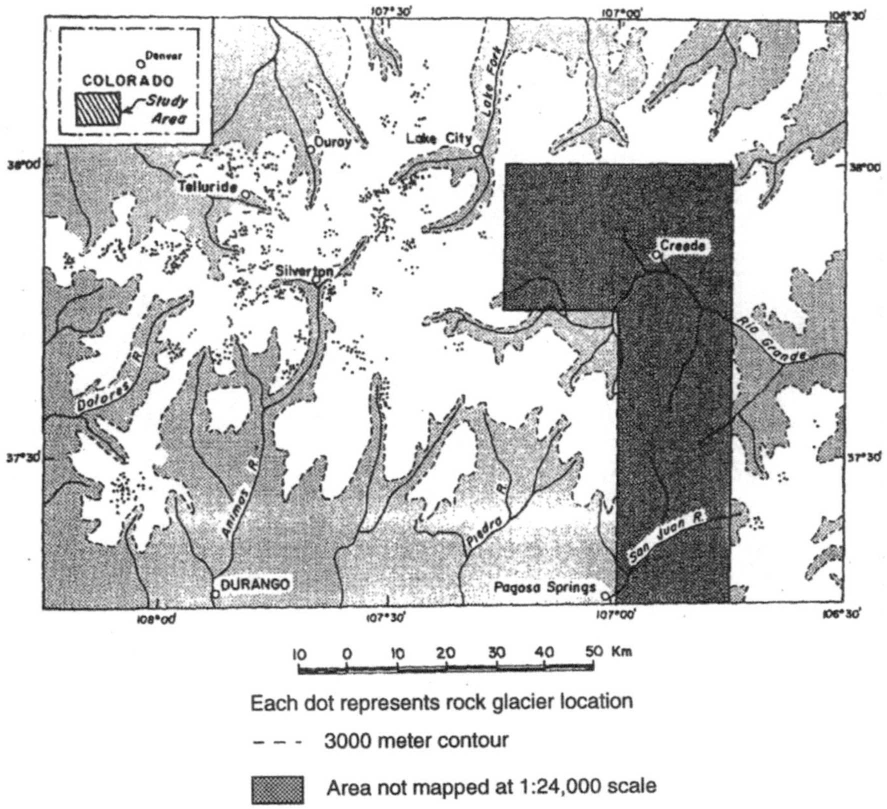

Fig. 5.16. Rockglacier distribution in the San Juan Mountains (Colorado) above 3000 m a.s.l. Each *dot* represents a rockglacier location. The map shows a very dense rockglacier population in this relatively continental high mountain system. (P.G. White 1979, p. 516)

extremely well developed rockglaciers in the Andes between Mendoza and Santiago de Chile (Corte 1976a; Marangunic 1976; Corte and Espizua 1981). In this semiarid to arid mountain environment, glaciers are not too well developed, but rockglaciers are numerous, well distributed and excitingly developed. A distribution map is presented as Fig. 5.19. The discussion in this area with a high-radiation climate is centered on the question as to whether the debris rockglaciers have been developed out of debris covered glaciers (Corte 1976, 1987b; Barsch 1983b; Barsch and King 1989; cf. Sect. 8.2).

In the northwest of Argentina, rockglaciers have been reported from the Sierra de Aguilar (Jujuy) at latitude 23° S by Rivelli (pers. comm.). It is very important to obtain more information on these supposedly active rockglaciers at 4650 m a.s.l.

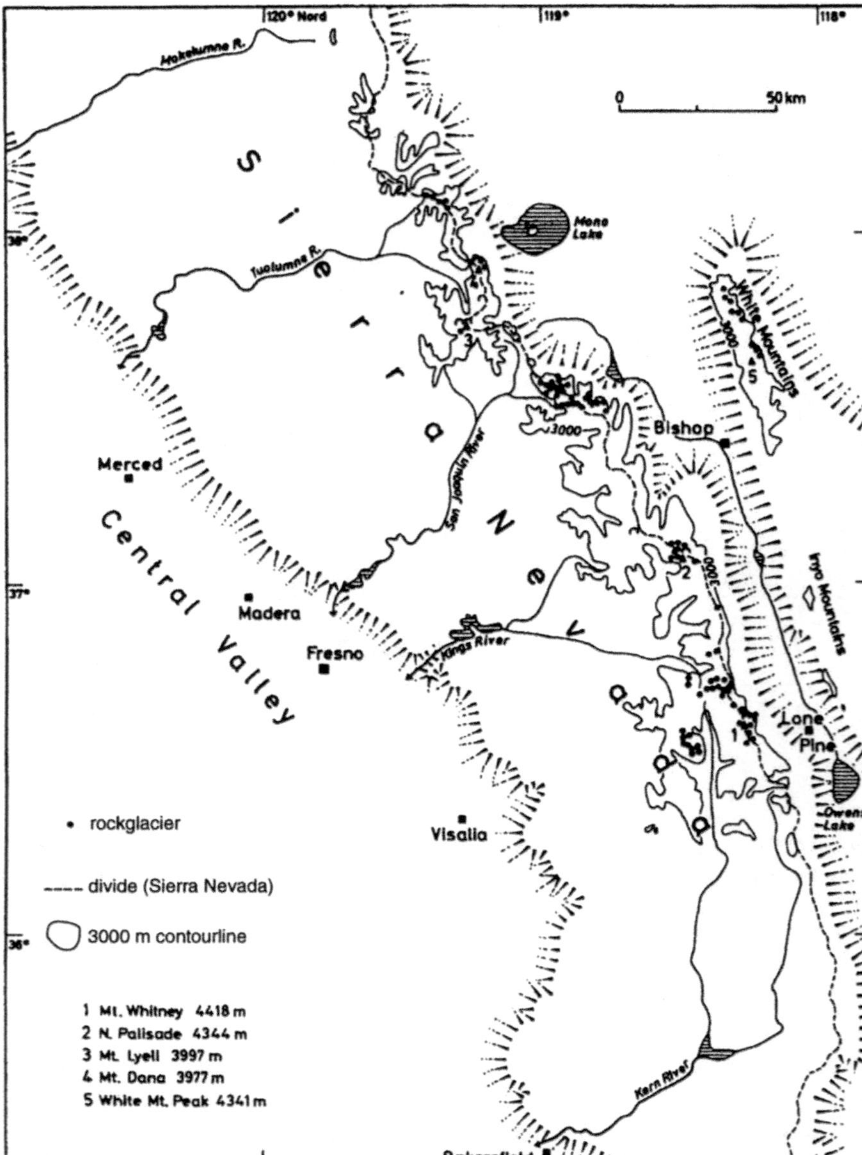

Fig. 5.17. Rockglacier distribution in the Sierra Nevada and the White Mountains of California (According to Höllermann 1983a)

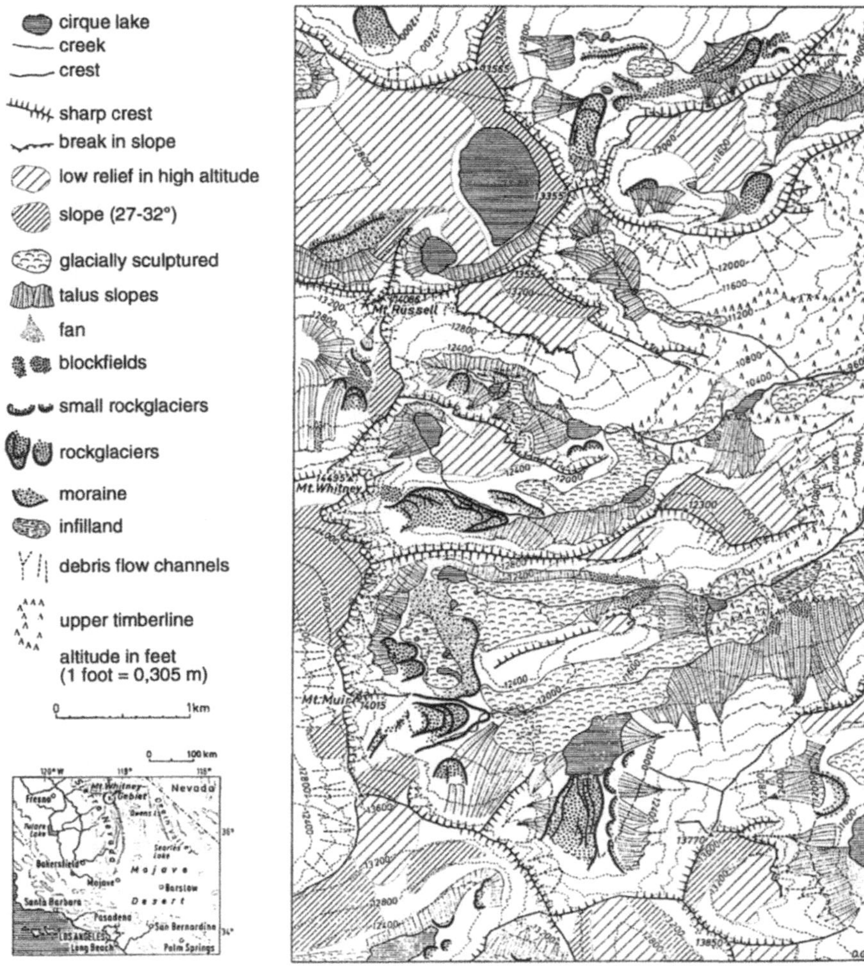

Fig. 5.18. Detailed map of rockglacier occurrence in the Mt. Whitney area, Sierra Nevada, California. (Höllerman 1983a)

5.2.6 The Mountains of Asia

This vast continent has a great number of high mountain systems; most of them contain suitable environments for rockglacier development. For large areas, the references are published in Russian. Outside the former Soviet Union rockglaciers have been reported from the Hindukush (Afghanistan) by Grötzbach (1965a), from the Kashmir Himalaya by Mayewski et al. (1981) or the Khumbu Himalaya (Nepal) by Higuchi (1978), Jakob (1992a,b), and

Fig. 5.19. Rockglacier distribution in part of the Andes of Mendoza. (According to Corte and Espizua 1981, maps in the annex)

Barsch and Jakob (1993). Information on the mountains of China has been given by Cui (1983) and by Cui and Zhu (1988).

The high mountains of central Asia, influenced by very continental climates with cold winters and pronounced aridity, ought to be favorable for rockglacier development. In the literature consulted, rockglaciers are described by Iveronova (1950) from the Tien Schan and especially the Altai Ranges (Krasnoslobodcev 1971; Tarakanov 1982; Gorbunov and Titov 1986), from the Pamir, the Altai, Transbaikalia (Chamar-Daban, Kodar) and eastern Sajan (Krasnoslobodcev 1971; Ivanovskij 1981) and from the Caucasus (Gorbunov and Titov 1986). Often, the possible existence of rockglaciers can be assumed. According to the discussion of geomorphic and geoecologic belts in the Changai Mountains (Mongolia) by Richter et al. (1961), it can be expected that rockglaciers occur above 2500 m a.s.l. Schröder (1992) describes the rockglaciers in the Tien Shan near Alma Ata. They occur down to 3000 m a.s.l.

Gorbunov (1983) summarizes the state of knowledge for central Asia. Over an area of 2000 km^2, ca. 10 000 rockglaciers are developed which transport ca. 20 to 30 km^3 of debris (total rockglacier volume is estimated to about 50 km^3). The rockglacier density is high (5 per km^2). Gorbunov proposes the calculation of a specific rockglacier density (SRD) which is defined as active rockglacier surface in m^2 (ARS) divided by basin area above the lower limit of rockglaciers in km^2 (ALR):

$$SRD = ARS/ALR\,(m^2 km^{-2})\ .$$

According to Gorbunov the average specific rockglacier density for the Zailijskiy Alatau is around 14 550 m^2/km^2; for the Swiss Alps ca. 3333, using the values given by Barsch (1978); for the Selwyn Mountains, Canada, 13 772, using values published by Kershaw (1978); and 1430 for the Khumbu Himalya, Nepal, following values published by Higuchi (1978). Regarding the Zailijskiy Alatau only, the differences in the specific rockglacier density are enormous. The reported maximum is found to be 41 610 m^2/km^2 in the Issy river basin, the minimum is 6500 m^2/km^2 in the Kargaul Dinka river basin. Even if the value of this index is demonstrated by the differences presented, it has practical limitations because the necessary data are not always available. It is probably better to relate the rockglacier surface to the total area of the mountain valley or the mountain system discussed.

From the Kunlun Shan, Cui (1983) describes active and inactive rockglaciers from the continuous mountain permafrost belt. They are well developed in northern to eastern exposures and terminate on average at 4680 m a.s.l. That is around 500 m lower than the present snowline (in northern exposure at 5100–5200 m a.s.l) and about 80–130 m above the assumed lower limit of continuous mountain permafrost.

Cui proposes that these slowly moving rockglaciers (estimated ca. 3 cm/a) be referred to as a special rockglacier type (the Kunlun Shan type). This does not seem to be correct. These rockglaciers are also due to the creep of mountain permafrost. Their movement is not very fast as a consequence of the low MAAT (ca. –4 °C), but steepness and size are similar to rockglaciers from other mountain ranges. The only difference is the gentle front slope angle (mainly 18–22°) which is very low for active forms. It may be that they are not climatic inactive but dynamic inactive (cf. Chap. 1).

Cheng (1983) publishes an empirical formula which describes a relationship between the lower limit of discontinuous mountain permafrost (H) and latitude (ψ):

$$H_{p\psi} = 3650 \cdot e^{-0.003 \cdot (\psi - 25.37)^2} + 1428\,[\mathrm{m}] \tag{5.1}$$

for latitudes between 20° and 60° N.

The equation gives a rough estimate of the lower boundery and has been mainly established for Asia. Apparently, it fits our still incomplete knowledge not too badly. It also roughly describes the lower limit of active rockglaciers, even if humidity, snowfall, continentality etc. are not considered, provided that the conditions of topography and material controls are fulfilled. The form of the equation is presented in Fig. 5.20.

Summarizing, the total number of reported rockglaciers in Asia seems to be comparatively small. Based on theoretical considerations, regarding climate and relief, it is obvious that the actual number of rockglaciers must be much higher. A systematic inventory would facilitate studies of high mountain belts and the comparison of different mountain systems. This would help towards a better geoecological assessment of these vast areas.

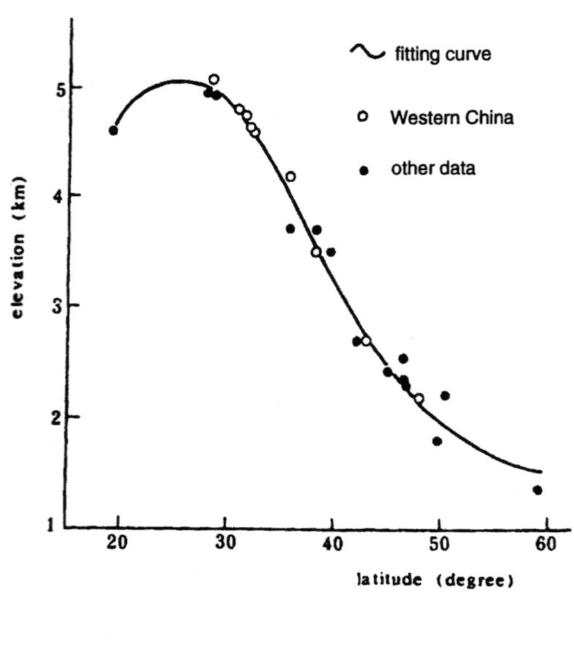

Fig. 5.20. Equation (5.1) describes the lower limit of active rockglaciers in mountains of the northern hemisphere between latitude 25 and 55. (Cheng 1983, p. 138). It presents still only a rough, but interesting estimate, because neither precipitation, relief, etc. are considered in the formula. In tropical mountains (roughly south of latitude 25), information on mountain permafrost and on rockglaciers is still extremely limited, therefore, the curve is not well defined in these areas

5.2.7 *The Southern Alps*

The first descriptions of rockglaciers in the Southern Alps of New Zealand were probably given by McGregor (1967). He mentions rockglaciers from the eastern side of the divide in the Ben Ohau Range (south of Mount Cook, 3766 m). Unfortunately, it seems that the author uses the terms rockglacier and ice-cored moraines as synonyms (cf. Sect. 8.2.4), by writing (p.745): "partly overridden by a large ice-cored moraine". The attached photograph displays a well-developed debris rockglacier. Further information on the existence of rockglaciers is given by Caine (1969) from the Two Thumbs range. Jeanneret (1975) supplies us with a systematic discussion of New Zealand's rockglaciers. He provides a map (p.85), which displays the possible range for rockglacier development on the southern island of New Zealand (Fig. 5.21).

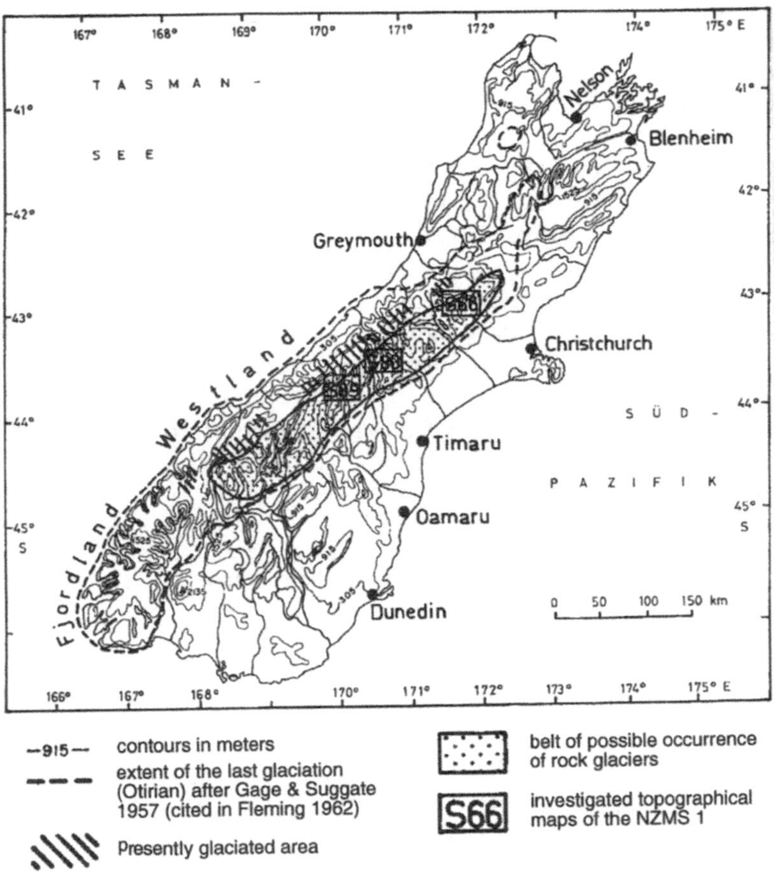

Fig. 5.21. The belt of possible rockglacier occurrence in the southern Alps (South Island, New Zealand) from Jeanneret (1975). Rockglaciers are especially developed on the relative dry eastern side of the main divide

Jeanneret studies three test regions using aerial photographs. These are shown on the following topographic maps (1:63 360) from south to north (cf. also Table 5.5):

S 89 Tasman: Burnett range (10 rockglaciers).
S 70 Macauley: Two Thumb range (8 rockglaciers).
S 66 Broken River: Craigieburn range (9 rockglaciers).

They are all located east of the divide.

All rockglaciers are found in cirques with southern aspects. Their mean length varies between 200 and 430 m. They are supposed to be active. It is surprising that their mean altitude decreases from the southern to the northern test area:

This may be explained by the small number of rockglaciers, and, therefore, it may be accidental. It may also be due to the fact that not all of the discussed rockglaciers are really active; in this case, a resurvey and a ground assessment is necessary. If this drop in altitude of the lower limit of active rockglaciers is verified, Jeanneret (1975) could be right by assuming that it is caused by topographic and local climatic effects. The topographic argument is based on the possibility that the appropriate forms (cirque floors etc.) are located only in these altitudes, and, therefore, the lower limit of rockglaciers is controlled by the overall topography and the geomorphic history of the Southern Alps.

Jeanneret believes the MAAT in his test area, Craigieburn, is about 5 °C higher than that in the permafrost belt of the Swiss Alps; i.e., the MAAT has to be assumed to +3 °C. In this case, these rockglaciers are probably either inactive or exist due to special climatic or topoclimatic conditions.

For a better geoecologic understanding of the Southern Alps, and for the distribution of alpine permafrost, more studies from these areas would be welcome. This is especially important because these often very maritime mountain climates and their periglacial/cryogenic environments in the southern hemisphere deserve a broader discussion.

Table 5.5. The lower limit of active rockglaciers in the southern Alps. (Jeanneret 1975)

Mountain range	Exposure	Altitude (m a.s.l.)	Length (av.) (m)
Burnett	S-SSE	2220	200
Two Thumbs	S-SSE	1940	430
Craigieburn	S-SSE	1700	270

The gradient Burnett Range to Two Thumbs Range is 80 m/60 km ~ 1.3 m/km.
The gradient Two Thumbs Range to Craigiebum Range is 240 m/100 km ~ 2.4 m/km.

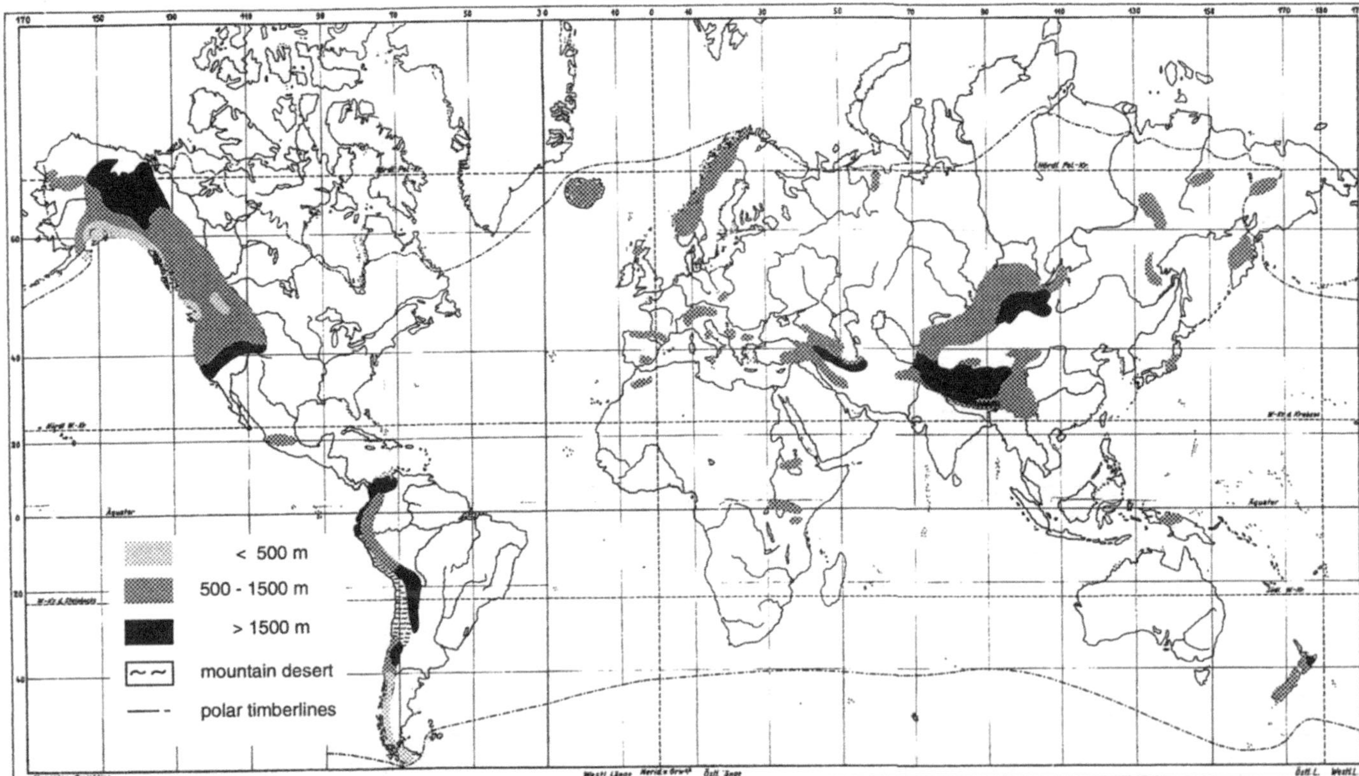

Fig. 5.22. The map displays the altitudinal difference between snowline and timberline (simplified from Hermes 1955). Rockglaciers are associated with mountain permafrost. They are most likely to occur in areas where the difference between snowline and timberline is greatest. The map indicates, therefore, possible places for rockglaciers

5.2.8 *Antarctic Mountains*

Only 2–5% of the land surface of Antarctica is not covered by glaciers. These small areas – mostly situated at the rim of the continent – are for the most part governed by a rigorous periglacial climate. They are places where rockglaciers are possible.

Barsch et al. (1985) report active talus rockglaciers from Punta Hennequin (Admiralty Bay, King George Island, South Shetland Islands). The rockglaciers are situated in the maritime part of the Antartic Peninsula, in areas with shallow, but continuous permafrost. They are moving with an average speed of 30 cm/a into the sea. Geoelectric measurements reveal a typical rockglacier structure (cf Chap. 6). The rockglaciers are crossing a raised beach, which was formed ca. 500–1000 years B.P. (Barsch and Mäusbacher 1986).

Mayewski (1979) and Mayewski and Hassinger (1980) report rockglaciers from the very continental southern Victoria Land.

Mjagkov (1980) studied rockglaciers in the Transantarctic Mountains. According to him rockglacier density is around 100–150 m^2/km^2.

5.2.9 *Conclusions*

Our knowledge of the regional distribution in the different mountain systems on Earth is still very incomplete. A possible summary can only try to display general rules. A world map of the possible distribution of active rockglaciers is also a distribution of the cryogenic belts on Earth. In a very rough approach, information about timberlines and snowlines (as indicators for high alpine climates) has been used to develop this general map (Fig. 5.22). It is based on the fact that in mountains with a continental climate the differences between timberline and firn line (as equilibrium line on glaciers) are greatest. Areas with a difference of more than 1500 m are more favorable for rockglaciers than those with a difference of less than 500 m. Thus, the map presents a rough estimate of where rockglaciers may be expected and where they are less likely on a world scale.

6 Rockglacier Material, Surficial Fabric and Internal Structure

Rockglaciers are two-layered phenomena according to texture and to temperature. The difference in texture is visible even at the front slopes where, under the bouldery upper layer, fine-grained material appears. Based on textures, it seems convenient to differentiate between the outer, bouldery *rockglacier mantle* and the inner *rockglacier core*. The distinction according to temperature is related to the active layer which is unfrozen in summer and the permanently frozen inner part. The two distinctions are normally not congruent, but most parts of the rockglacier core are frozen and contain ice. If only the "normal" pore volume of unconsolidated material in an unfrozen state of ca. 30–40 vol% is filled with ice, this part of the core is *ice-saturated.* If the whole core or parts contain more ice by volume than this "normal" pore space, then these parts are *ice-oversaturated* or *ice-supersaturated.*

Discussion of the rockglacier mantle includes information about the surface fabric; it could also includes surface relief. On the other hand, surface relief seems to be strongly related to the movement of rockglaciers, therefore it will be discussed in the next chapter on rockglacier rheology (Chap. 7).

6.1 Rock Type and Grain Size at and below the Surface

6.1.1 Rock Type

Rockglaciers can develop in almost any rock type which breaks down to particles of at least a few cubic centimeters (cm^3) and if other conditions (see Chap. 10) are favorable for rockglacier development. In the central Swiss Alps, they are well developed in all kinds of crystalline rock. According to Evin (1987b), most rockglaciers in the southeastern Alps (France) are built up of limestone and sandstone fragments. In the Andes of Mendoza, talus and debris of volcanic rocks make up the local rockglaciers. This shows that it is not possible to define a most-favored rock type for rockglacier development. If conditions become unfavorable for mountain permafrost and for rockglacier development, then rock types which break down to form coarse boulders are more suitable than those which break down only into fine grain sizes; the air-filled voids have an insulating effect (Balch ventilation, cf Sect. 10.1) and hamper the warming of the near-surface material. The nonrandom distribution of relict rockglaciers in the central Pyrenees is statistically explained in 56% of cases by the occurrence of crystalline rocks (Chueca 1992). In the context discussed here, these rockglaciers were probably close to the lower limit of discontinuous mountain permafrost and, thus, at an unfavorable altitude for the development of rockglaciers. Therefore, rock types presenting the right boulder size in sufficient quantities have been a necessary requirement.

6.1.2 Grain Sizes at the Surface

The most commonly observed grain sizes on a rockglacier are boulders at its surface. According to the statements above, all sizes occur. There are rockglaciers with house-sized boulders at the surface, which are difficult to cross. In cases like this (e.g., rockglacier Gigli, west of the Sustenpass, northern Swiss Alps), the voids between the boulders form real caves. But the opposite also exists as shown by some parts of the El Salto rockglacier (Argentinian Andes west of Mendoza) where only pebble-sized fragments occur at the surface. The "normal" rockglacier surface is characterized by boulders with an average length of between 0.6 and 1 m.

6.1.3 Development of the Bouldery Mantle

An explanation for the development of the bouldery top layer is not easy. On the talus slope or on the morainic ridge above the rockglacier all grain sizes can be found at the surface. At places where the rockglacier movement starts, a pure bouldery layer is developed, i.e., from here on downslope, only boulders form the surface (Barsch 1969a).

In general, two groups of processes seem to be responsible for the appearance of the coarse rockglacier surfaces:

- The first one is related to the boulder concentration and the separation of boulders and fines (material < 2 mm).
- The second causes the formation of the surface relief (cf. Chap. 7).

The separation of boulders and fines is partly due to downwashing of the finer grain sizes. However, in addition, an active process, the frost heave of boulders which has often been described in cold environments (Washburn 1979), is a principal factor in the separation of the bouldery layer. The downwashing of the fines seems to be only a secondary effect. Both processes together are responsible for the development of similar phenomena such as the two-layered block fields *(schwebende Blockmeere)* described from northern Scandinavia by Garleff (1970) or Meier (1987) which correspond to the main type 2 rock streams (Tyurin 1983): the "coarse-fragmented heaving material type". Frost heave also plays an important part in the formation of alpine boulder pavements (Pissart and Francou 1992).

Intense frost heave affects first the boulders in the active layer. Thus, changes in the active layer may lead to an increased supply of boulders. In most cases, it seems probable that the development of the bouldery mantle starts with the development of mountain permafrost. This is surely correct with talus rockglaciers, in which a first sorting is already given by the deposition of material on the talus slopes. In debris rockglaciers, this sorting is more difficult to understand, even if frost heave seems to be also the important factor. On the other hand, sorting shows up immediately at the starting point of the creep, despite the fact that those parts of the moraine that are not creeping are also frozen, but perhaps not ice-supersaturated. Barsch (1969a) observed this on small debris rockglaciers in Macun: the lateral moraine grading into the debris rockglacier displays all grain sizes at its surface. The lower part of the lateral moraine, which is incorporated into the debris rockglacier and its movement, shows the typical bouldery layer. The belt of transition between both parts is less than 5 m wide. Thus, it may be possible that in addition to frost heave the formation of the bouldery mantle is partly due to the creep and probably connected with the development of the very ice-rich core of the rockglacier.

Humlum (1988a) describes "cairns", i.e., boulders that are lifted above the "normal" rockglacier surface. The "rockglacier cairns" seem to be pushed by larger boulders from underneath. Because the bouldery mantle becomes thicker (as does the active layer; Sect. 6.4) towards the rockglacier terminus, these cairns are more frequent in the upper part of a rockglacier. Humlum believes that in the upper part of a rockglacier they are an indicator for a massive ice core. If we follow the definition in Chapter 1 then active rockglaciers are the visible expression of the creep of mountain permafrost. The occurrence of cairns is an indicator of a special activity in the active layer of the rockglacier. They seem to indicate that new boulders are freezing up to the surface. Three possible explanations exist:

1. The thickness of the bouldery mantle of the upper part of the rockglacier (or wherever these cairns occur) is not yet in equilibrium with the present climate.
2. The active layer (at these places) has been increased and new boulders are susceptible to frost heave, i.e., a new equilibrium has to develop.
3. As a consequence of the flow (e.g., compressive flow) boulders are pushed into the active layer of the rockglacier permafrost and are then susceptible to frost heave.

All explanations are related to frost heave in the active layer of mountain permafrost. The first and second are based on the information that the active layer of a rockglacier is passively riding on the rockglacier (cf Chap. 7). An influence of the movement of the rockglacier on the development of the cairns is, therefore, in these two cases not a necessary condition; however, it cannot be excluded, even if the cairns are by no means an indicator of a pushing ice core beneath the bouldery layer (cf. Chaps. 8, 10).

6.1.4 Surface Fabric

The boulders in the mantle of a rockglacier are to a great part not of the cubic form, naturally depending on the rock type. If they are elongated, their long axis displays a direction (trend) and a plunge. This can lead to a special fabric in which trend and plunge show a nonrandom distribution. Fabric studies have been done in a number of sedimentary environments. (cf. Caine 1968a,b for blockfields or Nicholas 1994 for additional references). Unfortunately, the number of fabric studies on true rockglaciers is very limited. This may be due to the fact that the formation of the bouldery mantle is not or only partly related to the movement (Sect. 7.7).

Evin and Assier (1983b) measured the orientations of oblong boulders on a number of active and fossil rockglaciers in the southern Alps (France). The measurements were done along different transects across the rockglacier. The long axes are classified into five groups. It seems that rockglaciers form a group separate from other bouldery deposits (moraines, etc.). It is difficult to evaluate the value of these measurements, because the dominant value seems to be influenced very much by position and local topography, the total field is not very distinct.

A methodologically interesting fabric study has been carried out by Nicholas (1994) on 13 more or less active rockglaciers in the La Sal mountains (Utah). In total at 24 sites (1 m^2 each) on longitudinal ridges, the trends and plunges of the long axes of the boulders at and immediately below the surface have been measured. Unfortunately, there is not a convincing, statistical preference. In 8 of the 24 sites, a random distribution is found. The weak fabric displays clusters as well as girdles. Besides a marked tendency for the boulders to lie parallel to the local slope, the fabrics resemble those found on lahars, blockfields and taluses. Thus, dry creep, passive transport and rockfall seem to be responsible for the orientation of the boulders. This confirms the hypothesis that the bouldery mantle on an active rockglacier is influenced more by secondary processes than by the interior creep of this mountain permafrost body.

Jakob (1992a) measured trends and plunges of the long axes of about 100 oblong boulders (length:width = 1:2, length > 30 cm) on 20 test plots on the rockglacier Nuptse (Khumbu, Nepal). He demonstrates again that trend and plunge are highly correlated with the local slope, i.e., with the microtopography of the present rockglacier surface.

6.1.5 Grain Sizes below the Bouldery Mantle

The boulders at the surface of a rockglacier form only a 2 to 5 m thick mantle. Beneath the bouldery layer finer-grained material is found. Sand and silt prevails, and coarser-grained fragments are "swimming" in the finer particles. This can be observed in boreholes (Barsch 1977b; Barsch et al. 1979; Haeberli et al. 1988; cf. also Figs. 6.1, 6.5, 6.6). This feature – as already stated – is visible at most front and side slopes: below the bouldery layer fine-grained material occurs.

Grain-size analyses of rockglacier core material collected either from boreholes or from excavations at the front and side slopes normally display maxima in the sand and silt fractions. Gravels and clay form – if they occur at all – a small part of the sample. As an example the grain sizes for some samples from the rockglacier Gruben are given in Table 6.1 and Fig. 6.1, according to

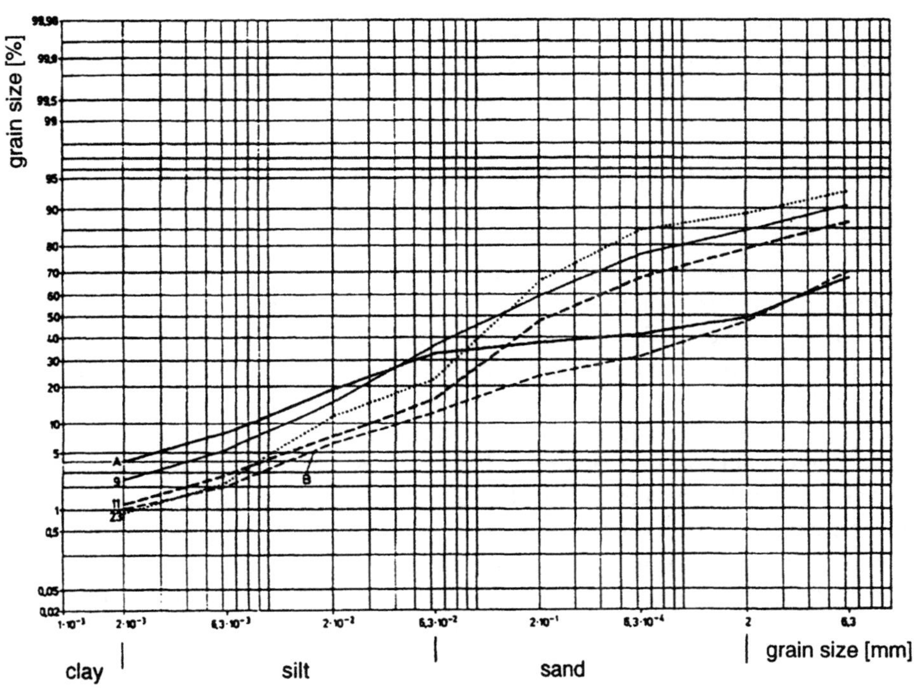

Fig. 6.1. Grain-size distribution of some samples (*9, 11, and 23*) from the core (drilling, June 1976) of the Gruben (Wallis) according to Barsch et al. (1979, p. 219). The three core samples contain mainly sand (2.0–0.0639 mm), and silt (0.063–0.002 mm). On the grid (horizontal, grain size in mm; vertical, %), a normal distribution (according to the Gauss integral) forms a straight line. The three samples of core (all taken from below 2 m depth) are very similar with a higher content of medium sand than a normal distribution. The samples *A* (excavation on the active rockglacier Gruben (depth 1 m) and especially *B* (surface of the 1850 moraine of the Gruben glacier) are much coarser

Table 6.1. Grain-size distribution (%) for some samples from the rockglacier Gruben, Wallis, Swiss Alps. (Barsch et al. 1979)

Material		ϕ(mm)	Sample				
			9	11	23	A	B
Gravel	Coarse	63–20	2.8	1.5	1.6	5.0	3.5
	Medium	20–6.3	8.1	11.3	7.5	29.1	26.7
	Fine	6.3–2.0	6.3	7.7	4.1	18.4	22.8
Sand	Coarse	2.0–0.63	7.8	11.8	4.1	7.7	15.3
	Medium	0.63-0.20	17.0	19.6	19.4	3.6	7.8
	Fine	0.20–0.063	23.4	31.8	42.9	4.6	11.1
Silt	Coarse	0.063–0.020	20.8	8.6	10.9	13.7	6.4
	Medium	0.020–0.0063	9.8	4.6	9.3	11.0	4.0
	Fine	0.0063–0.002	3.0	1.6	1.4	4.1	1.2
Clay		< 0.002	2.5	1.2	0.9	4.1	1.1
Total		10	1.5	99.7	102.1	101.3	99.9

Samples 9, 11, and 23 are from the borehole drilling into the rockglacier taken between 2 and 3 m depth. Sample A is from a small excavation near the drilling site and sample B is from the 1850 lateral moraine of the Gruben glacier

Barsch et al. (1979). In general, the material has to be classified as silty sand with ca. 60% sand and about 30% silt. This is about typical for most active rockglaciers.

Vere and Matthews (1985) carried out a sedimentological analysis on a debris rockglacier that flows out of a lateral moraine at Bukkeholbreen in Jotunheimen (Norway). Samples of the fraction less than 8 mm were taken at 50 cm depths and compared with material from the undisturbed lateral moraine and from subglacial sediments. They demonstrate that the rockglacier material shows a closer similarity to subglacial than to supraglacial material.

The conclusion that this rockglacier consists more of subglacial than of supraglacial material seems logical at the first view, since lateral moraines are often formed by glacial extrusions. On the other hand, if the lateral moraine contains supraglacial material in larger quantities, then the morainic material has probably been transported over short distances. That is due to the fact that a glacier overloaded with supraglacial material from lateral rock walls tries to deposit it at the first possible site. This has been proposed by Barsch (1977a) to explain the rockglacier Weissmies, Simplon, Switzerland (cf. Fig. 1.1) studied by Messerli and Zurbuchen (1968). Therefore, it seems to be possible that the similarity between subglacial and rockglacier material is not a consequence of its origin, but of the processes which are related to the sorting of the morainic material at the surface (forming the bouldery layer) and to the internal deformation during rockglacier development. Trombotto (1991) reports similar results. He compares quartz grains from glacial, cryogenic, and rockglacier deposits with the help of an electron microscope. The resemblance between the sand grains of glacier and rockglacier material is explained by the similar debris treatments inside glaciers and rockglaciers. In both cases, the mineral grains are subject to high pressures and, quite often, they will be moved against each other.

6.2 Internal Structure

The internal structure of rockglaciers is still poorly understood, but during the last 20 years an enormous amount of scientifically sound information has been collected mainly by geophysical means, by drilling investigations etc. Geophysical data and information from boreholes prove that ice lenses up to several meters thickness often with more or less well-developed silty-sandy layers are not rare in the frozen core of an active rockglacier. Additional field evidence of relict rockglaciers shows that the ice content varies between 50 and 80% of the total volume.

Despite their inaccessibility and despite the difficulties of direct excavations, some direct observations have been made. Unfortunately, their number is still sparse. In addition, these observations are confined to limited depths; however, they present first information at selected points on a geosystem which has been until now open only to speculation. Our knowledge has been greatly increased by the use of geophysical methods (seismic, geoelectric soundings) and by temperature measurements. Thus, today a first, still incomplete but scientifically sound picture of our present knowledge on the inner structure of rockglaciers is possible.

6.2.1 Direct Information

Direct access to the inner core of an active rockglacier is only possible at a very few places. The system is not only covered by a bouldery mantle with single elements often more than 1 Mg, but also the front and side slopes are so steep that sliding is always a danger for all attempts to dig into the system. The examples known hitherto are collected in the next chapter.

6.2.1.1 Excavations, Outcrops, Tunnels

First observations are made at the steep front and side slopes of active rockglaciers, where fine material (mostly sand and silt; c.f. Fig. 6.1) appears. Excavations, endangered by the toppling of fairly large boulders, reveal frozen material often at depths of less than 1 m at the upper side slopes (cf. Chap. 10 and Sect. 6.3). Meltwater can be heard flowing below the bouldery layer, often long after snowmelt during summer. In some cases, it is possible to move boulders, and it can be seen that water flows on frozen, but ice-rich, fine-grained material or even on apparently pure ice. This probably led to the "pure ice-core hypothesis" which assumes that a massive ice core exists in some rockglaciers (Potter 1972; cf. also Chap. 8). This conclusion is too far-reaching because this ice layer has never been dug up. Even at shallow depths, this layer is seen to be part of an extremely ice-rich (often ice-supersaturated) permafrost near the permafrost table. This is not unusual in permafrost areas, where meltwater normally flows on the permafrost table. As long as no more conclusive evidence is offered, the ice layer below the bouldery mantle can only be accepted as ice-supersaturated permafrost.

Dyke (1990) furnishes some photographs of apparently natural outcrops with well-developed permafrost. One example comprises ice from avalanche-snow deposits in the lower part of the talus slope, the other ice with clear dirt layers in a rockglacier. Both demonstrate again the existence of large ice bodies in high mountain environments and in rockglaciers.

Large excavations have been carried out in the course or road or building construction. One example is presented at the left side of the rockglacier in the combe de Prafleuri (Val d'Hérémence, Wallis). As Fisch et al. (1978) reports, the excavation revealed about 10 m of "angular debris cemented by interstitial ice beneath a thin layer of coarse debris". The ice content was very high. The ice crystals were very small and quite different from the nearby glacier ice. Another (and similar) example forms the excavation at the front slope of the rockglacier at the Flüelapass (Graubünden), which has been studied by King et al. (1987). Frozen, ice-rich material was found. Normally, this material is difficult to dig and people try to wait until the interstitial ice has melted. Being permafrost, all that happens is that a new active layer develops during summer, which is – depending on the local conditions – again 1 to 3 m thick. This method does not improve the availability of material for road construction.

A very exciting insight (Moore and Friedmann 1991; and Fig. 6.2) into a feature which is probably a rockglacier is given by a collapse above a mine in the Rocky Mountains (Red Mountain, 60 km west of Denver). The feature is found at 3475 m (11 400 ft) on a north facing slope; it has a surface gradient of 12°, and a front slope of 33°. It originates at the foot of a steep talus slope (29°–33°). The exposure along the axis of the feature is about 86 m wide. From the abstract, it is difficult to evaluate whether or not this is really a talus rockglacier. Its position, surface slope, bouldery mantle, and front slope can be regarded as typical; only the small thickness of 5–14 m is astonishing, but perhaps not unrealistic. If this permafrost body creeps (or has been creeping), it demonstrates perhaps that mountain permafrost deforms at very low shear stresses (Sect. 7.4). Unfortunately, there is not enough information on the length, the width, or on the activity of the feature.

Anyhow, the outcrop is exciting. It demonstrates at least a mountain permafrost ice body which is 3–10 m thick and which is mantled by a 2–3 m bouldery cover. The following observations seem to be important:

a

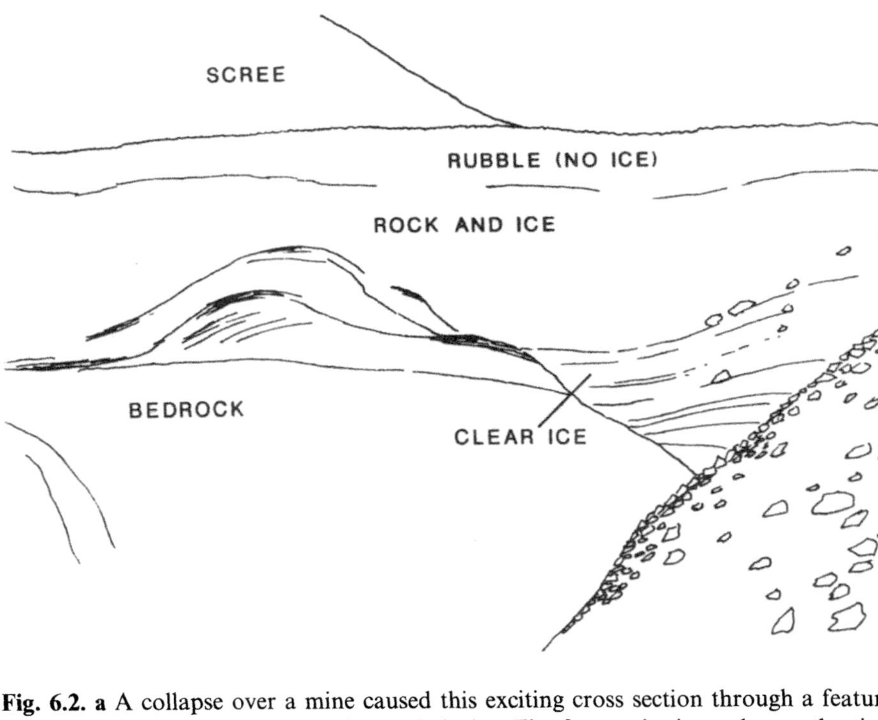

b

Fig. 6.2. a A collapse over a mine caused this exciting cross section through a feature which seems to be an inactive talus rockglacier. The feature is situated near the timberline at 3475 m a.s.l. On Red Mountain (60 km W of Denver, Colorado). It was discovered and photographed in August 1990 by David Moore, Denver. **b** The profile explains the photograph of **a**. It was produced by David Moore, Denver

- The ice body is free of angular boulders.
- The ice contains air bubbles.
- Laminae of silt and sand are reported.
- The sand and silt layers are not straight but undulating.
- The ice has not been dated yet, but the deuterium content is about 30‰ less negative than average modern snow packs and similar to that in streams (with summer precipitation and snow melt). The ice is, therefore, regarded as Holocene.
- The bouldery layer has a well-developed lichen cover, and thus creates the impression that it is "at least several thousand years old".

To summarize, even if the feature is not a rockglacier, it demonstrates the existence of mountain permafrost bodies with a very high ice content in relation to the mineral inclusions under the coarse bouldery mantle. In the reported case, ice may account for 90 vol% of the feature, but it seems that this ice – according to the mineral inclusions and to the air bubbles – is not glacier ice but apparently (old?) permafrost ice. It can be assumed that active rockglaciers may have a similar core perhaps with a more varibale, but also fine-grained mineral content.

Giardino and Vitek (1988a) did interesting fabric studies in a 15-m-deep, 100-m-long trench in a blocky deposit on the south slopes of Mt. Mestas (Colorado). The discussed landform ends at a very low altitude (2595 m, summit of Mt. Mestas is 3520 m at latitude 37° 30′). The deposit is stable at present. The authors conclude that there are three possible events depositing blocky debris, and indicating three distinct periods of movement (partly catastrophic ones). This opposes sharply the given definition of rockglaciers, which supposes the (slow) plastic deformation of the ice content in the frozen core as the main cause for movement. Thus, the inner fabric described by Giardino and Vitek does not allow the assumption that this feature is a rockglacier. This is further suggested by the unusually thick bouldery layer (more than 15 m), which is normally around 3 m. At this point, Wahrhaftig's comment (1987) has to be repeated that the discussed landforms around Mt. Mestas are not rockglaciers (cf. Chap. 2 and Sect. 8.2.1).

Tunnels excavated during mining activities have been of great importance for rockglacier research. The most frequently quoted is the one that was built through the Hurricane Basin rockglacier in the San Juan Mountains in Colorado (Brown 1925). Apparently, it was reopened in 1950, but in the early 1970s it was filled with ice (Carrara and Andrews 1975). This was also the case during summer 1980, when Paul Carrara, Nel Caine, and Jörg and Dietrich Barsch did a resurvey. According to the written communication by Nel Caine and Paul Carrara (cf. also Table 6.6 in Caine 1986), the rockglacier was still actively moving (4 cm a^{-1}) in the years up to 1980.

Unfortunately, the original description by Brown caused a terrible confusion, since the ice that was found was accepted without crystallographic studies as so-called glacier ice. This assumption was used as an indication that rockglaciers contain pure ice cores and are nothing else but debris covered glaciers (S.E. White 1971a,b, 1974b; Potter 1972; Corte 1978). Apparently, nobody who quoted this paper rechecked the locality. What looked like a longitudinal section in Brown's paper (1925 Fig. 1, p.465) is actually a cross section (Barsch 1983a, 1988). The resurveyed profile is presented in Fig. 6.3. It demonstrates that the left-hand side of the rockglacier is set against a rock wall, where meltwater seeps into the rockglacier. The massive ice at this place presented as a "probable fossil glacier" by Brown is probably nothing other than refrozen meltwater infiltrating over the rock wall into the rockglacier. The observation of massive ice in the tunnel cut through the lower end of the active Hurricane Basin rockglacier is of interest for mountain permafrost studies and for the genesis of massive underground ice-bodies, but it is not an evidence of a glacial origin of the ice or of the rockglacier itself, as frequently suggested (Giardino and Vick 1987). On the other hand, Brown describes the inner part of his "ice bound rock-stream" very exactly: the tunnel, which started at the right-hand side slope of the rockglacier, penetrated for the first few feet "typical rock-stream débris consisting of angular blocks. For the next three hundred feet the material was cemented by ice, but the quantity of rock in the ice varied as did the size of the angular blocks" (Brown 1925, p. 465).

This description seems to be typical: the cross section of about 100 m length gives an impression of the great variability of the rockglacier core. It reveals an apparently ice-rich (ice-saturated, partly ice-oversaturated) talus

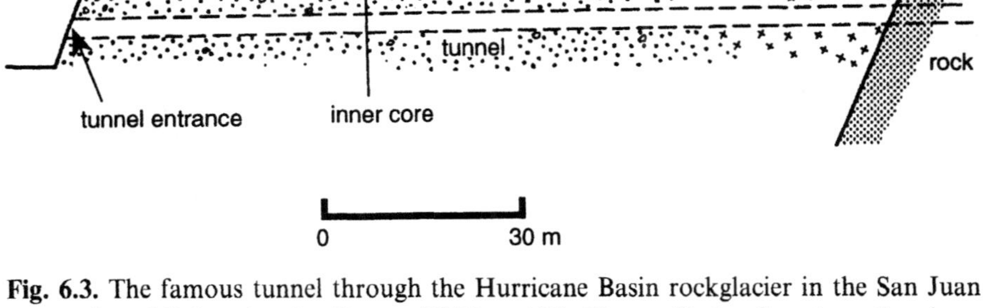

Fig. 6.3. The famous tunnel through the Hurricane Basin rockglacier in the San Juan mountains of Colorado after Brown (1925, p. 465). The profile has for a long time been accepted as a longitudinal profile. However, according to the situation at the Hurricane Basin rockglacier, it is a cross-section profile. This changes all possible intrepretations which believe in a buried glacier in the upper part of a still active rockglacier. According to Barsch (1983a, p.136), the ice is refrozen meltwater which infiltrated the rockglacier below the rock wall at its left side

with a very variable boulder content and a predominance of fine-grained material.

6.2.1.2 Smaller Boreholes

The boreholes drilled so far confirm the direct observations (cf. Table 6.2). The results of the borehole through the rockglacier Murtèl I (Barsch 1977b) are presented in Table 6.3 and Fig. 6.4. In the upper 10 m of the rockglacier, which is around 45–50 m thick at this point (Barsch and Hell 1975; Haeberli et al. 1988) only interstitial ice or some decimeter thick ice lenses are encountered. Boulders are found down to a depth of 3 m below the surface. The first ice lenses were hit below 3.2 m. The average ice volume varies between 30 and 100%. The borehole was sealed; water circulated for cooling purposes was never lost into the rockglacier.

The cored borehole into the Gruben rockglacier (Barsch et al. 1979; Haeberli et al. 1979; cf. Fig. 6.5) was drilled because this debris rockglacier had been described as a debris-covered glacier by Whalley (1974a,b). This has been questioned (Haeberli 1985a) using all information available. The borehole should have revealed if there was a pure "glacier" core existing directly below the bouldery mantle. As Fig. 6.5 displays this was not the case. In addition, chemical analysis of the ice, by Roland Souchez, Bruxelles, revealed that the ion content of the ice was very high in relation to the ice of the nearby Gruben

Table 6.2. Drillings into and through rockglacier permafrost

Name	Type	Status	Altitude (m a.s.l)	Year	Depth (m)	Core (mm)	Reference
RG II Yukon	Talus	Active	1090	1969	21.5[a]	89	Johnson and Nickling (1979)
Murtèl I GR	Talus	Active		1975	10.4	40	Barsch (1977b)
Gruben VS	Talus/ debris	Active		1976	7.0	40	Barsch et al. (1979)
Murtèl I GR	Talus	Active	2.670	1/1987	21	70	Haeberli (1989b)
				2/1987	62.5[b]	70	Haeberli et al (1988)
Las Sours[c]	Talus	Active?	2.755	1/1990	67.0[d]		Vonder Mühll (1993)
				2/1990	37.0[d]		VDM 93 Vonder Mühll (1993)

GR, Graubünden/Swiss Alps; VS, Wallis/Swiss Alps
[a]Measurements were only possible to a depth of 16 m.
[b]Active permafrost thickness is assumed down to 52 m.
[c]Study area Pontresina-Schafsberg in Vonder Mühll (1993).
[d]There was no core sampling during drilling.

Table 6.3. Results of the first borehole into the rockglacier Murtèl I. (According to Barsch 1977b)

Depth (cm)	Thickness (cm)	Material
0–160	160	Boulders
–260	100	Boulders, permafrost table
–320	60	Frozen gravel, sand, silt
–365	45	Ice lenses, segregated ice
–590	225	Frozen sand and silt with a high ice content
660	70	Ice lenses with many bubbles and silty layers
790	135	Ice lenses
875	85	Ice lenses with many bubbles and silty layers
1040	165	Frozen sand and silt with a high ice content

The ice content below the permafrost layer is estimated to be around 50–60 vol%.

glacier. This demonstrates that the water had been in contact with rock material (as groundwater?) for a long time before draining into the rockglacier and freezing.

The borehole through the talus rockglacier RG II in the Kluane Range at mile 1061 of the Alaskan highway (Yukon) was drilled in 1969 (Johnson 1972; Johnson and Nickling 1979). The lower end of the rockglacier lies at 1090 m a.s.l. The borehole was situated on a prominent lobe with a relative height of 6 m. A depth of 21.5 m was reached, but 2 m was lost due to collapse. Frozen ground was reached at 2.2 m below the bouldery mantle. In 1970 and in 1971, temperature measurements again confirmed the frozen core, but in 1976 the whole length of the borehole (except a layer between 11 and 12.5 m depth) displayed positive temperatures. The results of this borehole regarding temperature and inclinination measurements are very important; therefore, they are discussed later (Sect. 6.2.2.5).

6.2.1.3 The Deep Borehole through the Rockglacier Murtèl I

Wilfried Haeberli took the initiative to start a drilling project through an active rockglacier (Haeberli et al. 1988). The already well-known and easily accessible talus rockglacier Murtèl I was chosen. Field operations were carried out between April and early July 1987. After an experimental borehole was drilled down to about 20 m, the main borehole (about 2 m away from the first borehole) was sunk and cored (core diameter 70 mm). Drilling was stopped in solid bedrock at a depth of 62.5 m. It was carried out in order to obtain information about the development of the Holocene cryogenic environment that is contained in alpine permafrost. In addition, it revealed a number of very

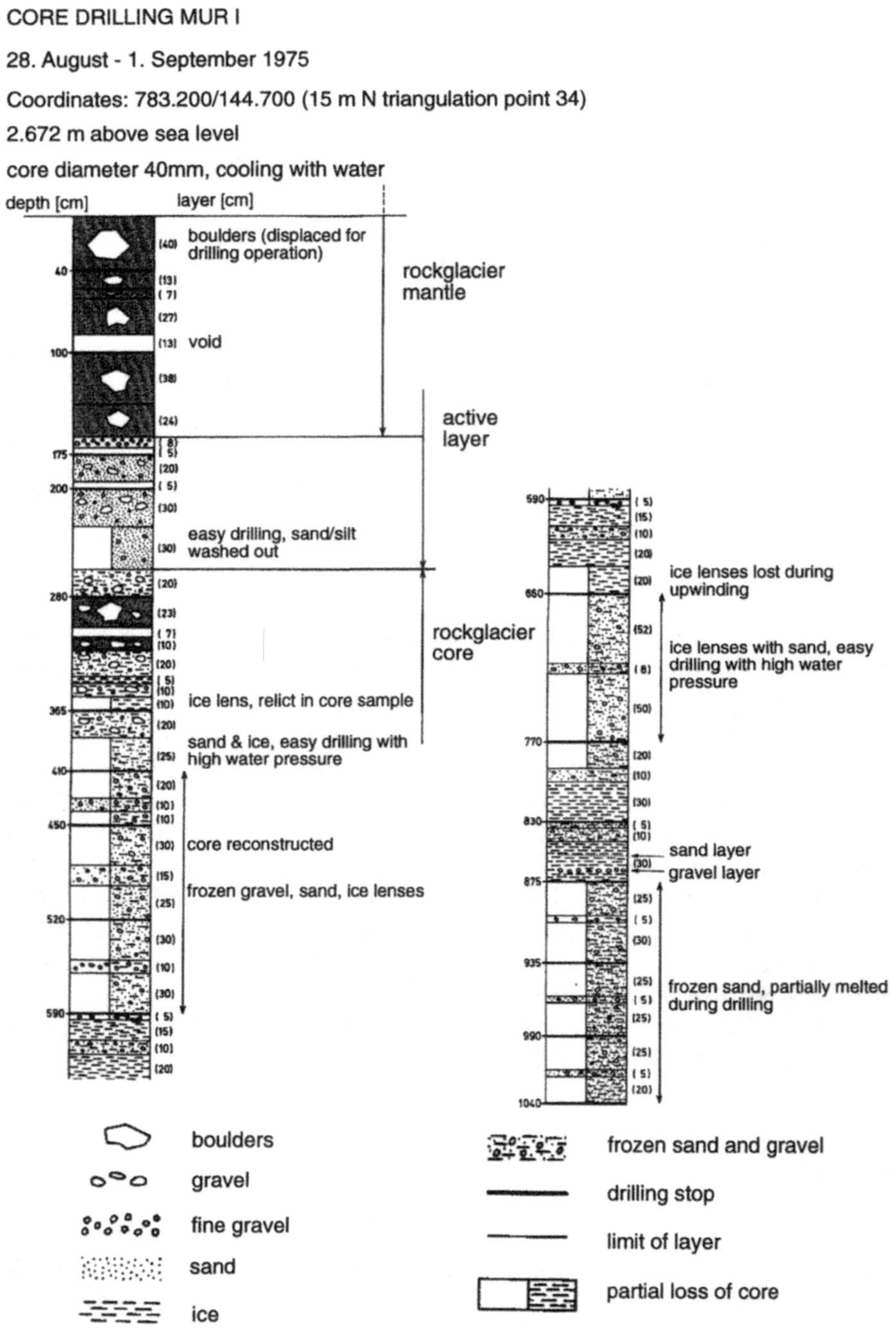

Fig. 6.4. Core drilling MUR 1 (1975) on the rockglacier Murtel I. (Barsch 1977b). The profile gives the visual stratigraphy of the core, which was described during the drilling operation (28 August–1 September 1975). Below a depth of 3m, no boulders were encountered. The ice content is estimated to be 50–60 vol%

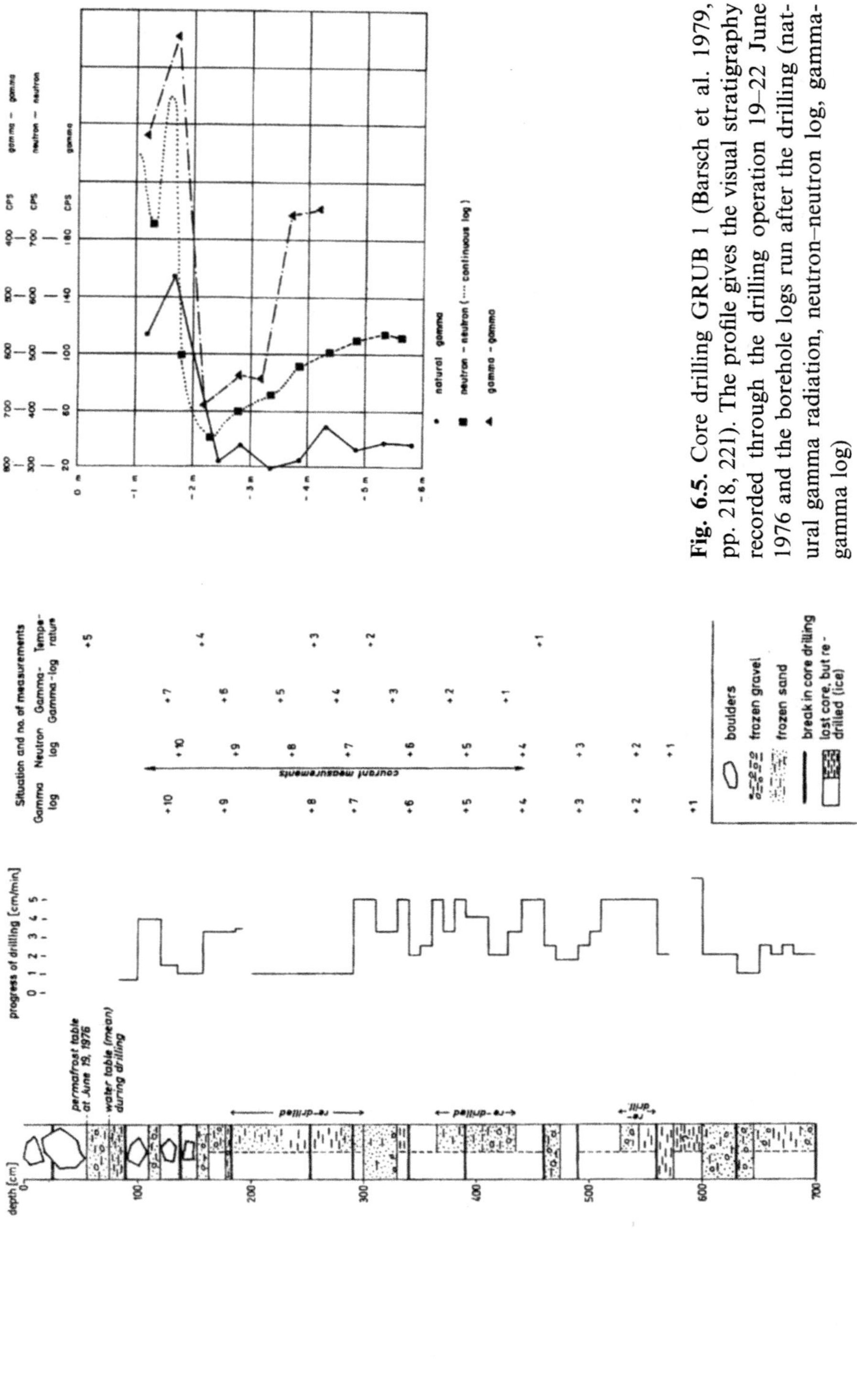

Fig. 6.5. Core drilling GRUB 1 (Barsch et al. 1979, pp. 218, 221). The profile gives the visual stratigraphy recorded through the drilling operation 19–22 June 1976 and the borehole logs run after the drilling (natural gamma radiation, neutron–neutron log, gamma-gamma log)

important details about active rockglaciers. The first results, including the shallow borehole (Barsch 1977b), indicate, according to Haeberli et al. (1988) and Haeberli (1989b), the following (Fig. 6.6):

1. The borehole revealed that the rockglacier thickness above bedrock is around 50 m, which confirms the seismic (Barsch and Hell 1975) and geo-electric measurements (King et al. 1987).
2. The stratigraphy below the active layer (ca. 3 m) is described by Haeberli et al. (1988, Table. 6.4)
3. The amount of ice (90 to 100% of volume) between 3 and 15 m is surprisingly high. It is higher than encountered in the borehole drilled by Barsch (1977b) about 100 m to the east and more or less at the right-hand side of the rockglacier. Incorporation of avalanche snow, segregation ice formation, frost heave and self-purification may be the principal causes for this large amount of ice. More information will probably be obtained from chemical analyses.
4. Temperature at the bottom is at melting point, which seems rather high. The temperature gradient also seems to be very high.
5. The bedrock is, in its uppermost part, heavily fissured and thus permeable. This conflicts sharply with the bedrock observed directly in front of the rockglacier.

The deep borehole at Murtèl reveals that the internal structure of an active talus rockglacier is very complex, and that no glacial ice core exists beneath the talus that fills this cirque as a talus rockglacier.

6.2.2 Indirect Information

As already mentioned, the direct insight into the core of an active rockglacier is seldom obtained and is not always unambiguous. Therefore, it is necessary to use indirect and scientifically sound methods. From the field of geophysics, a number of helpful methods are available that can be used in rockglacier as well as in mountain permafrost research. A valuable summary of applied methods has just been published in connection with the Interlaken International Workshop on Mountain Permafrost by King et al. (1992).

6.2.2.1 Seismic Information

There are two forms of seismic rockglacier exploration. First, the whole rockglacier is investigated to gain information on its total thickness and on its internal structure. This is normally done with a multichannel apparatus and the shocks are generated by explosions or by vibration. The velocity of the shock waves (P-waves) is registered by 12 or 24 geophones along profile lines.

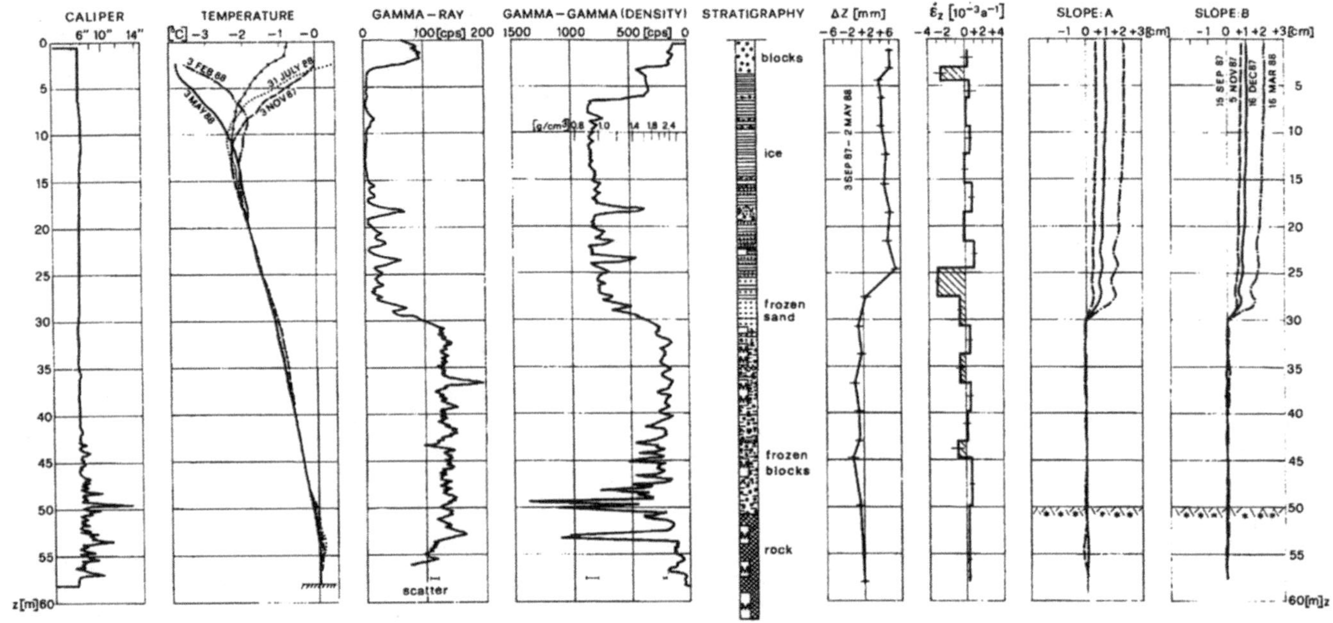

Fig. 6.6. Core drilling (2/1987) through the rockglacier Murtél I from Haeberli (1989b). The stratigraphy recorded visually during drilling is given in the *center*. The logs refer to borehole diameter immediately after drilling (caliper), horizontal borehole deformation (*slopes A* and *B* differ by 90°, main flow direction is between the two), reflected radiation from an artificial source (gamma–gamma), indicating density, vertical strain rate, and slope of the borehole at different times

Table 6.4. Stratigraphy of the active talus rockglacier Murtèl I according to the core drilling 2/1987. (Haeberli 1989b)

Depth (m)	Material[a]	Density[b] ($Mg\ m^{-3}$)	Temperature[c] (°C)	Shear Stress[d] (kPa)
0–3	Boulders (mantle)	–	–	–
3–15	Massive ice	1	–2/–3	~25
15–28	Frozen silt, sand, gravel highly ice-supersaturated	1–1.5	–1/–2	~50
28–32	Frozen silt, sand, gravel ice-saturated	2	–1	~60
32–50	Coarse boulders ice-saturated layers of frozen sand	2–2.5		~150
50	Bedrock			

[a]Material is given by visual description of the core.
[b]Density is determined by weighing.
[c]Temperature is based on measurements in fall; it has been corrected for thermal disturbance due to drilling operation. Error is assumed to ±0.2 °C (cf. also Sect. 6.2.2.5).
[d]Shear stress is calculated from $\tau = \delta \cdot g \cdot h \cdot \sin \alpha$
with: δ = density, g = acceleration due to gravity, h = thickness, α = average surface slope over 250 m (10°).

These are the deep seismic soundings which may determine the thickness of the rockglacier permafrost over bedrock.

Second, information is gained on the upper 20 m of a rockglacier by sledge-hammer seismic traverses along profiles of, normally, 30 to 50 m length. The shock waves are triggered by hammer blows to the ground. The apparatus is in most cases a single channel unit, often furnished with a switch to use two geophones (at both ends of the profile). This device can readily be operated by two people.

Deep Seismic Soundings

The first deep seismic soundings on rockglaciers were carried out in Graubünden: the first one in 1968 on the rockglacier Macun I (Lower Engadin/ Swiss Alps; Barsch 1971b); the second in 1971 on the rockglacier Murtèl I (Upper Engadin/Swiss Alps; Barsch and Hell 1975). In both cases, longitudinal and cross-sectional profiles were measured. Recently, Vonder Mühll (1993) organized further deep seismic investigations on the rockglaciers Murtèl I, Muragl, Suvretta and in the region of Pontresina-Schafsberg in Graubünden; Gerber (1994) did the same in the southern Wallis.

The deep seismic soundings allow the detection of subsurface layers of different densities, because the travel velocity of the shock waves is directly correlated with the density of the material traversed. Thus, neglecting temperature dependencies, on active rockglaciers the following P-wave velocities are normally registered:

- Outer mantle, 250–1500 m s^{-1}.
- Inner core, 2400–4000 m s^{-1}.
- Bedrock, 5000 m s^{-1}.

The main requirement for the application of the method is that the density of the different layers increases with depth. An unfrozen layer between the rockglacier permafrost and the underlying bedrock, for instance, causes problems, because it will not be detected by this geophysical method. The measured travel times for the signals will be too high, and a wrong thickness of the permafrost will be calculated. The same error occurs if the underlying bedrock has a seismic velocity around 4000 m s^{-1} (due to shattering or structure).

Other problems are connected with the irregularity of the blocky surface or with the fact that the different layers have undulating boundaries. Nevertheless, deep seismic soundings reveal a large amount of very valuable information on the inner structure of active rockglaciers. They furnish scientifically sound data, and – used correctly – end or may end speculation in rockglacier research.

Financed by the Schweizerische Nationalfonds and organized by D. Barsch, the deep seismic soundings on the rockglacier Macun I (Lower Engadin, Swiss Alps) were carried out between 12 and 17 August 1968 by A. Schneider (GEOTEST AG Bern) in cooperation with the Department of Geography of the University of Basel. It was the first deep seismic sounding on an active rockglacier with no previous experience to aid interpretation. Nevertheless, exciting information was collected:

1. The coarse rockglacier mantle, with seismic velocities (V_1) around 800 m s^{-1}, is shallow.
2. The rockglacier core, with seismic velocities (V_2) between 3000 and 3800 m s^{-1}, is a compact body.
3. The underlying bedrock reveals velocities (V_3) around 4500 to 4600 m s^{-1}.
4. Thus, assuming that no low-velocity layer has been overlooked, the velocity differences allow a clear interpretation.
5. The rockglacier hides an overdeepened basin. Its maximum thickness is close to 100 m, i.e., much greater that that assumed from the height of the front slope.

The velocities of the core are extremely interesting. They differ from those of pure glacier ice(ca. 3600±200 m s^{-1}). Their high variability allows the conclusion that this inner core is not just simple glacier ice, but of a very complex structure (Haeberli and Patzelt 1982). The existence of ice-rich debris has to be assumed, because the velocities measured display values which are typical of frozen ground (Röthlisberger 1972). The interpretation is displayed in Figs. 6.7 and 6.8.

These results were confirmed by the deep seismic soundings on the rockglacier Murtèl I, which were carried out in the frame work of the same project during September 1971. In particular the velocity variations in the core were again registered:

$$3000 < V_2 < 4000 \text{ m s}^{-1}.$$

The bedrock velocity (granite) varies between 2600 m s^{-1} north of the rockglacier, 4700 m s^{-1} below the front and 5200 m s^{-1} below the main part of the rockglacier. This may be explained by jointing or by the fact that the joints in the bedrock under the rockglacier are filled with ice, and that the permafrost is thicker than the rockglacier.

In addition to the longitudinal profile (Fig. 6.9), contour lines of the bedrock surface have been constructed (Fig. 6.10). Again, the rockglacier fills a closed depression. The maximum thickness of the rockglacier was calculated to be about 50 m ± 5 m. This value was confirmed by the drilling organized by Wilfried Haeberli (Sect. 6.2.1) and by further deep seismic soundings done in 1990 by Vonder Mühll (1993). In addition, the continuity of the first refractor indicates a thermal control in the position of the permafrost table (Fig. 6.11).

The deep seismic soundings reported by Vonder Mühll (1993) confirm the hitherto known information; in addition, they furnish some more interesting details. On the same rockglacier, the seismic velocity (V_2) of the rockglacier core decreases from 3600 m s^{-1} in the upper part towards 3000 m s^{-1} in the lower (ca. 2300 m a.s.l.). This is probably caused by an increased water content in the permafrost due to the increasing temperature in this low altitude for active mountain permafrost.

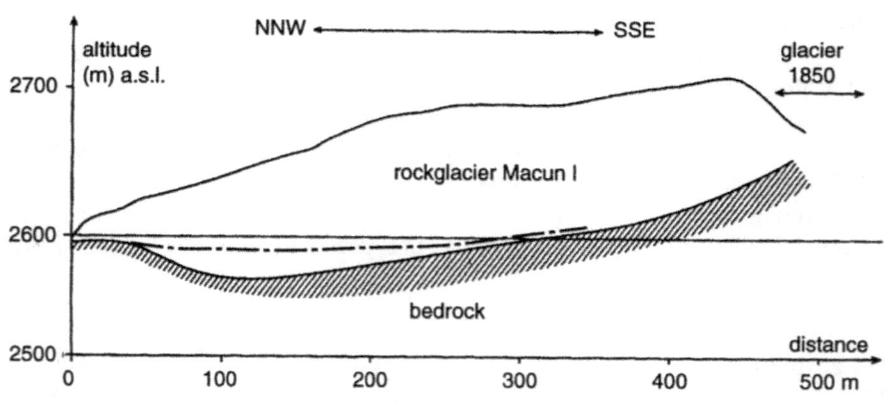

Fig. 6.7. From the deep seismic soundings in 1968, the cross section through the rockglacier Macun I has been developed. The profile follows the stromstrich (main flow direction). The rockglacier covers a bedrock depression. According to the results of the drilling 2/1987 through the rockglacier Murtél I, the possible high deformation zone (shear zone) in the rockglacier has been marked (cf. Fig. 6.8)

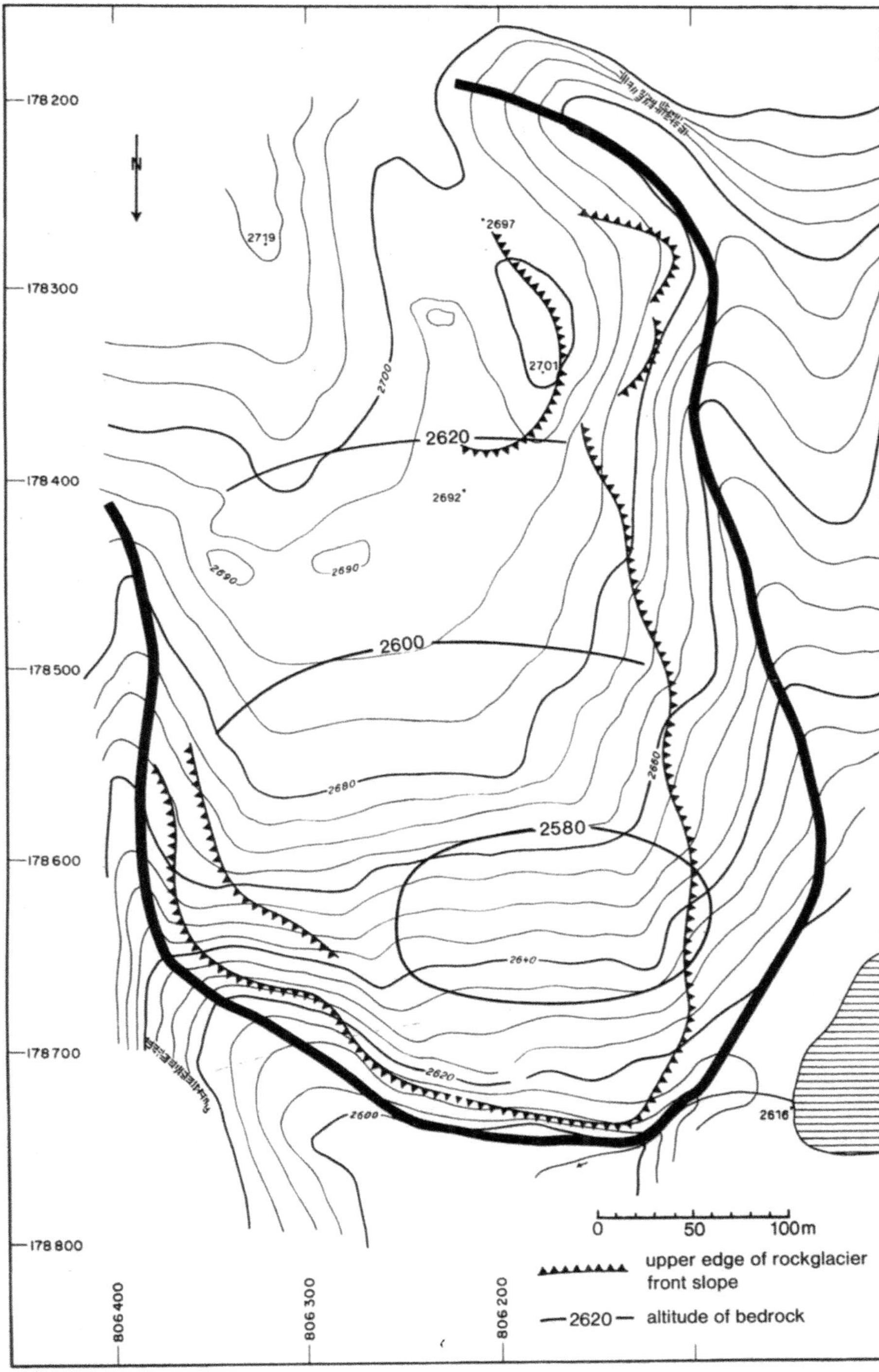

Fig. 6.8. Bedrock altitude beneath the rockglacier Macun I based on the deep seismic sounding in 1968. In the contourline plan, the foot and the upper edge of the rockglacier are marked

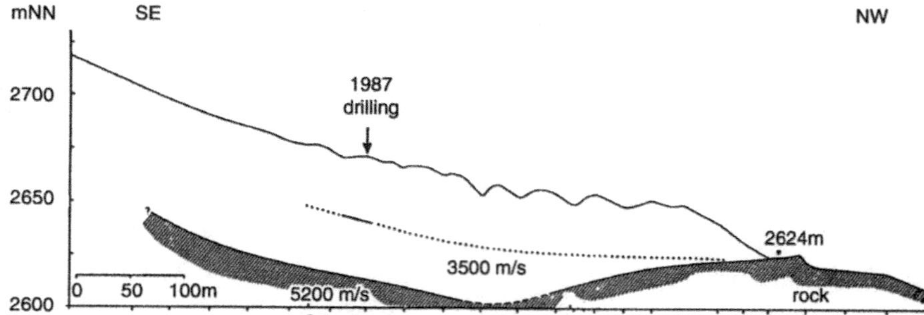

Fig. 6.9

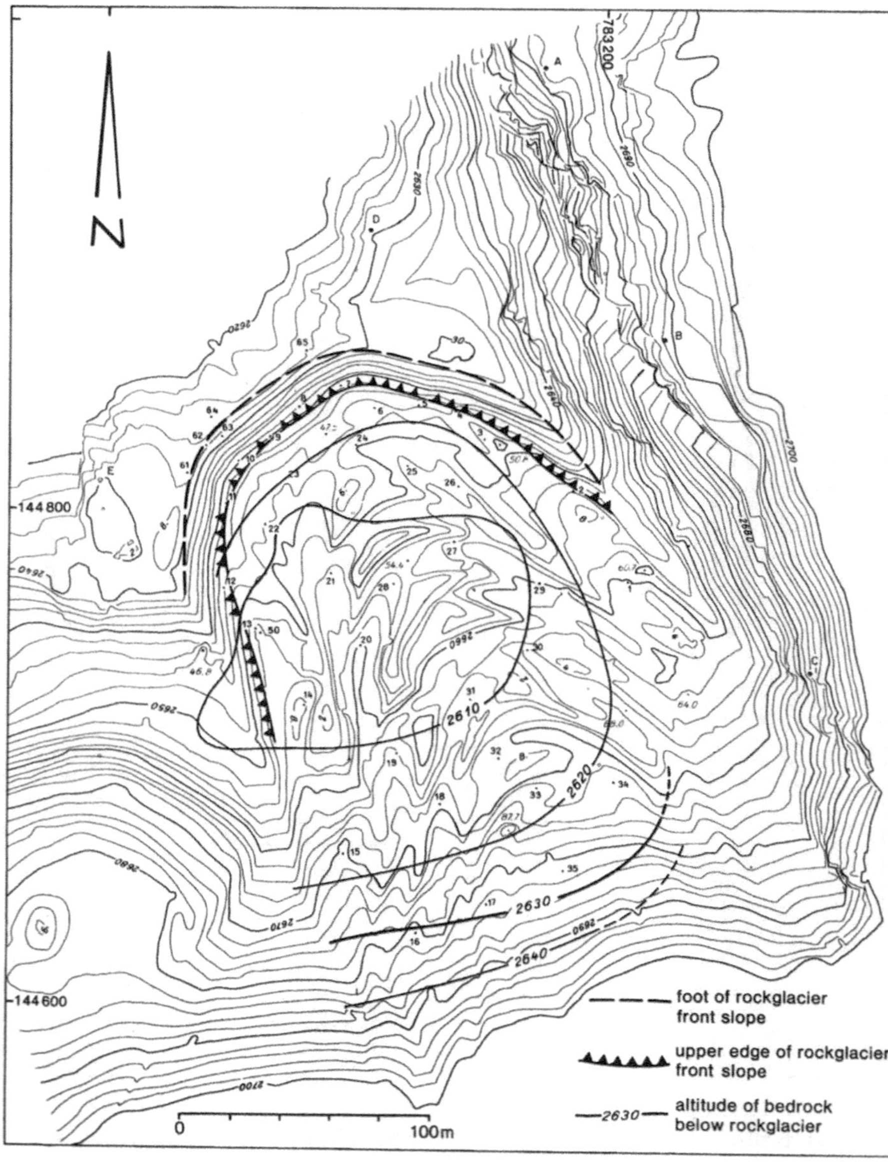

Fig. 6.10

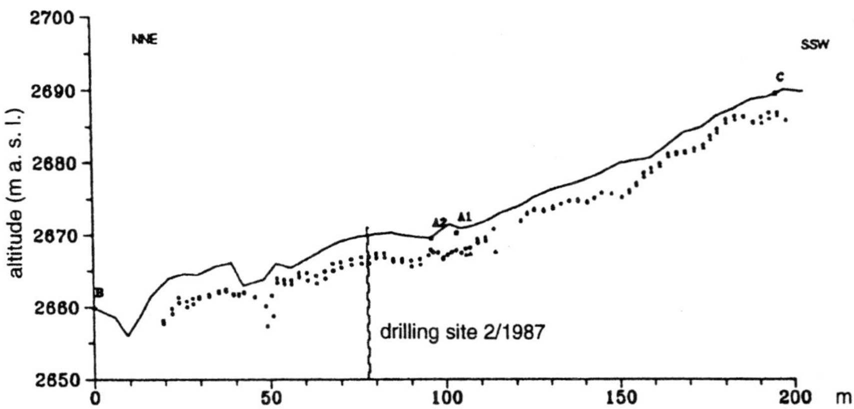

Fig. 6.11. Seismically determined thickness of the active layer of the rockglacier Murtél I from Vonder Mühll (1993)

Sledge Hammer Shallow Seismic Soundings

Shallow seismic studies have been used in permafrost research since the early sixties; sledge hammer seismic measurements on rockglaciers were done almost simultaneously by Potter (1972) and Barsch (1973). More information has been added by King (1976), Haeberli (1979), Haeberli and Patzelt (1982). For the internal structure of rockglaciers, these studies reveal information on the "active layer" (cf. Sect. 6.3) and the uppermost parts of rockglacier permafrost. The active layer shows P-wave velocities of 300–800 m s^{-1} (Table 6.5), the upper rockglacier permafrost displays P-wave velocities between 2800 and 4000 m s^{-1}, which can be sometimes as low as 2400 m s^{-1} depending on temperature and water content.

The large variation in P-wave velocities again demonstrates that active rockglaciers are of a more diverse structure than a glacial ice core could account for. It indicates frozen material with different ice contents and perhaps different temperatures and water contents as well, even if a quantitative evaluation is at the moment not possible.

Typical results of sledge-hammer seismic soundings are presented in Fig. 6.12. The profile has been measured in both directions. Profiles all around the world measured on active rockglaciers display similar curves. A number of measurements have been collected together in Table. 6.5. They display typical velocities of the rockglacier mantle (v_1) and the rockglacier core (v_2). Ac-

Fig. 6.9. The cross section through the rockglacier Murtél I along the stromstrich has been developed from deep seismic soundings done September 1971 by A. Schneider and D. Barsch (Barsch and Hell 1975). Following the results of the drilling 2/1987 through this rockglacier, the possible position of the low viscosity layer has been marked

Fig. 6.10. In the contourline plan of the rockglacier Murtél I, the altitude of bedrock underneath the rockglacier is presented according to the deep seismic sounding in 1971

Table 6.5. P-wave velocities, active layer thickness and ice content from sledge-hammer seismic measurements on active rockglaciers

Rockglacier	Region	Length (m)	V_1 (ms^{-1})	V_2 (ms^{-1})	Error (±%)	D_1 (m)	Sin φ	Ice Content (vol %)
Murtèl I[a]	Engadin	35.0	550	2900	~10	2.0	0.190	77
		35.0	590	3000	~10	2.5	0.197	72
		54.7	900	3330	~10	3.5	0.270	59
Muragl[a]	Engadin	62.0	590	3200	~10	5.3	0.184	63
		60.0	670	3300	~10	5.0	0.203	59
Graben[b]	Wallis	100.0	500	3105	4	1.9	0.161	68
		100.0	700	3905	5	1.2	0.179	40
		100.0	725	3375	4	0.8	0.215	57
Äuss. Hochebenkar[c]	Austria	100.0	425	3630	5	2.0	0.117	48
		50.0	600	3145	10	2.3	0.191	66
		40.0	700	2560	11	3.5	0.273	96
Lintgen	Khumba Nepal	40.0	500	3330	~10	1.6	0.150	59
Nuptse[d]	Khumba Nepal	40.0	804	3000	~10	2.8	0.268	72
Dugla[d]	Khumba Nepal	40.0	510	2750	~10	2.6	0.185	85
Galena Creek[e]	Rocky Mountains	50.0	240	2440	~10		0.098	
		50.0	340	3350	~10		0.102	57

Remarks: φ = critical angle of refraction, $\sin\varphi = \frac{V_1}{V_2}$,
icecontent $= \left[\frac{4167}{V_2} - 0.67\right]$ [vol%].
The Müller formula (Röthlisberger 1972; Haeberli 1985a) gives a rough estimate of the ice content.
[a]Barsch (1973 pp.).
[b]Haeberli (1985a).
[c]Haeberli and Patzelt (1982).
[d]Jakob (1992a).
[e]Potter (1972), Barsch (1987b).

cording to the proposition by Haeberli (1985a; following the empiric formula of Müller fide Röthlisberger 1972), values for the ice content (n_i) (despite the uncertainty of the formula) and for the sine of the angle of refraction have been added.

$$\frac{4167}{v_2} - 0.7[\text{vol\%}] \tag{6.1}$$

$$\phi = \frac{v_1}{v_2} \ . \tag{6.2}$$

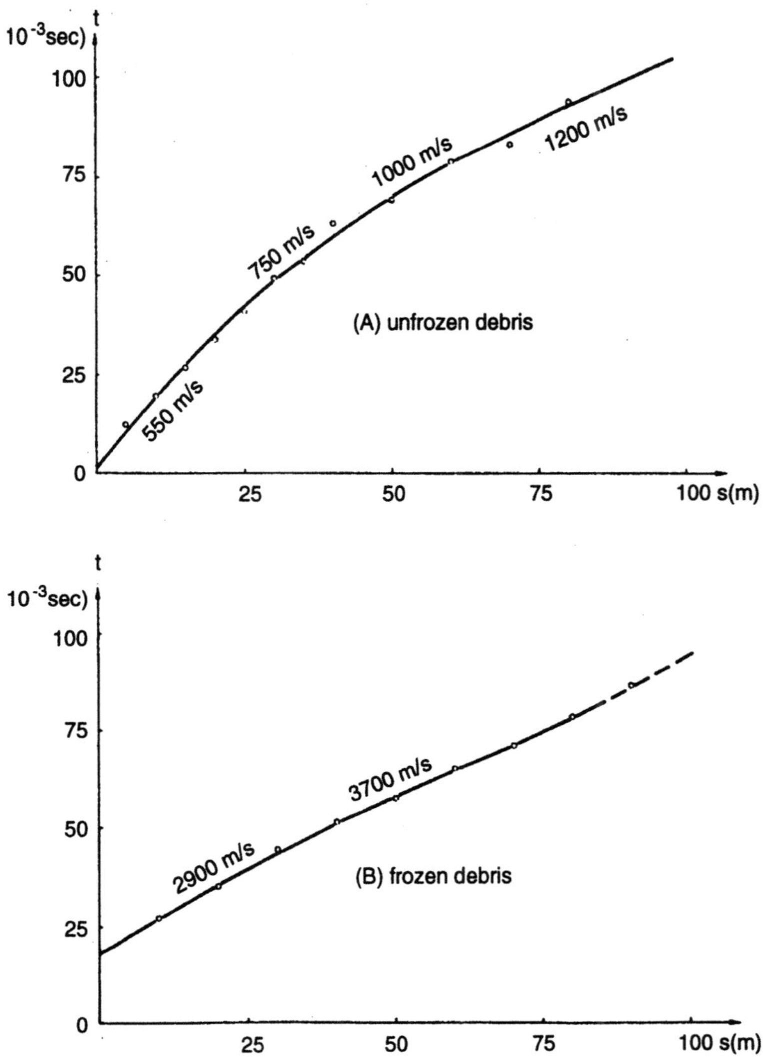

Fig. 6.12. Seismic velocities (P-wave), according to sledge-hammer seismic studies in unfrozen, water-unsaturated debris and in frozen debris, combined into time-distance graphs (Haeberli et al. 1979). The diagram displays the difference between frozen and unfrozen debris as well as increasing density with depth. In the case of frozen debris, it may demonstrate a decrease in ice content

Sledge-hammer seismic soundings are easily done and reveal a lot of information. As with every seismic method it provides some problems. One is connected with the calculation of the data. Normally, the depth of the upper layer (D_1) and of the second layer (D_2) are calculated according to the following formula (Barsch 1973):

$$D_1 = \frac{x_1}{2}\sqrt{\frac{v_2 - v_1}{v_2 + v_1}}\ [\mathrm{m}] \tag{6.3}$$

$$D_2 = \frac{x_2}{2}\sqrt{\frac{v_3 - v_2}{v_3 + v_2}}\ [\mathrm{m}]\ . \tag{6.4}$$

This "nick point" method needs profiles in both directions, because otherwise it cannot be controlled if the refractor (V_2 layer) is about horizontal.

According to Hagedoorn (1959; cf. Haeberli 1985a; Vonder Mühll (1993), it is possible to determine wave fronts in a two dimensional diagram. From this, the depth (D) below every geophone of the profile to the discussed layer can be calculated

$$d = \frac{v_1 \cdot t_+}{2\cos(i_c)}\ [\mathrm{m}] \tag{6.5}$$

$$i_c = \arcsin\frac{v_2}{v_1} \tag{6.6}$$

$$t_+ = t_{1A} + t_{1B} - t_{AB}, \tag{6.7}$$

where t_{1A} : time from the shot A to geophone 1;
t_{1B} : time from the shot B to geophone 1;
t_{AB} : time from shot A to the place of shot B.

This plus-minus method is time-consuming, but with good data it is worthwhile to use it.

In addition, King et al. (1992) point out that the careful evaluation of the diagram is also essential for a good interpretation. In Fig. 6.13, two possibi-

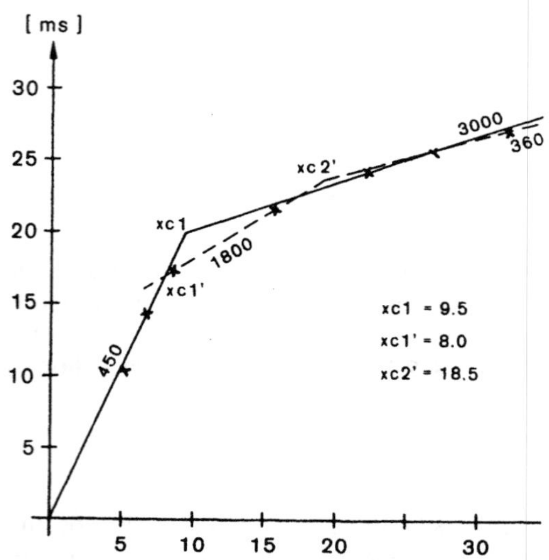

Fig. 6.13. Possible interpretation of sledge–hammer seismic measurements according to King et al. (1992). The thickness of the active layer is 4 m (*solid line*) and 8 m (*dashed line*)

Table 6.6. Two possible interpretations of sledge-hammer seismic soundings (Fig. 6.13). King et al. (1992)

		Interpretation 1	Interpretation 2
V_1	[ms^{-1}]	450	450
V_2	[ms^{-1}]	3000	1800
V_3	[ms^{-1}]	5500	3600
X_1	[m]	9.5	8.0
X_2	[m]	~35	18.5
D_1	[m]	~4.1	~3.1
D_2	[m]	–	~5.3
D_3	[m]	~9.5	–

lities are presented. The first interpretation (solid line) gives an active layer of about 4 m thickness over frozen material. The second reveals probably inactive permafrost (active layer about 8 m thick) over frozen debris (Table 6.6).

This illustrates that:

- Interpretation has to be done extremely carefully.
- Every indirect method has always to be checked by another.

Sledge-hammer seismic soundings provide considerable information on the active layer to be obtained (Sect. 6.3). Figure 6.14 displays the thickness of the summer time unfrozen rockglacier mantle of the rockglacier Murtèl I. In addition, the method may be used to differentiate between active and inactive rockglaciers. Barsch (1973) demonstrated that on climatic inactive rockglaciers the unfrozen mantle is more than 10 m thick, i.e., the rockglacier permafrost is inactive, because the mantle will not freeze completely in winter. Table 6.7 summarizes some of the relevant values.

6.2.2.2 Geoelectric Soundings

Geoelectric soundings have been used in permafrost research for quite a time (Barnes 1965; Hoekstra and McNeil 1973). Although never published, the first geoelectric resistivity measurements on rockglaciers were carried out by Barsch and Schneider in 1972 on a number of rockglaciers in the Upper Engadin, Graubünden. Fisch et al. (1978) independently used this method in the mountains of the Wallis where he had already worked between 1952 and 1956 on rockglaciers. Evin (1983b, 1987b) applied geoelectric measurements to rockglaciers in the French Alps, Barsch and King (1989) to different types of rockglaciers in the Andes of Mendoza. In particular King (1982) has worked on the differentiation of ice-rich permafrost and clear underground ice bodies. King et al. (1987) also compared geoelectric measurements with radio echo

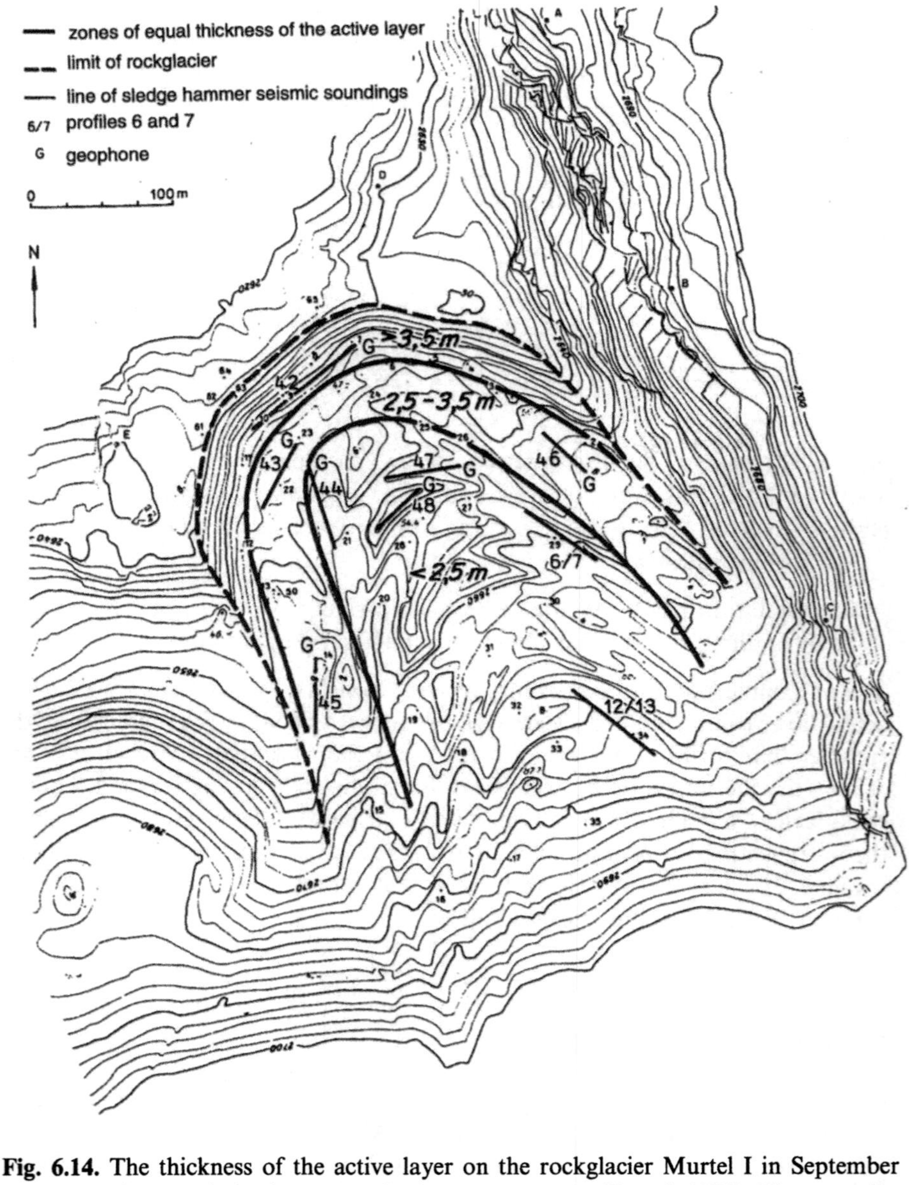

Fig. 6.14. The thickness of the active layer on the rockglacier Murtel I in September 1972 according to sledge-hammer seismic measurements (Barsch 1973). The map displays the decreasing thickness of the active layer from the toe to the rooting zone of the rockglacier

Table 6.7. Data of sledge-hammer seismic soundings on inactive rockglaciers. (Barsch 1973, unpubl. data)

Rockglacier	Region	Length (m)	V_1 (ms^{-1})	V_2 (ms^{-1})	X_C (m)	D_1 (m)
Albana	Engadin	50	810	3200	31.0	11.9
Val Sassa	Engadin	30	330	1250	6.2	10.3[a]
Val Sassa	Engadin	60	370	3320	15.4	7.1
Val Sassa	Engadin	30	360	3160	11.2	5.1
Flüelastraße	Engadin	44	940	3200	25.0	9.25

All profiles in the Engadin were measured in September 1972
Profil Albana was measured directly above the front slope
The three profiles on the rockglacier Val Sassa were measured directly above the front slope, 45 m, and 100 m above the front slope, respectively.
[a]This is the minimum depth above an assumed frozen material in greater depth.

soundings. Lately, Vonder Mühll (1993) published his results in relation to the hitherto gained knowledge.

Geoelectric resistivity measurement is today a very effective field method in permafrost and rockglacier research (Fisch et al. 1978; King 1982; Evin and Assier 1984; King et al. 1987; Barsch and King 1989; Vonder Mühll 1993). It is based on the fact that subsurface materials display different resistivity values which are normally measured between a fixed array and variably spaced electrodes.

Electric resistivity of a body of permafrost is a function of:

- Temperature (decreasing with increasing temperature),
- (Unfrozen) water content
- Grain size,
- Lithology of the debris.

The relationship between temperature and resistivity is displayed in Fig. 6.15.

Fortunately unfrozen sediments, frozen debris, and glacier ice show quite different resistivities. Problems are connected with:

- The difficulties of connecting the electrodes with the blocky terrain,
- Lateral effects due to the sounding on often narrow bodies of permafrost (oversteepened sounding graphs),
- The necessity for a minimum thickness of near-surface layers to be detectable,
- The temperature dependence of the resistivity (permafrost resistivity increases with decreasing temperatures), e.g., in an active rockglacier from bottom to top,

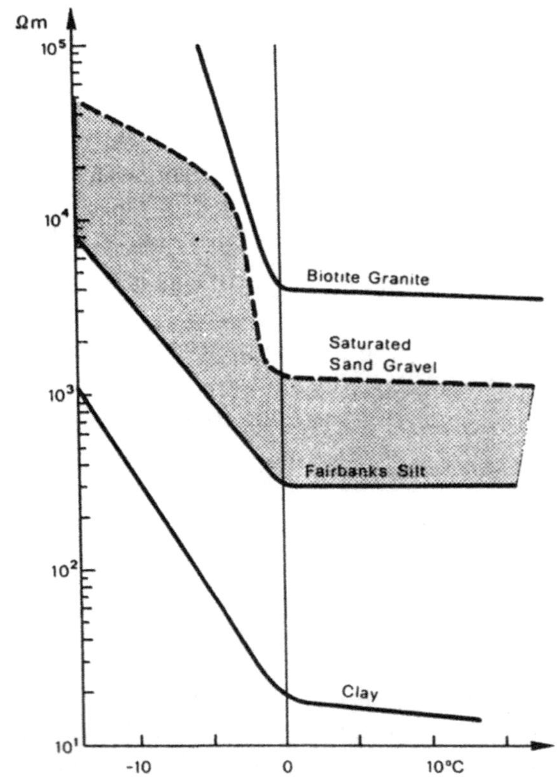

Fig. 6.15. Relationship between temperature, grain size, and resistivity from Haeberli (1985a; after Hoekstra and McNeill 1973). The *shaded part* indicates grain sizes which occur in rockglacier cores. Note the marked increase in resistivity below –1 °C

- The fact that the resistivity curves can be moved parallel. Therefore, other information (e.g., the depth of the active layer by seismic refraction) is necessary to stabilize the interpretation.

Resistivity is measured in Ω m (1000 Ωm = 1 kΩm = 0.001 MΩm). As a basis for the interpretation of geoelectric soundings, three values according to Fisch et al. (1978) shall be given:

Unfrozen debris	ca..1000 Ωm
Frozen debris	10 to > 100 kΩm
Temperate glacier ice	> 10 MΩm

Lower resistivities for glacier ice are only found under very cold arctic conditions. Resistivity values for nonpolar glacier ice have been complied by Rothlisberger and Vögtli (1967). They range between 10 and 600 M Ωm, whereas Östrem (1967) reports somewhat lower values for snowbank ice measured in laboratory tests. Similar values for unfrozen and frozen ground are given by Hoekstra and McNeil (1973) as 1000 Ωm (unfrozen, but saturated sand and gravel) and 10 000 to 100 000 Ωm (frozen material).

In any case, it is necessary to calculate model curves with existing computer programs (Koefoed 1979; Mundry et al. 1985). In addition, one should keep in mind that the conceptual models used have to be simple. In most cases, it is not worthwhile discussing too sophisticated models. On active rockglaciers three-layered models are normally assumed if the layout is long enough, although four-to five-layered systems are more realistic. The resistivity values indicated are typical but often the vertical sequence of the geoelectric resistivity is more important than the absolute figures. Some characteristic sounding graphs are presented in Fig. 6.16.

The uppermost layer on an active rockglacier normally displays a high resistivity of 10 000 Ωm or more (Table 6.8 and Fig. 6.16). Even 15 000 to 40 000 Ωm (King et al. 1987) are possible and often characteristic for the dry, coarse upper layer. The observation by Fisch et al. (1978) that a value of 10 000 Ωm indicates already frozen material is only acceptable in fine-grained frozen deposits.

The second layer, which represents the mountain permafrost in an active rockglacier, shows a still higher resistivity of 85 000 to 250 000 Ωm, which excludes temperate glacier ice. The values in Table 6.6 indicate that an ice core can be excluded. A calculation with an equivalent model indicates that the measured resistivity curve can only be obtained by the substitution of a 3m thick ice core.

The third layer has a resistivity of about 20 to 25 kΩm. It is normally moist unfrozen or frozen bedrock or wet unconsolidated and unfrozen material (cf. the data from the El Salto rockglacier; (Barsch and King 1989; Fig. 6.17).

Similar values have been measured on the Gruben rockglacier (King et al. 1987). The second layer displays apparent resistivities of more than 2 MΩm. This is probably due to lateral effects. The authors assume the "true" specific resistivity to be 500 to 600 kΩm. In a model calculation, the effects of a near-surface layer of ice (15 MΩm) has been tested. Even if this layer is assumed to be only 4 m thick, the apparent resistivities are still too high. Thus, a complete ice core can be excluded, in contrast to the assumption by Whalley (1974a), even if at some restricted places in the upper part of the Gruben rockglacier dead ice of the glacier advance of 1856 has been found.

An instructive example is presented by Fisch et al. (1978) from the rockglacier in the Combe de Prafleuri (Fig. 6.18). It presents a long profile over the rockglacier which creeps over unfrozen debris. The thickness given for the rockglacier permafrost is a minimum value because depending on the increasing temperature (with depth) the lower boundary of the frozen material is difficult to determine. Thus, the rockglacier may be 1.5 to 2 times thicker than shown. Even if this is not the case, it moves with a minimum thickness of about 40 m.

Summarizing, resistivity soundings are a valuable method for determining frozen ground. The resistivity ratio between the mantle and the inner core of an active rockglacier is 10 to 200. This allows a clear differentiation. In addition, the resistivity of frozen ground ("congelation" ice according to Shumskii 1964)

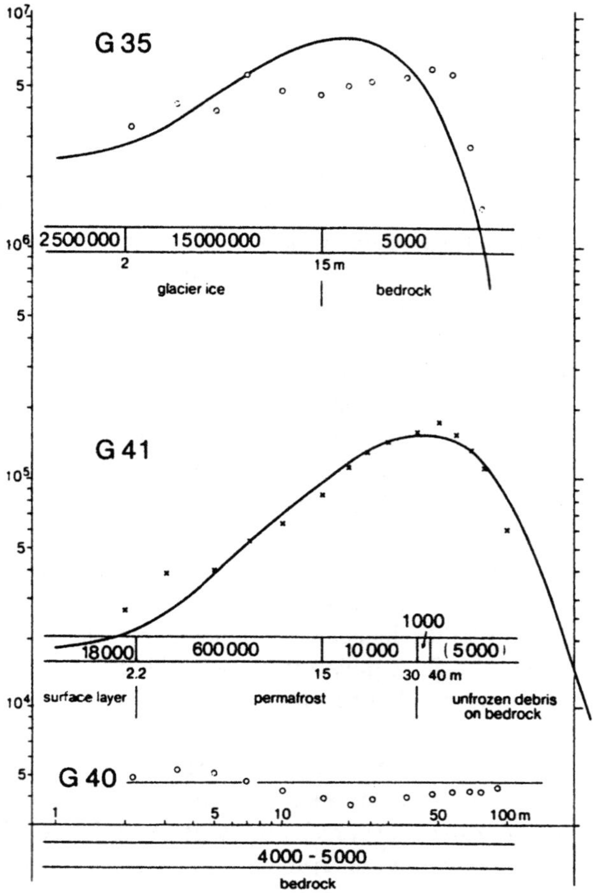

Fig. 6.16. Geoelectric sounding graphs (resistivity) on bedrock (*G40*), on rockglacier permafrost (*G41*) and on a glacier in the Corvatsch/Murtél area. The corresponding numerical models are given in Ohm-m below the graphs, the resulting depths are in meters. (King et al. 1987)

Table 6.8. Resistivity data measured on different rockglaciers

Rockglacier	region	Mantle	Core (k W m)	Base	Depth (m)	Reference
Gruben	Wallis	4–28	520–2500[a]	2–6.8	70–110	King et al. (1987)
El Salto	Argent.	12–50	85–250	20–25	65–110	Barsch and King (1989)
Morenas Coloradas	Argent.	7–12.5	30–50	?	> 50	Barsch and King (1989)
Murtèl I	Graub.	5–15	50–2000	ca.5	max. 100	Mühll (1993)
Muragl	Graub	15–30	80–3000	2–3	30–50	Mühll (1993)
Suvretta	Graub.	5–20	50–1000	ca. 1	?	Mühll (1993)

[a]According to the authors, this is probably due to lateral effects.

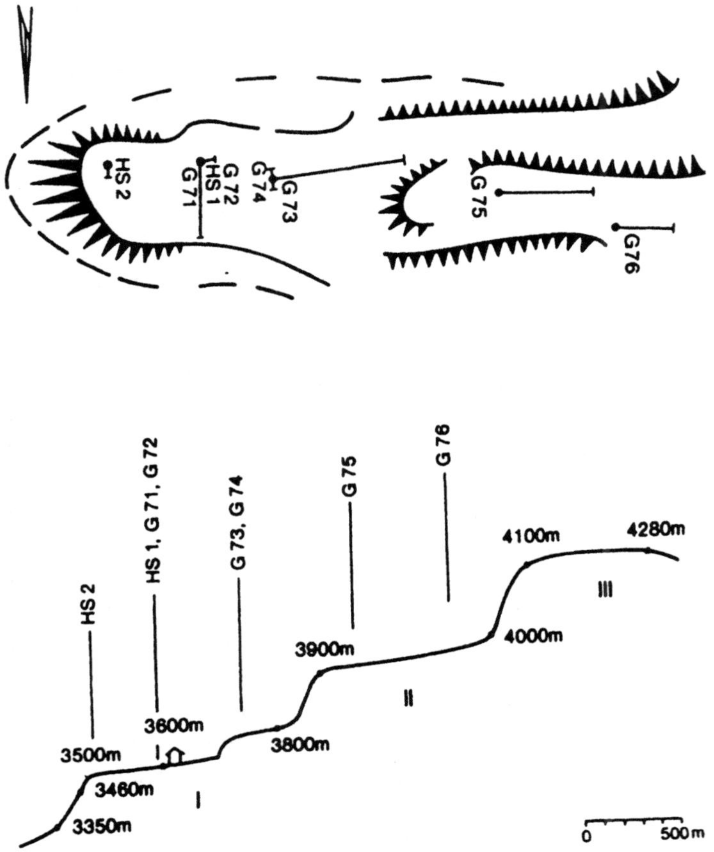

Fig. 6.17. Sketch map and long profile of the rockglacier El Salto (W of Mendoza, Argentinian Andes)

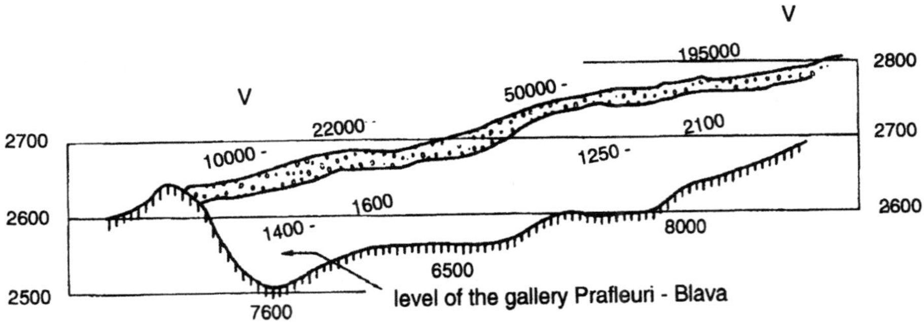

Fig. 6.18. Results of geoelectric soundings on a longitudinal profile over the rockglacier in the Combe de Prafleuri (Val d' Heremence, Wallis) (Fisch et al. 1978): Rockglacier permafrost thickness is given as minimum value, because the increasing temperature with depth makes a determination of the lower boundary of the rockglacier permafrost difficult. The rockglacier is moving with a minimum thickness of 40 m. Vertical and horizontal scales are equal

is much lower than that of glacier ice ("sedimentary" ice). Thus, all hitherto known resistivity measurements reveal that rockglaciers have a permafrost core and not an ice core derived from a glacier. Geoelectric sounding also allows the mapping of the thickness of the rockglacier mantle (active layer).

6.2.2.3 Radio-Echo Soundings

Radio-echo sounding or impulse radar ("georadar") are used for measuring ice thickness. Especially with low-frequency instruments (Watts and Wright 1981), thickness determination of temperate and cold glacier ice has become very easy. This well-developed technique has been used on rockglaciers too. Haeberli (1985a; cf also King et al. 1987), reports his experience with the US Geological Survey Monopulse ice radar (Hodge 1978), which uses a frequency of 2.5–5 MHz. On the debris-covered temperate Rossbodengletscher, the reflected signals are strong and clear; the unfrozen debris cover has no effects on the technique. The thickness of the glacier has been determined to be between 25 and 70 m. On the rockglaciers Gruben (Wallis) and Murtèl I (Graubünden) the return signal is undetectable or very small. Absorbtion and scattering are so great that a determination of rockglacier permafrost seems impossible if thickness is greater than 30 m.

These experiences indicate that the use of radio echo sounding in rockglacier permafrost research is very limited. The only important information which can be taken from these measurements is the fact that rockglacier permafrost, on the one hand, and glaciers as well as debris covered glaciers, on the other, display quite different results, probably due to very different internal structures.

Measurements with a georadar using a frequency of 50–300 MHz (maximum energy around 65 MHz) have also been undertaken in recent years (Vonder Mühll 1993; pp. 111–114) on the rockglacier Murtèl I. Unfortunately, the wavelength of 65 MHz frequence corresponds to the size of the larger boulders. The results are again ambiguous and not very promising. Vonder Mühll (1993) believes that perhaps lower frequencies (< 50 MHz) would give better results. At the moment, georadar is not helpful in rockglacier research.

6.2.2.4 Gravimetry

The gravity field on Earth is influenced by latitude, altitude, topography, tidal effects, and the subsurface density distribution. If the first four parameters are known (this is normally the case) then the difference between the measured and the theoretical gravity at each point of the Earth's surface is a function of the subsurface density distribution. Gravimetry has already been used in permafrost research (Kawasaki et al. 1983). It was first used on a rockglacier by

Vonder Mühll (1993, pp. 115–126; cf. Vonder Mühll and Kingelè 1994). During summer 1991, three profiles on the rockglacier Murtèl I were measured. The regional anomaly is around –156 mgal; the local anomaly (on the rockglacier) is up to 1.5 mgal lower. The difference between measured and calculated gravity is ca. 0.1 mgal which represents ca. 1.3 to 2 m of ice depending on the basal topography (Vonder Mühll and Kingelè 1994, p. 14). For the theoretical model, it is assumed that the rockglacier consists of four layers:

1. The mantle with a density of 2.47 Mg m^{-3}, thickness 3 m;
2. The upper core with a density of 0.95 Mg m^{-3}, depth 3–20 m;
3. The lower core with a density of 1.00 Mg m^{-3}, depth 20–30 m;
4. The lower permafrost with 2.20 Mg m^{-3} density, depth 30–55 m.

This is reasonable except for the mantle where the density is too high (92.5% of granite). The voids between the boulders in the mantle normally have an estimated volume of at least 30–40%. Thus, the density would be around 1.60 to 1.90 Mg m^{-3}. Also for layer 4 (with an ice content of 30–40 vol%) the density is, from 2.00 to 2.20 Mg m^{-3}, slightly lower than assumed in the model.

Together with the density log from the borehole Murtèl I (2/1987), it is possible to display a model of the thickness of the four layers (Fig. 6.19). The results confirm the interpretation based on the deep seismic sounding (Figs. 6.8, 6.10). In particular the overdeepening of the bedrock in the central part is well visible.

The thicknesses of the four layers reveal additional information:

1. The active layer (Fig. 6.11) corresponds generally to Fig. 6.14, which is constructed from data gained by sledge-hammer seismic survey, although, the newly calculated figure seems to present slightly higher values (1–2 m more) than the data from 1972 (Fig. 6.14). This may be due either to the warming during the almost 20 years between the two measurements (1972/1991), or to the combined error of both methods, or to the high bulk density assumed for this layer.
2. The second layer, the ice-supersaturated upper core, is thickest in the central part. If this is generally the case, it would explain the central depressions (the central collapse features) of fossil rockglaciers, whereby the outer parts always stand high as ridges above the central depression. It may also explain the central zone of high velocity found on many rockglaciers (cf Chap. 7; Fig. 7.5).
3. The lower core displays an asymmetric form; the maximum thickness is found to the right (Fig. 6.19c). This may be caused by bedrock topography.
4. The permafrost layer below the low-viscosity layer (below the rockglacier proper) is still more asymmetrically developed (Fig. 6.19d); this is certainly due to the bedrock topography.

It may, therefore, be summarized that gravimetry allows quite a number of additional insights that are (in this example) partly based on the large amount of additional geophysical data on the rockglacier Murtèl I. This helps to

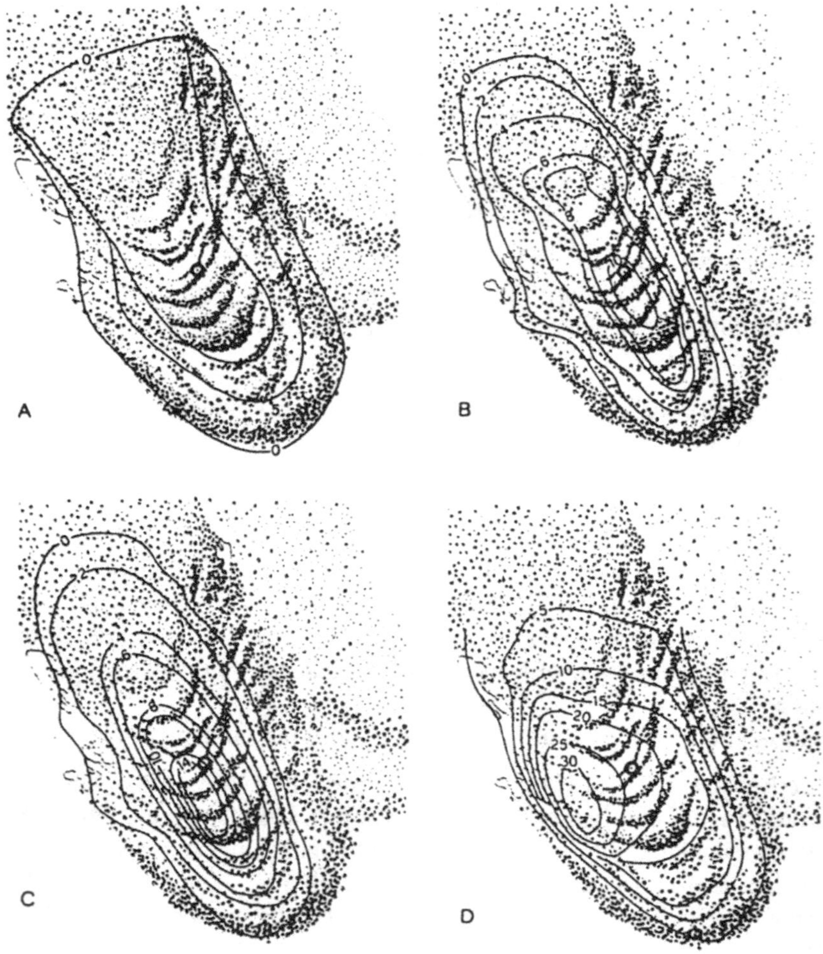

Fig. 6.19A-D. Using a four-layer model, gravimetry allows a determination of the thickness of each layer of the rockglacier Murtél I (Vonder Mühll 1993). The structure of the rockglacier is taken from a photograph for better representation of the results

minimize interpretation errors. It cannot be recommended for use as the only geophysical method on a rockglacier.

6.2.2.5 Borehole Geophysics and Related Measurements

A cored borehole gives an enormous amount of information about the substratum in which or through which it is drilled Sect. 6.2.1). The stratigraphy is represented by the log and the more or less detailed profiles. The information can be greatly enhanced by use of borehole geophysics, i.e., by logs which are run more or less continuously in the borehole. Coring during drilling is very expensive; borehole geophysics is a suitable but indirect means of obtaining information on the subsurface material. The method itself is well established in oil and gas prospecting but has hitherto been very seldom used in rockglacier research.

Normally, the following logs have been run (Barsch et al. 1979; Haeberli et al. 1979; Haeberli 1985a; Vonder Mühll 1993):

1. The caliper log; information on the diameter of the drillhole (which can be quite variable in frozen material).
2. The gamma log; measurement of the natural gamma radiation (indirect information on the ice content because minerals contain more radioactive elements than ice).
3. The gamma–gamma log; measurement of the reflected radiation from an artificial source in the borehole (information on the density of the material up to about 15 cm around the borehole).
4. The neutron–neutron log; measurement of the hydrogen content stimulated by an artificial source (indicating the water content in the material around the borehole).
5. The resistivity log; detailed measurements of geoelectric resistivity along the borehole indicating also the ice content in layers with high resistivity (interesting in comparison with surface geoelectric measurements).
6. The sonic log; data on seismic velocities in the material around the borehole.

In the borehole on the Gruben rockglacier (19–24 June 1976), three logs were run (gamma, gamma–gamma, neutron–neutron). The results and the interpretation are presented in Fig. 6.5 (for details, see Barsch et al. 1979; Haeberli 1985a). Under the permafrost table (55 cm below ground level on 19 June 1976), the ice content is less than 50%. After a sudden rise between 1.8 and 2.3 m, it increases to nearly 100%. Below 4 m, it is slightly higher than 50%. The suggested interpretation seems reasonable. It demonstrates an average ice content of ca. 50% and relatively clear ice lenses near the permafrost table.

In the two boreholes on the rockglacier Murtèl I (1/1987 and 2/1987), a more sophisticated use was made of borehole geophysics. All logs were run, and the results are presented in Fig. 6.6 (Haeberli et al. 1988; Vonder Mühll 1993)

The ice content lies between 35 and 80 vol%. Ice-supersaturated permafrost has a seismic velocity of 3700–3800 ms^{-1} which is somewhat higher, but still close to that of glacier ice. Resistivity in the supersaturated permafrost is low (15 kΩm) in relation to that of glacier ice and actually decreases with depth (due to the temperature increase). Freezing-point depression is smaller than 0.1 °C. Correlation between different borehole measurements (Vonder Mühll 1993, p. 151, Table 5) in the frozen parts of the borehole confirms some interesting relationships:

1. A high ice content (neutron–neutron) reduces seismic velocity.
2. In zones with a high density (gamma–gamma), the radioactive counts are also high (radiation is activated from the rock material and not from the ice); i.e, in dense parts (gamma–gamma), the ice content is small.
3. Resistivity is higher in zones with low density (gamma–gamma).
4. Seismic velocity is higher in zones with more rock material (gamma as well as gamma–gamma).

Seismic Measurements

In addition, Vonder Mühll (1993) reports seismic measurements made in the borehole (geophones in the borehole, triggering of the shock waves on the ground or at the permafrost table). This is interesting because more information on the horizontal and vertical velocities of P- and S-waves in rockglaciers is required. Unfortunately the results are not totally clear, except that there are no P-wave velocity differences between ice-supersaturated and ice-saturated permafrost. Also, the signals caused by hammer strokes and by explosions are, for the short distances involved, of the same quality.

Inclination Measurements

Measurements of the borehole deformation or of the changing inclination in tubes (or drillholes) over time are standard methods for determining small, three-dimensional movements of quite different features. The method is seldom used, on rockglaciers because the number of boreholes is quite limited. (Girsperger 1973; Johnson and Nickling 1979; Wagner 1992).

Johnson and Nickling (1979) report two measurements from a borehole into the rockglacier RGII (Yukon) that were carried out immediately after the drilling (1969) and seven years later. In 1969, average inclination was about 4° towards an azimuth of 300° (i.e., towards the head of the rockglacier). In 1976, the average inclination was higher (6° towards 308°). Accuracy of the measurements is assumed to be ±0.5°.

The results, corrected for an apparent azimuth of 304°, are presented in Fig. 6.20. Horizontal movement at the top of the borehole was 51 cm a^{-1} (3.6 m) and at the bottom 46 cm a^{-1} (3.2 m) towards 160°. The graph shows on the left-hand side the movement of the tube in the rockglacier. The right-hand side of Fig. 6.20 displays a uniform deformation (0 to 40 cm in 7 years) with depth. It demonstrates that the deformation rate with depth is about zero in the top 3 m. Below this depth, it is more or less constant:

$$y = 37\,x - 16 \qquad (6.8)$$

where y = depth (m), x = deformation (m).

Small variations in the deformation rate are probably within the limits of accuracy. They should not be overinterpreted even though the authors point out that these variations are situated at depths where during drilling operation differences in the subsurface material were encountered.

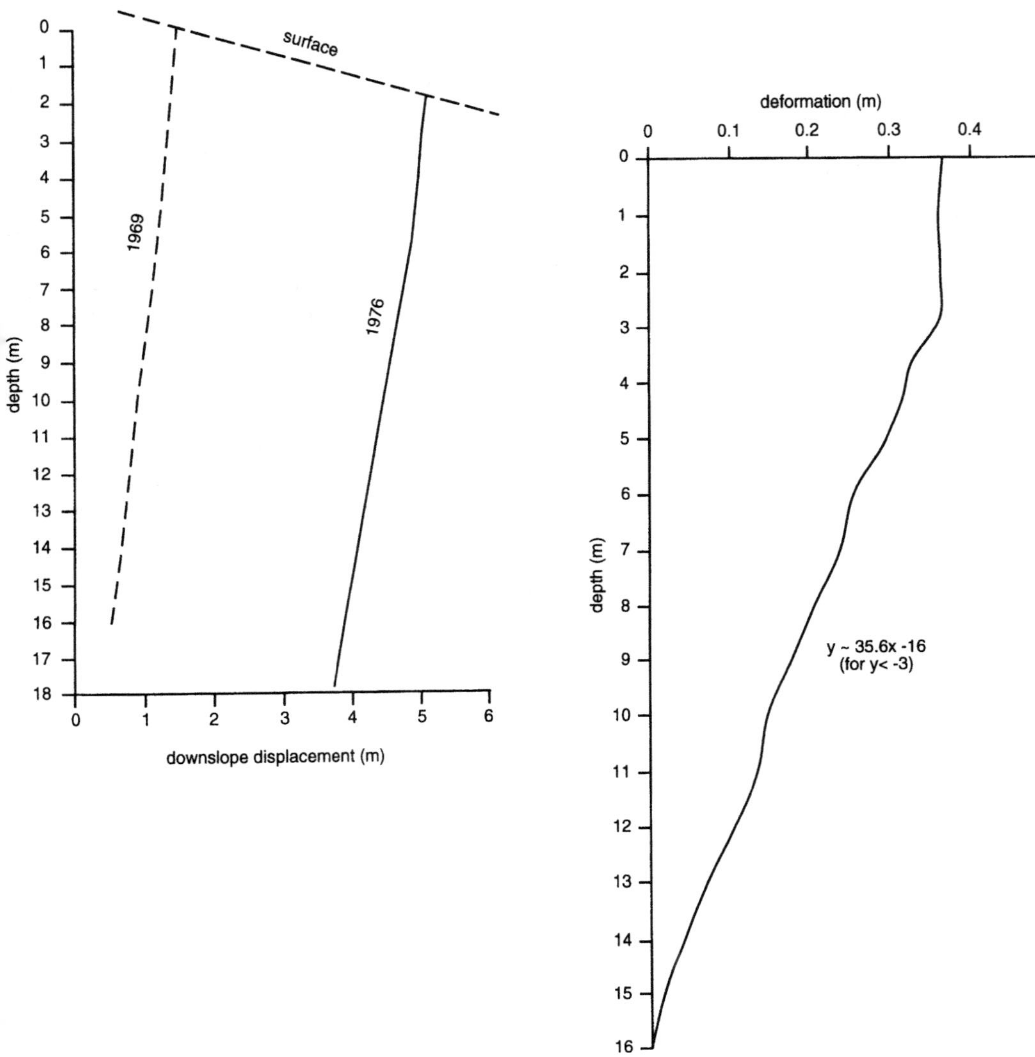

Fig. 6.20. Inclination measurements done in the borehole into the rockglacier RGII (Yukon) according to Johnson and Nickling (1979). The *left* part displays the downslope displacement of the borehole between 1969 and 1976; the *right* part represents the deformation in relation to depth

In general, it can be stated that it appears that the bouldery mantle (top 3 m) is passively riding on the rockglacier core. Also, even if the rockglacier is, according to the temperature measurements, at the edge of inactivity, the lower of the surface by 13 cm a^{-1} is due to the gradient ($14 \pm 1°$, assuming a horizontal displacement of about 51 cm a^{-1}). Assuming Eq. (6.8) to be correct and assuming no shear planes below 16 m depth, the maximum deformation depth of the rockglacier can be extrapolated from the difference in horizontal displacements to be about 120 to 130 m. This is probably much too high because the linear gradient can only be the upper part of an exponentially changing gradient as in the Murtèl I borehole (see below).

The inclination measurements by Girsperger (1973) were carried out for the scientific commission of the Swiss National Park on the rockglacier Val Sassa (Graubünden). They were done in small tubes drilled into the rockglacier mantle. The data suggested a horizontal displacement of 20 cm a^{-1} at the time of measurement (1972). Unfortunately, the apparent displacement was due to subsidence (cf. Sect. 7.1.3). This demonstrates again that these measurements have to be done in accordance with a theoretical setup based on the best possible knowledge on the geosystem which is studied. The application of sophisticated methods alone does not guarantee success.

Very sophisticated and very careful inclination measurements have been done in borehole 2/1987 in the rockglacier Murtèl I (Haeberli et al. 1988; Haeberli 1990a). Inclination was measured by slope inclinometers (horizontal deformation) and magnetic rings (vertical deformation) according to Wagner (1992, p.158). The measurements done in November 1987 demonstrated that the inclinometers did not have enough contact to the rockglacier after re-freezing. Thus, reliable data were not recorded until 9 months after drilling, i.e., since December 1987 (Figs. 6.21, 6.22). Both figures display clearly the

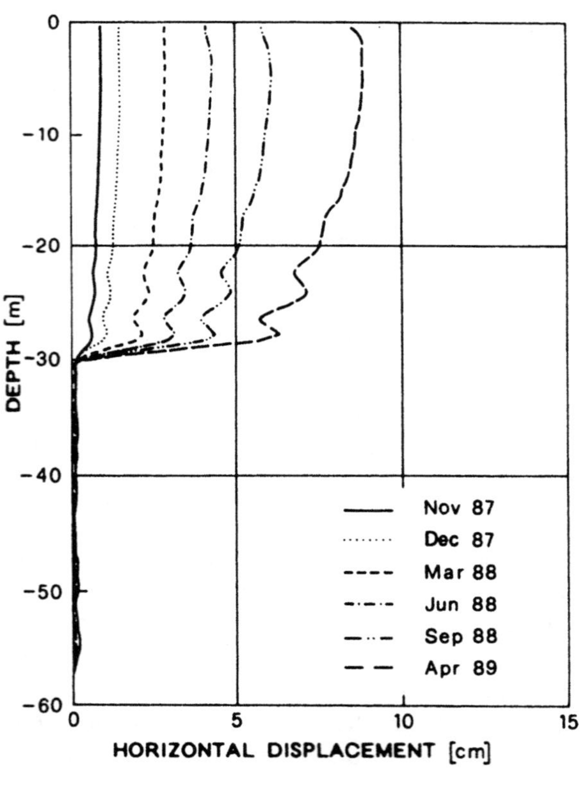

Fig. 6.21. Horizontal displacement in the borehole 2/1987 on the rockglacier Murtèl I from Wagner (1992).The displacement is calculated from the slope measurements; a normalization to the same temperature has not been done

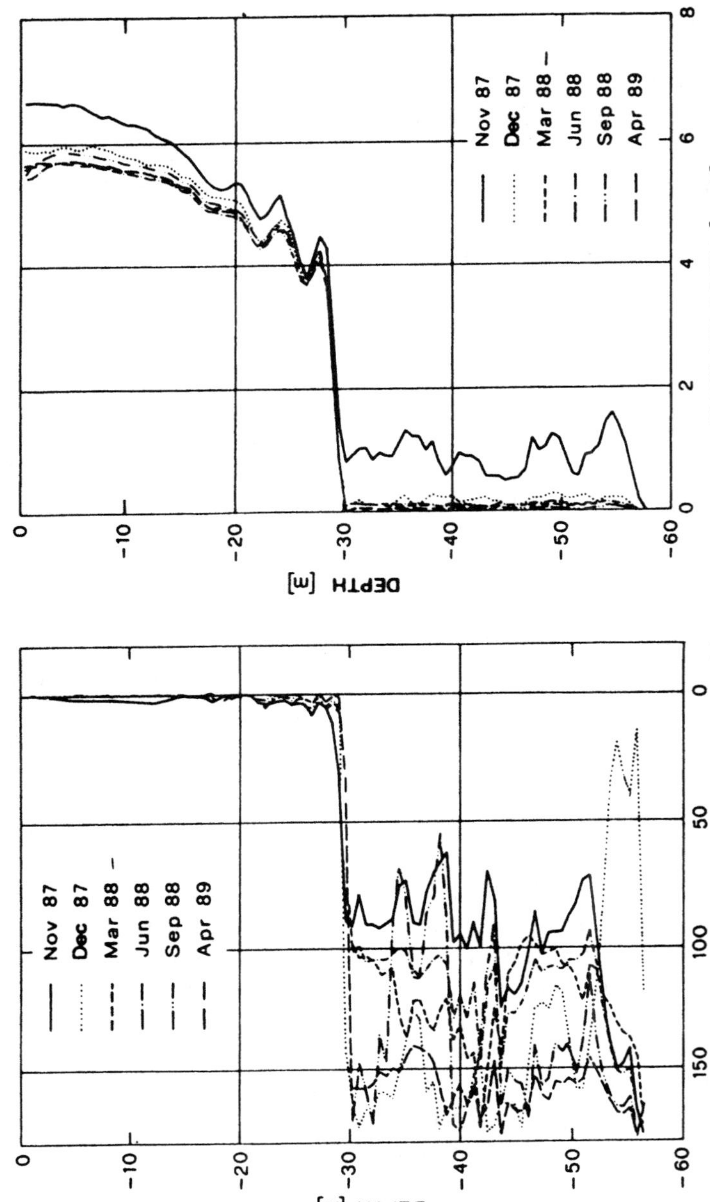

Fig. 6.22. Flow direction in a horizontal plane and flow velocity (cm a^{-1}) calculated from the horizontal displacement in the drillhole 2/1987 through the rockglacier Murtèl (I Wagner 1992)

two-layered structure: the upper part shows a remarkable deformation which is concentrated at the depths of 28–32 m where 75% of the deformation occurs. The lower part (32–58 m) moves with a velocity which is an order of magnitude smaller than that of the upper part. Similar behaviour is also confirmed by the vertical deformation. Following horizontal and vertical deformation in the upper ice-supersaturated layer, a flow angle of ca. 10° can be calculated which corresponds to the surface gradient of the lower part of the rockglacier.

From the results, the following points are of interest:

1. The flow velocity calculated from the horizontal displacement (Fig. 6.22) does not show the velocity variations in time which are found by precise geodetic surveys at the surface. There is no explanation for this behaviour at the moment.
2. Glen's flow law can be applied under a number of conditions to describe the relationship between the effective strain rate and the effective shear stress. The results are shown in Fig. 6.23.
3. The "weak" layer (28–32 m) is about four times less viscous than the overlying section.
4. The ice with the low debris content in the upper part (below the active layer) is weaker than glacier ice under comparable conditions.

These are the first very comprehensive results recording the deformation in an active rockglacier. Even if it is necessary to be cautious with any generalizations, it can be said that they give very interesting new insights into a very complex system.

Temperature Measurements

Temperature measurements in boreholes in permafrost areas have been used for a long time to obtain better information on these geosystems (Gold and Lachenbruch 1973; Lachenbruch et al. 1988). There are a number of problems involved regarding thermal conductivity along the measuring chain, the sensors used, calibration and drift of sensors after installation etc. Nevertheless, it is a possible to gain a better understanding of a very complex system. Vonder Mühll and Haeberli (1990) discussed heat flow and temperature influences on active rockglaciers (Fig. 6.24). Following their approach, present-day temperatures in an active rockglacier are a function of:

1. External factors such as
   - climate (e.g., radiation, advective heat, etc.)
   - paleoclimatic influences
   - sedimentation influences (e.g., input of snow by avalanches)
2. Boundary layer factors such as
   - heat-flow transition between the atmosphere and the rough and irregular rockglacier surface
3. Internal factors such as
   - phase transitions in the active layer

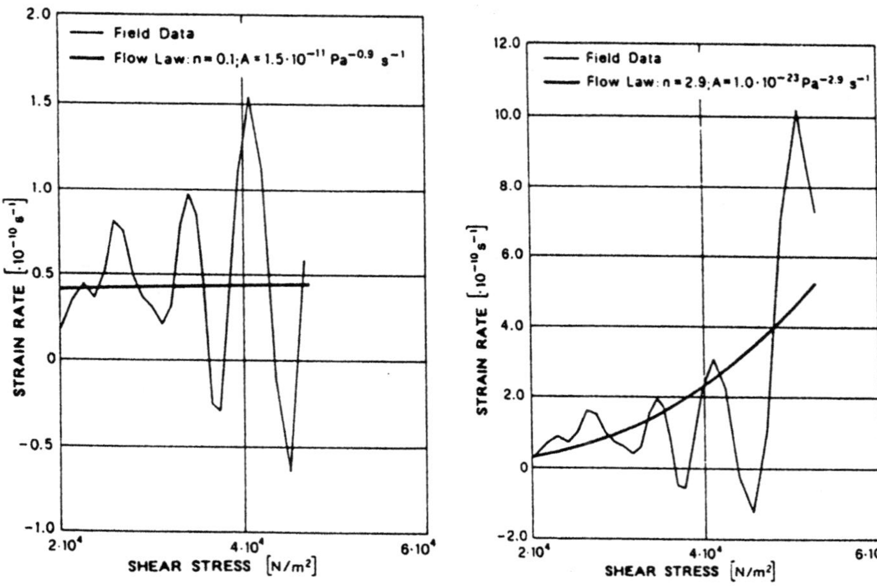

Fig. 6.23. Strain rate and shear stress calculated (least square interpolation) from the deformation data in the drillhole 2/1987 through the rockglacier Murtél following Glen's flow law (Wagner 1992). *Left* The upper layer without the weak layer (12–27 m); *right* the upper part including the weak layer (12–29 m)

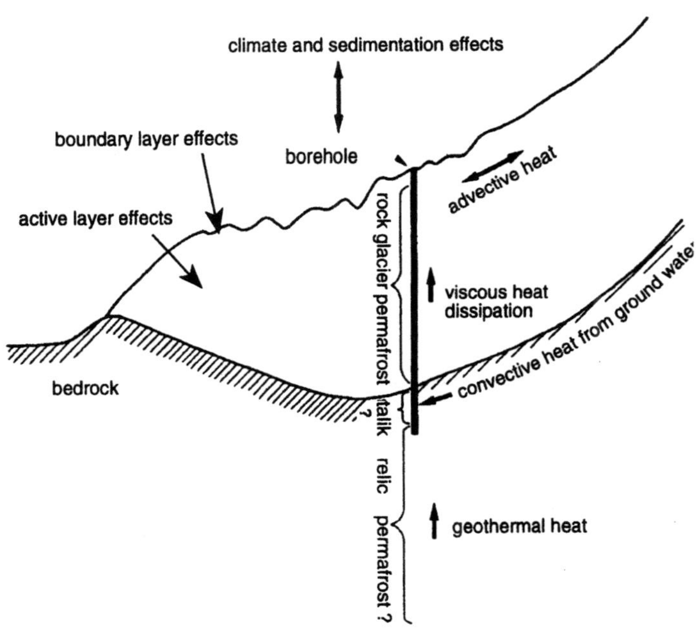

Fig. 6.24. Conceptual model of the heat flow in an active rockglacier. (After Vonder Mühll and Haeberli 1990)

  - advective heat (e.g., by water on the permafrost table etc.)
  - viscous heat dissipation (frictional heat)
  - convective heat (e.g., groundwater transport)
4. Interior factors such as
  - geothermal heat flow

The heat flow (q) along a vertical profile through an active rockglacier is a function of the temperature gradient and the thermal conductivity. In a simple approach, the latter can be assumed to be constant with depth because most crystalline rocks and ice have similar thermal conductivites. Variations in the calculated values of q are, therefore, directly correlated with the temperature gradient. Unfortunately, the amount of temperature data from rockglacier permafrost is rather limited; however, the information collected hitherto is sufficient for an initial evaluation.

Temperature measurements were made in the borehole located about 100 m above the front slope of the rockglacier RGII in the Yukon (Johnson and Nickling 1979) on 15 July 1970 (about 1 year after the drilling), 7 August 1971 and 5 August 1976. The northerly exposed talus rockglacier ends at 1090 m a.s.l (300 m above the valley floor). The active layer on the rockglacier supports stands of trees and seems not to have be in equilibrium with the climate in the early 1970s in 1970 it was about 4 m thick, in 1971 only 2 m thick, and in 1976 8 to 11 m thick. At the time of the measurements in 1970 and 1971, the temperature below the permafrost table was slightly negative. Under the presumption that there are no perturbations, the linear extrapolation of the temperature profile of August 1971 allows the depth of the permafrost body to be estimated at 27 to 28 m. In 1979 only a thin layer between 11 and 12.5 m displayed a temperature just below zero. Below 15 m, a temperature of around +0.8 °C was recorded at this time. This probably indicates that heat is transported by flowing water which seems reasonable because during drilling water losses were recorded. If these temperature measurements are correct and if they indicate a trend then this rockglacier is at the edge of inactivity. It would be worthwhile repeating the measurements on this rockglacier again today.

After a number of installations on several rockglaciers in Graubünden, which encountered some of the difficulties described above, very intense measurements were carried out in the borehole on the Gruben rockglacier (Barsch et al. 1979; Haeberli 1985a). They are discussed in detail in Haeberli (1985a) and presented in Figs. 6.25 and 6.26.

During the observation period (1976–1982), the MAAT at nearby meteorologic stations was close to the mean of the period 1901–1960 (summer temperatures slightly low, winter temperatures slightly high). Figure 6.25 shows that the 3 m thermistor is situated below the permafrost table, which lies at ca. 2.3 m depth, during these years. The mean temperature at 3 m depth is –1.1 °C, the extremes are recorded as –4 °C and close to zero. According to Haeberli (1985a), the mean temperature in February and May t_3(F,M) is well

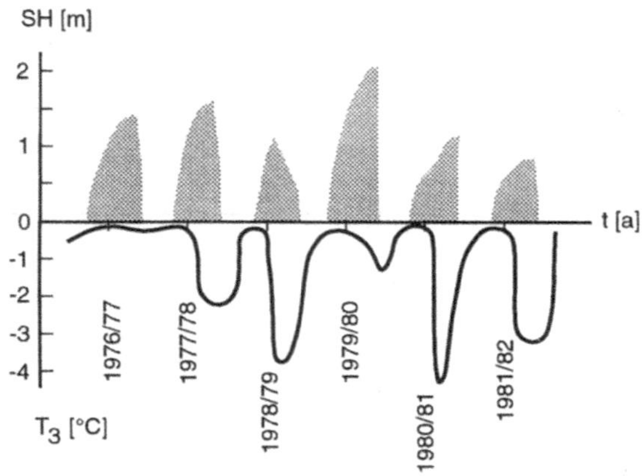

Fig. 6.25. Borehole temperature at 3 m depth (T_3) and snow thickness *(SH)* during winter on the rockglacier Gruben. (Haeberli 1985a)

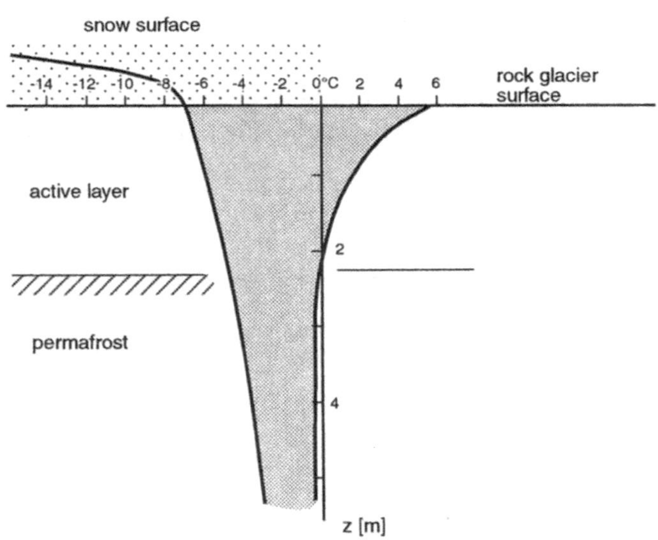

Fig. 6.26. Temperature extremes in the borehole and snow cover thickness on the rockglacier Gruben 1976 to 1982. (Haeberli 1985a)

correlated with the mean thickness of the snow cover in November and February d_s(N,F) in meters (cf. Fig. 6.26):

$$t3(F, M) = -5.195 + 3.893\, ds(N, F) \; r = 0.9797 \; . \qquad (6.9)$$

Table 6.9. First results from the temperature measurements in the rockglacier Murtèl I, Graubünden, Swiss Alps. (According to Vonder Mühll and Haeberli 1990; Vonder Mühll 1993

Borehole	:	2/1987, 62.5 m depth
Altitude	:	2670 m

Temperature disturbance by the drilling	Mid-1988
Mean annual surface temp.	–3 °C
Permafrost table	ca. 3 m depth
Annual amplitude < 1 °C	ca. 10 m depth (at 11.6 m: 0.6 °C)
Phase difference at 11.6 m	ca. 0.5 a
Zero annual amplitude	ca. 20 m depth
Temperature gradient	
20–30 m depth	69 ± 11 °C km^{-1}
30–50 m depth	47 ± 5 °C km^{-1}
Thermal conductivity	2.5–3.0 W m^{-1} $°C^{-1}$
Heat flow	150 W m^{-2} (av. Switzerland: 85 W m^{-2})

Table 6.10. Mean seasonal temperature amplitude 1987–1992 in the borehole 2/1987 on the rockglacier Murtèl I, Graubünden, Swiss Alps according to the data in Vonder Mühll (1993)

Depth[a] (m)	Mean fluctuation (°C)
1.6	9
2.6	5
7.6	1.5
11.6	0.6
15.6	0.2
21.6	< 0.1

[a]Depth is measured from the top of the liner tube which is 0.5 m above the ground.

A thicker average snow cover during these months (e.g., of 0.1 m) increases the mean monthly temperature of February and May at this depth by 0.39 °C. This also affects the mean annual permafrost temperature at this depth. Haeberli (1985a) cites an example: an increase in the mean snow cover (during these months) of just 8 cm increases the mean annual permafrost temperature by 0.15 °C.

On the rockglacier Murtèl I (borehole 2/1987), temperature measurements have been carried out since 1987 (Vonder Mühll and Haeberli 1990; Vonder

Mühll 1993). Measurements are done with fixed thermistors (NTC) and with a movable thermistor chain. The absolute error is assumed to be smaller than ±0.05 °C (Vonder Mühll 1993, p.165). Because the relative error is four times smaller than the absolute error (Haeberli and Funk 1991), the measurements are certainly correct in the range of ±0.02 °C. The most important information obtained is summarized in Table 6.9. The decreasing seasonal temperature fluctuations with depth are presented in Table 6.10. It demonstrates a known effect that has not been demonstrated before in rockglacier permafrost.

The temperature series for the base of the active layer is given in Fig. 6.27 and that for two selected depths (1.6 and 11.6 m) is given in Figs. 6.28 and 6.29. The three figures indicate a marked increase in the temperatures since 1989. This is also clearly visible in the summer maximum temperatures in the rockglacier mantle (active layer, thermistor 1.6 m), despite the fact that the temperature is not recorded continuously. At this depth the increase in winter minima is less pronounced probably due to special conditions in this system. At 11.6 m depth, the increase (1988–1992) amounts to 0.3 °C. A more subdued, but still clear, increase in temperature is found at the thermistors at greater depths. The data are summarized in two profiles (Fig. 6.30) which also show

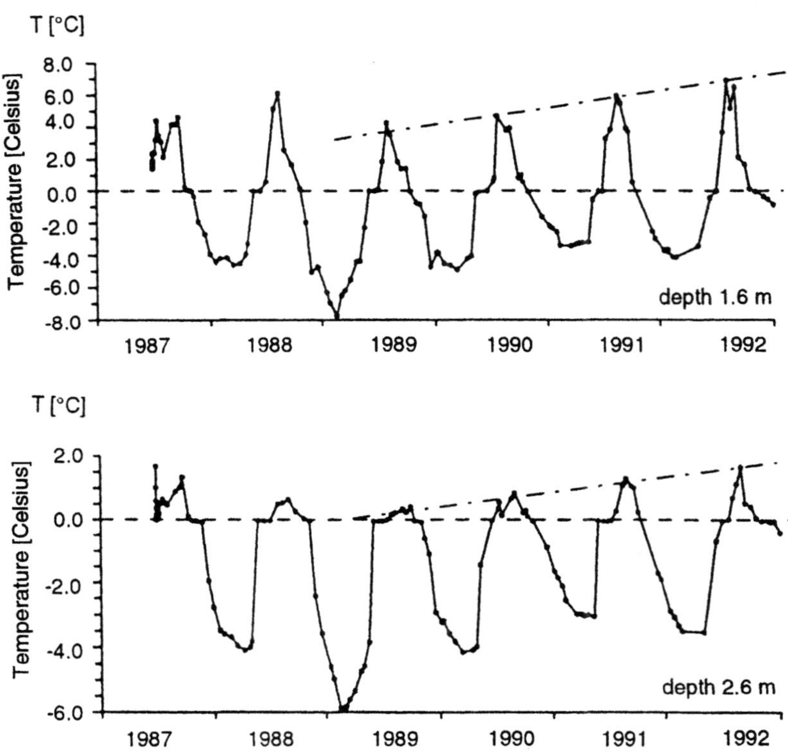

Fig. 6.27. Borehole temperatures in and near the base of the active layer on the rockglacier Murtél I, (Vonder Mühll 1993)

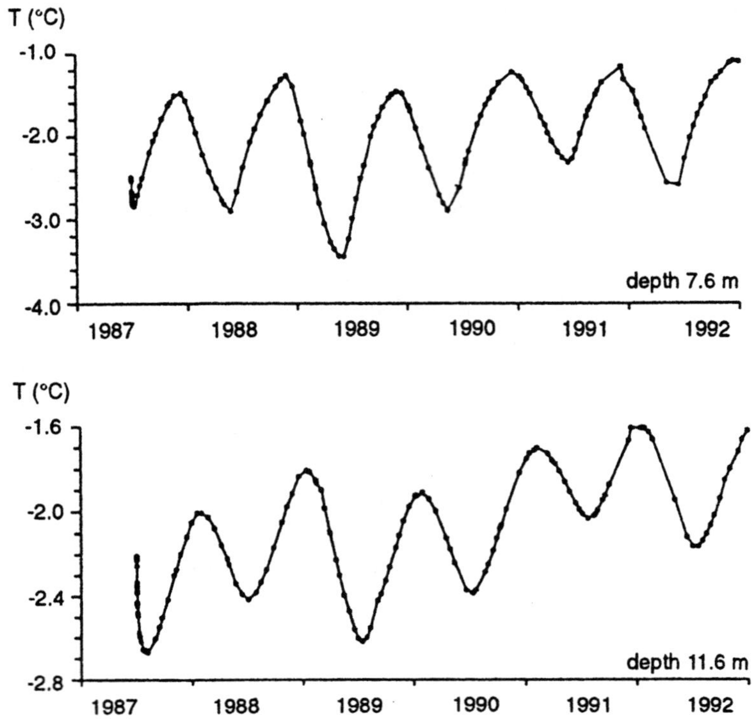

Fig. 6.28. Borehole temperatures on the rockglacier Murtél I (Feb. 1987) in 7.6 and 11.6 m depth (Vonder Mühll 1993). The warming in the 5 years at 11.6 m depth amounts to 0.3 °C

the warming trend, probably due to the warm summers since 1980 and a change in winter snow cover (see above).

Greatest interest is caused by the observation that at 52–56 m depth temperatures around and slightly above 0 °C are encountered during summer. The positive temperature far below the zero annual temperature fluctuation can only be explained by water circulating in the permafrost. It is assumed, therefore, that during summer water circulates in a talik and is responsible for the heat transport at this depth. Information on this phenomenon is still too incomplete to allow a final explanation. In any case with regard to the hydrology of rockglaciers (cf. Sect. 10.4), this observation is of the greatest importance.

6.2.2.6 BTS Measurements

The thermal balance of rockglacier or mountain permafrost and its active layer has in general a great influence on the wintertime surface temperature below an insulating snow cover. This fact has been used by Haeberli (1973) to measure

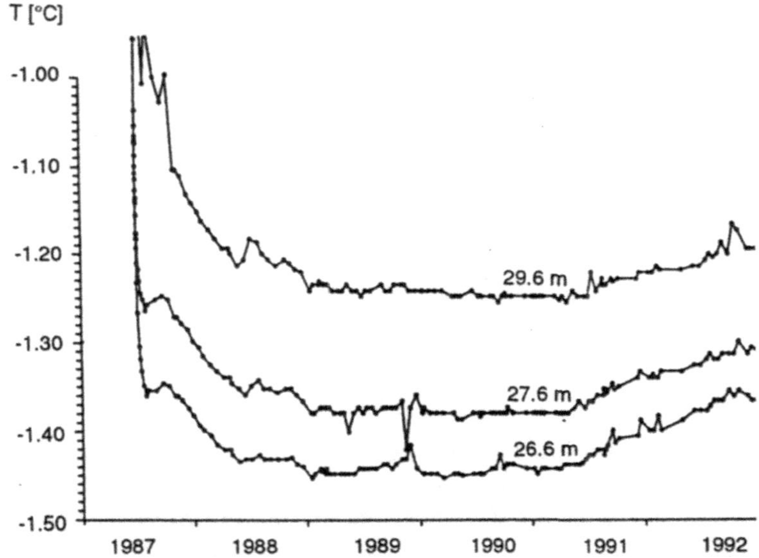

Fig. 6.29. Borehole temperatures on the rockglacier Murtél I (Feb. 1987 at greater depths (26.6, 27.6, and 29.6 m) (Vonder Mühll 1993). The heat disturbance by the drilling ends in mid-1988. The warming after 1990 is due to the warm temperatures in the 1980s

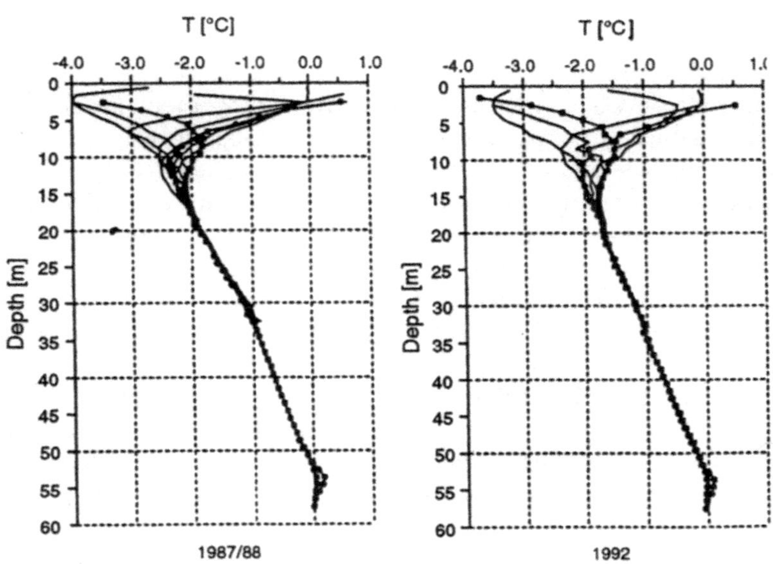

Fig. 6.30. Temperature profiles through the rockglacier Murtél I. (Vonder Müll 1993)

Table 6.11. BTS temperatures as mountain permafrost indicators according to Haeberli (1973, 1975) and Haeberli and Patzelt (1982)

BTS (°C)	Mountain permafrost
<–2	Not probable
–2 to –3	Possible
<–3	Probable

The temperature at the base of a snow cover of at least 0.8 to 1 m is measured at the end of winter (February/March).

soil-surface temperatures below a snow cover of at least 0.8 to 1 m at the end of winter. The insulation of this snow cover is thick enough to prevent daily temperature variations influencing the temperature of the soil surface and to allow an equilibrium (steady state) to be established that is controlled by the heat flow from the ground. Thus, the temperature at the base of the wintertime snow cover (BTS, *Basistemperatur der winterlichen Schneedecke*) is an indicator of the thermal balance in the ground (cf. Table 6.11). It can, therefore, also be utilized as an indicator for the existence of rockglacier permafrost.

BTS was introduced in the Grialetsch area (Graubünden) by Haeberli (1975) in the early 1970s. Since then, a number of studies (Haeberli and Patzelt 1982; King 1984; Haeberli 1985a; Hoelzle 1992) have collected more information on this method. Critical comments by Harris (1986) have been rejected (King et al. 1992).

According to many measurements, a BTS below −2 °C is an indicator for the probable existence of permafrost at the test site; a BTS between −2 and −1 °C does not preclude mountain permafrost, and a BTS above –1 °C makes mountain permafrost unlikely. The lower boundary value was originally set at −2 °C but Haeberli and Patzelt (1982) selected −2.3 °C (due to a corresponding active layer of 5 m on the rockglacier Äußeres Hochebenkar). On active rockglaciers permafrost BTS is often as low as −5 or even −7 °C, i.e., in most cases it is clearly below −3 °C (cf. Table 6.11). An example is presented in Fig. 6.31.

According to Haeberli and Patzelt (1982) a strong correlation exists between BTS and the thickness of the active layer in summer (Fig 6.32). This relationship has been established on the rockglacier Äußeres Hochebenkar, but it should hold for similar rockglaciers (in terms of exposure, altitude) in the neighborhood.

The hitherto collected BTS measurements are practically the same from year to year. Thus, it may also be concluded that the active layer is constant on active rockglaciers from year to year (Haeberli and Patzelt 1982). On the other hand, the thickness of the active layer seems to change at least in decades. It

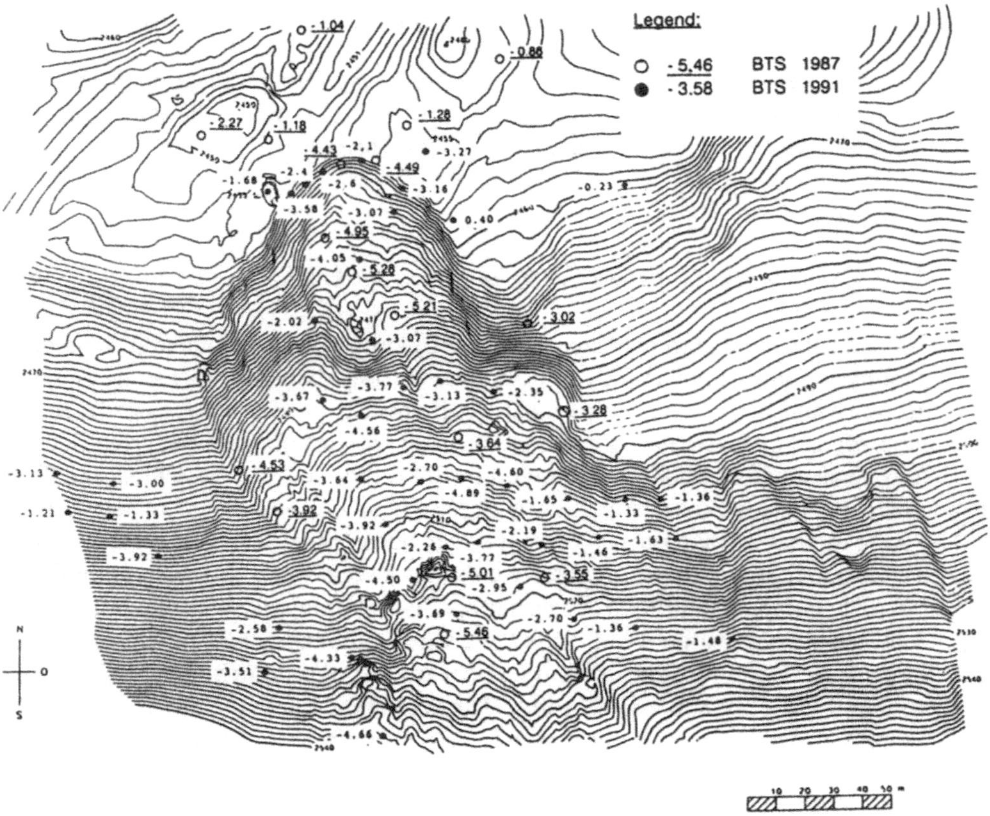

Fig. 6.31. BTS (temperatures) in 1987 and 1991 on the rockglacier in the Furggentäli. (Krummennacher and Budmiger 1992)

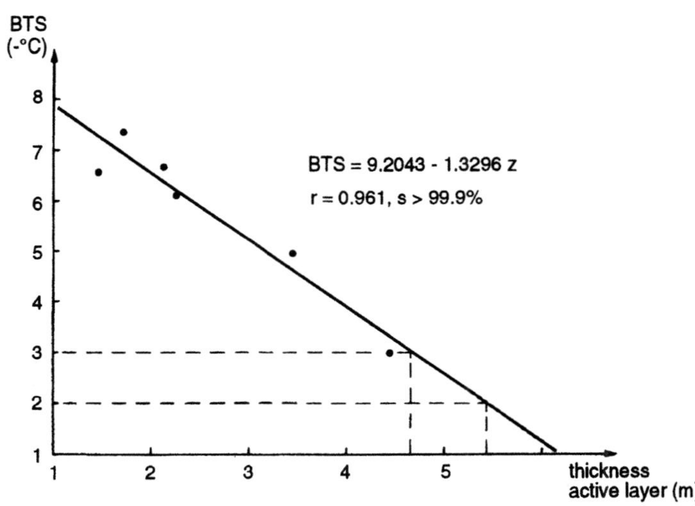

Fig. 6.32. Relationship between BTS (temperatures) and thickness of the active layer on the rockglacier Äußeres Hochebenkar. (Haeberli and Patzelt 1982)

would, therefore, be very interesting to set up a long-term monitoring program to check BTS from year to year to study probable changes in the thickness of the active layer that may occur due to global warming.

6.2.2.7 Summary: The Inner Core of an Active Rockglacier

All scientifically sound evidence collected by direct observations and indirect geophysical measurements permits the conclusion that the inner core of an active rockglacier is of a complex nature. It consists of frozen debris or frozen talus often supersaturated or at least saturated with ice. By volume, ground ice may be as much as 80–90% or even up to 100% [boreholes at Mutèl I (1 and 2/1987); on average it is estimated to be 40–60%. In any case, it is definitely wrong to assume that glaciers form the rockglacier cores (cf. Sect. 8.2.2), even if occasional observations of massive ground ice seem to prove this hypothesis.

The complexity of the rockglacier core is not only related to the different contents of rock material but also to the variation of the grain sizes involved. In contrast to the rockglacier mantle, finer fractions (sand, silt) are dominant; boulders are only singular features. The rock type encountered reflects the bedrock distribution in the area which feeds material into the rockglacier.

Unfortunately, the influence of the rock type on the geophysical signals from the rockglacier core have not yet been studied.

Geophysical methods allow information on the bedrock surface beneath the rockglacier or beneath the rockglacier permafrost to be obtained. In the cases studied, overdeepened basins are hidden under the rockglaciers Macun I and Murtel I. Under the rockglacier Suvretta (Vonder Mühll 1993), the bedrock surface has not been found. Deep seismic soundings therefore allow very good approximations of the rockfloor. Geoelectric and gravimetric measurements confirm these conclusions and provide additional information. Final proof of the interpretation of the deep seismic data on the rockglacier Murtèl I, which is today the best-known rockglacier, is given by the Haeberli drilling (2/1987).

The studied rockglaciers fill the overdeepened basins. They do not move directly on the bedrock/permafrost interface. Maximum movement occurs in a low-viscosity layer which can extend to the top of the rock threshold over which the rockglacier moves (cf. Fig. 6.9). This layer is not a single shear plane or a group of shear planes, but a low-viscosity, highly deformable layer in which 75% of the deformation responsible for the movement occurs.

Questions which have still not been satisfactorily answered relate to the existence of reflectors, to other inner deformation planes, or to stratigraphic bounderies in the rockglacier core. The aforementioned low-viscosity layer has been found only by the deformation of the 2/1987 borehole, even if its position seems mechanically logical (cf. Fig. 6.8). It did not show up in any geophysical measurement before or even after it was known. Also, apparent reflectors were found in the early deep seismic soundings on the rockglaciers Macun I and

Murtel I. They could not be explained or connected with special features of these rockglaciers.

According to the big variations in seismic velocities of the rockglacier core (V_2), it seems probable that water content and temperature increase in the rockglacier core from the root (Fig. 3.1) towards the tongue. This is not surprising if we consider the decrease in altitude, the increase in summer meltwater transport on the permafrost table below the mantle, and probably the warming in the Alps since 1856, which has increased since 1980, and which affects the lower parts more than the usually shaded upper parts. It is also not surprising that temperature increases below the depth of the zero annual temperature variation due to geothermal heat flow.

6.3 The Active Layer on Rockglaciers

Active rockglaciers are bodies of mountain permafrost. They are more or less in equilibrium with the present climate in their surroundings. As temperature measurements reveal, they adjust to climatic changes. Thus, heat flow is either from the surface down during climatic warming or up during climatic cooling, provided that the situation is not masked by other influences. A good indicator for the present heat balance on active rockglaciers is the thickness of the active layer, i.e., the summertime–unfrozen part of the mantle of an active rockglacier. Information on the active layer can be gained by sledge-hammer seismic soundings (cf. Fig. 6.14). These show that the thickness of the active layer is thickest above the toe (normally 3–5 m). It decreases towards the rooting zone where it is often less than 1 m thick. In addition, the active layer is well developed on the transverse ridges; in the furrows it is often near zero probably due to long-lying snow at these places (Barsch 1973; Vonder Mühll 1993).

It seems that during climate warming the active layer increases in thickness. First indications of this development are given by a comparison of the measurements by Barsch (1973) and Vonder Mühll (1993) which were made 20 years apart. Based on the new deep seismic sounding on the rockglacier Murtèl I, Vonder Mühll constructed a long profile which displays the thickness of the active layer (Fig. 6.11). It demonstrates again the expected increasing depth of the unfrozen layer from the higher to the lower part. It increases from 0.5 m (at 2700 m a.s.l) to 5 m (at 2650 m a.s.l). This is slightly more than the values published by Barsch (1973). At the moment, it is uncertain whether this is due to the different methods used or to a real change in the active layer during the 20 years between the two studies.

In relation to the early detection of the effect of global warming, the development of the active layer can form a very helpful indicator. Thus, the monitoring of the BTS from year to year (cf. Sect. 6.2.2.6) could be a good method.

On the cross-sectional profile S3 of the rockglacier Suvretta (2435–2445 m a.s.l.), the active layer is 5–7 m thick beneath fine-grained material, and 3–4 m thick beneath the blocky surface. The difference is greater than the error (±1 m). Thus, this may be explained by different thermal conductivities. Probably, the well-known "Balch" ventilation (cf. Sect. 10.1) is responsible for the difference.

In the development from activity towards inactivity, the increase in thickness of the unfrozen upper layer at the end of summer is remarkable. According to the sledge-hammer soundings by Barsch (1973), the unfrozen mantle on the toe of inactive rockglaciers is more than 10 m. In the case of the Val Sassa rockglacier (Lower Engadin, Swiss Alps: unfrozen mantle 1972>10 m), this development took place in less than 50 years. On the Albana rockglacier (Upper Engadin, Swiss Alps: unfrozen mantle 1972, ca. 11–12 m), the change was completed in probably no more than about 25 years. During this time, a layer of about 6–8 m of ice-supersaturated permafrost must have been melted. Assuming a linear development, a minimum of about 12 cm a^{-1} (Val Sassa) and a maximum of ca. 25 cm a^{-1} (allowing 750 cm in 30 years) was melted. For this, an energy input of about 35 to 75 × 10^3 $Jm^{-2}a^{-1}$ was necessary, which seems possible and reasonable. With precise measurements, developments like this should already be detectable after a very short time if temperature measurements of the quality of those from the borehole 2/1987 through the Murtèl I rockglacier are available. On the Murtèl I rockglacier (cf. Sect. 6.3.6), the increase in temperature in the upper 10 m from 1988 to 1992 was recorded as 0.3 ° C, i.e., 0.06 °C a^{-1}. By linear extrapolation from Fig. 6.28, the mean annual temperature at 11.6 m depth will be around 0 °C in about 30 years.

7 Rockglacier Movement, Velocity, and Rheology

The movement of rockglaciers has caused quite a lot of speculation. Theoretically, a huge mass of debris can be moved by a number of different processes (Johnson 1987). A discussion of these various theoretical possibilities has to be based on information of measured velocities and on the geometry of the rockglacier movement. This is possible today, because measurements have been carried out for more than 70 years on different rockglaciers around the world. Even if the methods used are extremely different, quite a good picture of the geometry of the surface movement of rockglaciers has been developed.

The rate of motion is recorded naturally on active rockglaciers. Inactive and relict (fossil) forms no longer display any horizontal movement, according to the definitions in Chapter 1, but their behavior during their change from activity to inactivity is of interest for all rheologic considerations. Unfortunately, the transition from activity to inactivity (and vice versa) as well as to the relict status has never been studied systematically; information was always gained accidentally (cf. the record of the Val Sassa rockglacier in the Swiss National Park, which had to be compiled from various sources; Sect. 7.1.5).

For convenience, the discussion of annual rockglacier movement may be differentiated into:

- Short-term measurements over some years that display the individual and the internal velocity variations on the rockglacier surface.
- Long-term annual averages (often >10 a) based on direct measurements.
- Long-term estimates derived by indirect means.
- Long-term series that display velocity variations in time or long-term steady creep.

Seasonal fluctuations as well as the vertical components of the movement are recorded only in a very limited number of cases.

7.1 The Horizontal and Vertical Movement of Active Rockglaciers

7.1.1 Measurement Methods

Numerous methods have been applied. They begin with very simple approaches such as trying to record the advances of rockglaciers in relation to a certain reference point. Normally, the distance between the foot of the front slope and the reference point is measured with a tape every few years. In this case, accuracy is limited, not only because the measurement is imprecise, but also because the foot of the rockglacier front slope is not well enough defined. Boulders which have fallen from the upper front slope frequently form a diffuse transition. Nevertheless, this method applied over years gives at least the order of magnitude of the movement (Chaix 1919, 1923, 1943).

Much better results are gained by marking profile lines across the rockglacier, especially if the movement is checked from nearby solid bedrock by means of a theodolite. This method has been applied in a number of cases (e.g., Wahrhaftig and Cox 1959; S.E. White 1987; Vietoris 1972). Abundant good data have been gathered with this procedure.

More information on the geometry of movement is gained if points can be selected more freely. In this case, a network has to be established. Normally, this is only possible in cooperation with a geodetic engineer. Within the net, a (more or less) free number of points can be marked and checked annually (Barsch 1969a; Barsch and Zick 1991). If the measurement is accurate (at least ±1 cm or better) and if the rockglacier movement is fast enough, then measurements over shorter periods (e.g., monthly measurements) are possible (cf. Sect. 7.1.5).

Terrestrial photogrammetric measurements were introduced as early as 1936 in rockglacier research (Pillewizer 1957). They are especially valuable if the photos are still available today. Photogrammetric measurements on air photographs which have been taken in different years and which allow the identification of individual boulders, can provide complete pictures of the movement geometry. Messerli and Zurbuchen (1968) were the first to use this method in rockglacier research. The procedure is especially helpful if older aerial photographs exist (Barsch and Hell 1975). In these cases, long-term trends in the behavior of rockglaciers may be determined. This method also allows monitoring of the present-day behavior of rockglaciers and rockglacier permafrost (Haeberli and Schmid 1988) if the photos are taken just for the monitoring purpose at regular time intervals and thus, are on a large scale.

The main problem of all velocity measurements on rockglaciers is connected with the fact that the control points are marked on boulders of the upper surface. Unfortunately, the individual movement of boulders may be accepted as the movement of the entire feature. The geodetic accuracy forms, therefore, only part of the error in the data. It is vital for the quality of the measurement to select large boulders which may be partly frozen into the rockglacier permafrost and to mark the measured points as precisely as possible. The hitherto published data series are consistent and seem to represent the rockglacier movement correctly, i.e., the recorded movement is due to the rockglacier permafrost creep. Extreme accuracy is expensive; it can only be financed in connection with big construction works (Keusen and Haeberli 1983).

The movement of an active rockglacier has, like all linear geomorphic processes, a horizontal and a vertical component. Both should be discussed separately because they may reveal quite different information about the rockglacier behavior. In the following sections the horizontal displacement will be treated first. The horizontal movement of an active rockglacier is often assumed to be constant over decades and during different seasons. It is, therefore, necessary to discuss short-and long-term annual behavior and seasonal measurements separately. Long-term measurements (as well as long-term estimates) may give interesting averages of rockglacier velocities which have to be compared with the short-term information. In addition, they may also help to provide a better understanding of long-term rockglacier velocity fluctuations. The discussion on rockglacier movement will be summarized as movement geometry, and with respect to the velocity changes, over cross-sectional and longitudinal profiles.

7.1.2 Annual Horizontal Displacement

Measurements of the annual horizontal displacement of active rockglaciers indicate the individual and the internal variations of the creep of rockglacier permafrost. The variability of velocity with time is normally not considered, because most measurement series are too short. The different velocities measured at different rockglaciers during different times may be used only as a first indication that there are also velocity differences in time.

The first measurements were probably done in Graubünden by both Chaix from Geneva. Chaix (1919, 1923) published the first results (cf. Sect. 7.1.3). The marked points on the rockglacier Val Sassa (Fig. 7.1) indicated an annual downvalley movement of between 0 and 2 m (average 168 cm a^{-1}). The nearby Val dell'Acqua (today, Val da l'Acqua) rockglacier displayed movement rates between 65 and 155 cm a^{-1} (average: 127 cm a^{-1}).

Almost 40 years later, Wahrhaftig and Cox (1959) reported much lower velocities measured on rockglaciers in the Alaska Range. Their results from Clear Creek rockglacier (No. 51 in their inventory) are presented in Table 7.1 (cf. also Fig. 7.2). During the 8-year period (1949-1957), this rockglacier moved in its lower part with an average velocity of 57 cm a^{-1} and in its upper part with 64 cm a^{-1}.

Since the papers by Chaix and Wahrhaftig and Cox, more and more measurements have been published. Thus, White (1971b) reported measure-

Table 7.1. Horizontal displacement of the Clear Creek rockglacier (Alaska Range) according to Wahrhaftig and Cox (1959)

	Displacement (cm)			Velocity (cm a^{-1})			
	1949–1951	1949–1952	1949–1957	1949–1951	1951–1952	1952–1957	1949–1957
Front	106.7	170.7	396.2	56.0	64.0	45.1	50.1
I 1	97.5	155.5	350.5	51.2	57.9	39.0	44.3
2	131.1	195.1	478.5	68.7	64.0	56.7	60.5
3	146.3	210.3	518.2	76.7	64.0	61.1	65.5
4	112.8	182.9	484.6	59.1	70.1	60.4	61.3
5	64.0	109.7	–	33.6	45.7	–	
ϕ				57.9	60.3	(54.4)	(57.9)
II 1	103.6	149.4	390.1	54.4	45.7	48.2	49.3
2	131.1	201.2	506.0	68.7	70.1	61.0	64.0
3	143.3	225.6	539.5	75.1	82.3	62.8	68.2
4	176.8	225.6	570.0	92.7	48.8	68.9	72.1
ϕ				72.7	61.2	60.2	63.4

Point 1 is on the right side, points 5 and 4 are on the left of the rockglacier.
The original values were reported in feet (1 ft. = 30.48 cm).
The error is assumed by the authors to 0.2 ft. (= ± 6 cm).
Velocity at the front was measured as distance to a reference point.
Surveys: 18 September 1949, 16 August 1951, 15 August 1952, 15 August 1957.
Abbreviations: I: profile on the lower part, II: profile on the upper part, ϕ: average, 1: point 1, etc.

Fig. 7.1. Val Sassa rockglacier, August 1972. In the *foreground*, the boulder that was used by Chaix as reference point

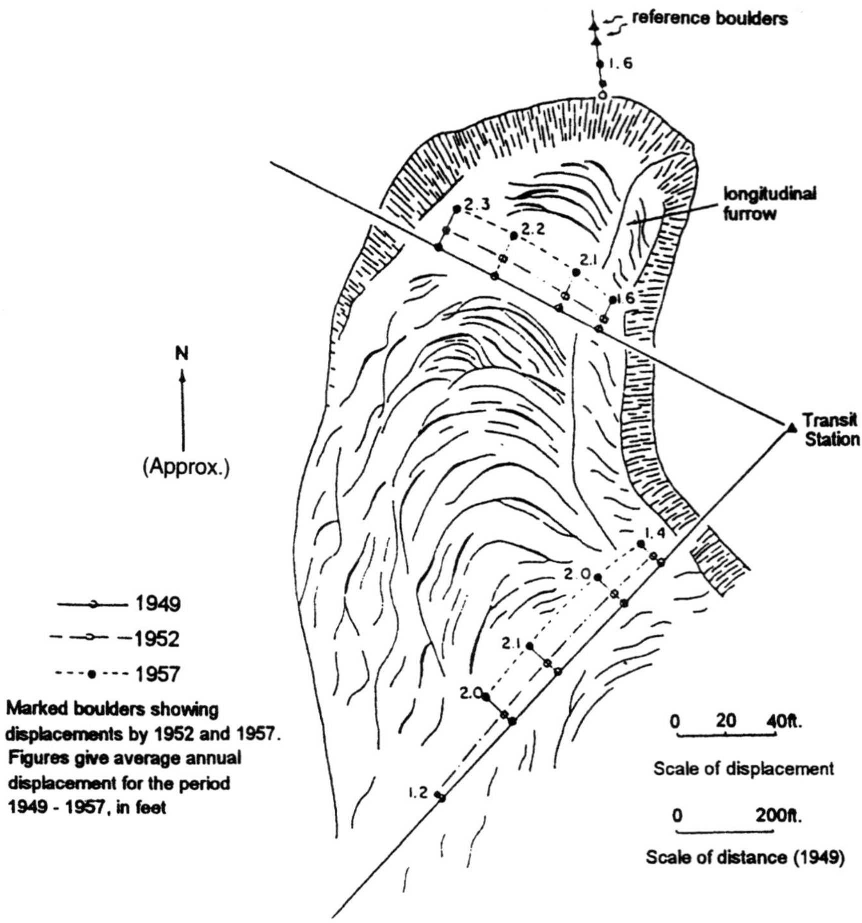

Fig. 7.2. Horizontal movement of the Clear Creek rockglacier 1952–1957. (Wahrhaftig and Cox 1959) 1 ft. equals 30.48 cm

ments from the lower part of the Arapaho rockglacier (Colorado). For the 5-year period (1961-1966), the data are summarized in Table 7.2. The highest velocity is found (as normal) in the center; however, comparison to the hitherto reported values this rockglacier moves slowly with only 5.9 cm a^{-1}. Despite the slow movement rates, the variability at the same point is high from period to period. Point T III-l displays, for instance, an increase of more than 250% (5.3 cm a^{-1} in 1961-1964 against 13.7 cm a^{-1} in 1964-1966). A similar increase for the same periods (240%, 2.5 against 6.0 cm a^{-1}) is found for the point T2-2 as well as for others. This leads to the assumption that special circumstances are responsible for this behavior.

Calkin et al. (1987) measured surface velocities on two tongue-shaped rockglaciers near Atigun Pass (central Brooks Range, Alaska). The measure-

Table 7.2. Horizontal movement rates on the Arapaho rockglacier, Front Range, Colorado, USA. (According to S.E. White 1971b)

Profile	Point	Movement (cm a^{-1})		
		1961–1964	1964–1966	1961–1966
I	1	0.0	17.0	6.8
	2	2.5	6.0	3.9
	3	0.0	22.5	9.0
	4	4.6	2.0	3.6
	5		17.7	–
		(8.6)[a]		
II	1	5.5	6.2	5.8
	2	0.0	8.7	3.5
	3		5.5	–
		(1.5)[a]		
III	1		(13.7)[a]	–
		(5.3)[a]		
	2		(0.5)[a]	–
		(2.6)[a]		
	3	3.0	2.0	2.6
	4	16.0	6.2	12.1
				Average = 5.9

Points are counted from right to left on the rockglacier.
[a]Apparent upvalley movement.

ments cover a 5-year period (1978/1982/1983). On each rockglacier, a profile just above the front was measured. The average horizontal movement is about 40 cm a^{-1} on the central lobe of the rockglacier Jaeger and ca. 10 cm a^{-1} on the rockglacier Pika.

Short-term measurements over 1 or 2 years already reveal that active rockglaciers move downslope with a horizontal velocity of between a few centimeters to 1 or 2 year (2–200 cm a^{-1}). Higher velocities are rare and are associated with special conditions as in the case of the rockglacier Äußeres Hochebenkar (Obergurgl/Tirol/Austria). It displays velocities up to ca. 500 cm a^{-1} (Pillewizer 1957; Vietoris 1972). The high-velocity zone is situated below a sudden break in slope. Thus, the high gradient certainly causes the marked increase in the speed of movement. The high stress at the toe also causes the formation of deep crevasses (Haeberli and Patzelt 1982). Another example of high creep velocities has been published by Haeberli et al. (1979) from the Gruben rockglacier (Sect. 7.2), which displays maximum velocities of more than 3 m a^{-1}. Here, the high annual movement rates are due to the fact that the rockglacier is moving into the basin which became ice-free after the retreat of the Gruben glacier since 1850.

7.1.3 Long-Term Annual Averages

Pillewizer (1957) reported measurements from the Ölgruben rockglacier (Kaunertal, Tirol, Austria). The surveys were done in 1936 (17 September) and in 1953 (16 September). The rockglacier displays different behaviors in its right and in its left parts (cf. Table 7.3). The faster movement on the left-hand side is probably assisted by topography, whereas rockglacier flow on the right is hampered by relief. The reported increase in thickness of 6–7 m since 1922 (ca. +19 to 23 cm a^{-1}) at the left part is difficult to explain.

A long period (1963–1980) elapsed between the establishment of five profile lines on the Tungsten rockglacier (N.W.T. Canada) and the resurvey (Jackson and McDonald 1980). Even if the assumed error is high (ca. 3m = ±20 cm a^{-1}), the absolute movement is big enough to overcome this problem. The front of this talus rockglacier moves at about 40 cm a^{-1}, its central parts (200 m above the front) at about 250 cm a^{-1}, and even higher up (ca. 400 m above the front) at 210 cm a^{-1}. The points at the side, especially on the lateral ridges display almost no movement. The rockglacier, thus, demonstrates a normal behavior (cf. Sect. 7.2) even if it moves into the forest with a well-developed surface relief of furrows and ridges in the center (compressive flow) and a trend to a laterally extending flow.

Messerli and Zurbuchen (1968) report the movement behaviors of the rockglaciers Weissmies (S Wallis) and Grosses Gufer (N Wallis) over periods of 6 years (Weissmies: 1958–1964) and 12 years (Grosses Gufer: 1950–1962). They used air photogrammetry for the first time in rockglacier research. Horizontal displacement generally increased downslope (from the rooting zone) towards the front, and from the lateral ridges to the center. An interesting profile across the tongue of the rockglacier Weissmies is given in Fig. 7.3. It demonstrates that the velocity is about twice as high in the center as at the sides. Despite an average velocity of 60 cm a^{-1} at the surface of the center of

Table 7.3. Long-term averages of the horizontal surface velocity (cm a^{-1}) of the Ölgruben rockglacier, Kaunsertal, Tirol, Austria. (Pillewizer 1957)

	Altitude (m a.s.l.)	Left side (cm a^{-1})	Right side (cm a^{-1})
Upper profile	2660	75	35
Lower profile	2580	–	55
Front		97	–

The 17–years apart measurements were done on 17 September 1936 and 16 September 1953.
The movement of the front is an average for 1922–1953 (31 a).
The left side of the rockglacier was in 1953 about 6 to 7 m thicker than in 1922.

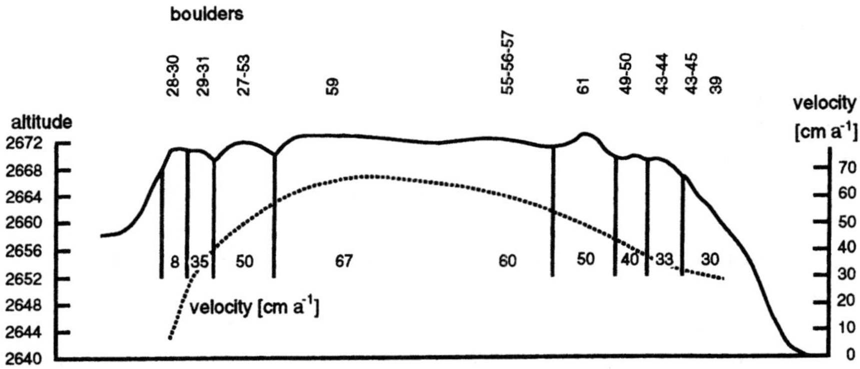

Fig. 7.3. Horizontal movement of different points of a cross-sectional profile over the tongue of the rockglacier Weissmies (1958–1964) (modified from Messerli and Zurbuchen 1968). The velocity has been taken from nearby points

the tongue, the rockglacier front seems to be stable during the 6-year period: on the aerial photographs, neither rockglacier shows a marked advance of the front slope. It seems that this fact can be, at least partly, explained by a compressive flow at the left-hand side and at the right front part of the rockglacier Weissmies.

A very interesting data series has been presented by Francou and Reynaud (1992). During the period 1979–1991, about seven lines were repeatedly monitored by theodolite survey on a large rockglacier in the Combe de Laurichard in the southern French Alps. The results (Table 7.4, Figs 7.4, 7.5) can be summarized as follows:

1. Directly beneath the talus slope at the rooting zone of the rockglacier the average velocity amounts to 25 cm a^{-1} over 12 years. From 1979 to 1985, all seven points of this line displayed a constant velocity with relatively little

Table 7.4. Horizontal velocities (cm a^{-1}) on a longitudinal profile on the rockglacier "Laurichard" in the southern French Alps according to Francou and Reynaud (1992)

Points	1	2	3	4	5	6	7	8	9	10	11	12	13
1983–1986	35	45	42	55	63	85	81	72	56	51	50	39	28
1986–1989	42	39	48	63	67	93	86	70	52	51	53	37	32
1989–1991	34	45	89	67	69	92	85	70	54	47	49	33	28
1983–1991	39	43	55	61	66	90	84	70	55	50	51	36	30

Velocities are presented as averages in cm for three periods: 1983–1986, 1986–1989 and 1989–1991.
Points are counted from the top to the front.

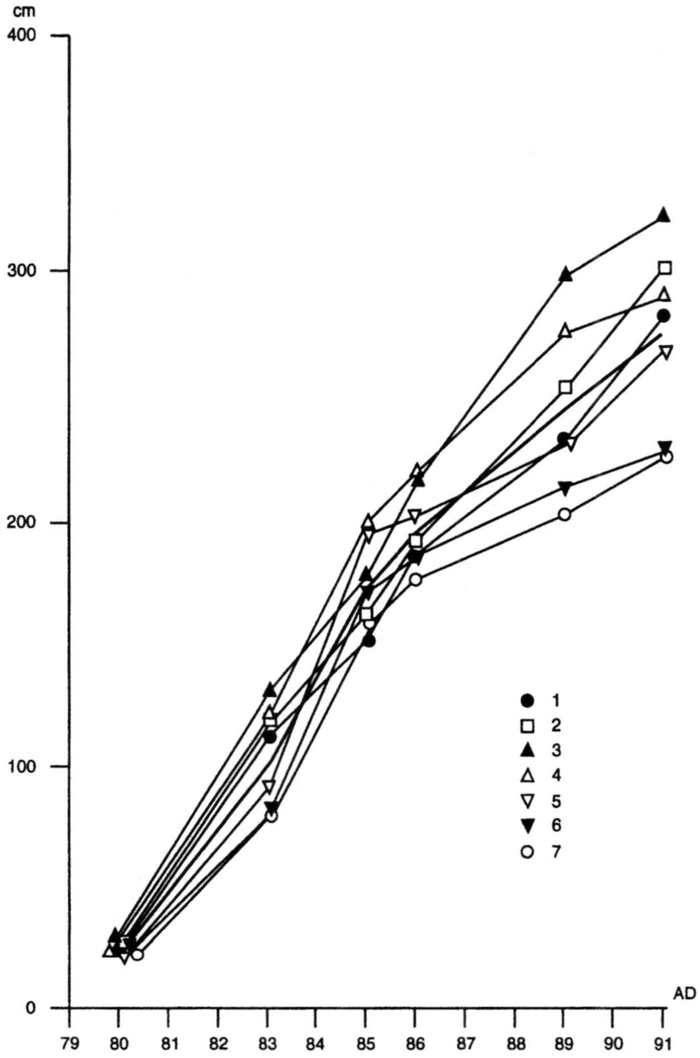

Fig. 7.4. Velocity changes on the rockglacier in the Combe de Laurichard. The *symbols* and *numbers* represent the points on cross-profile A in the rooting zone of the rockglacier. Numbering is from *right* to *left*. (Francou and Reynaud 1992)

variation. Since 1986, a decrease by 10% in velocity has occurred; the variation between the points has increased.

2. The middle part of the rockglacier moves faster than the upper part. The longitudinal profile (cf. Fig. 7.5) displays a low velocity (40 cm a^{-1}) in the upper part, a maximum velocity (90 cm a^{-1}) in the middle part, and a minimum velocity (ca. 30 cm a^{-1} at the toe.

The decrease in velocity in the upper cross-sectional profile may indicate that a change in input (snow, talus, or both) occurred below the talus slopes,

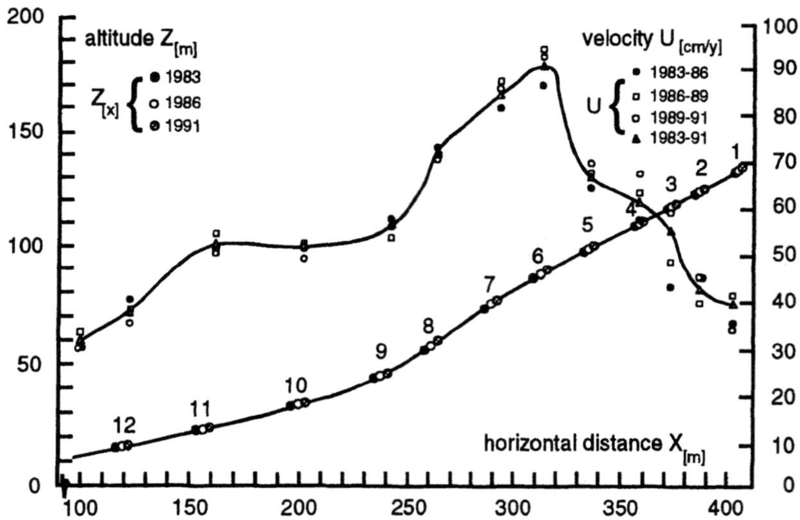

Fig. 7.5. Horizontal velocity of different points on the longitudinal profile of the rockglacier in the Combe de Laurichard. (Francou and Renaud 1992)

because a melting caused by increased warming should influence first the lower parts. Thus, a direct climatic influence cannot be assumed in this case. The velocity distribution on the central longitudinal profile is not clearly related to gradient. There is probably some compressive flow in the lower part involved. The high velocity in the middle part may be explained by former higher inputs. In addition, the explanation by the authors, who believe that the high-velocity zone is related to the downslope increase in gradient, may not be completely disregarded.

7.1.4 Long-Term Estimates

The Glockturm rockglacier (Ötztal, Tyrol, Austria) has a length of about 1 km and a total width of 500 m. It starts in a cirque on the north side of the Glockturm mountain (3355 m). Following a discussion of late glacial moraines, Gerhold (1970) believes it to be at least 7000 years old. However, an even greater age of about 10,000 years cannot be precluded. In any case, its mean velocity has been 10–15 cm a^{-1} over a very long period of several thousand years.

7.1.5 Longer Time Series

Long-term measurements are rare on rockglaciers. This is unfortunate, because only long time series allow a discussion of velocity variation with time, i.e., a discussion of whether the creep of active rockglaciers is time or climate-dependent.

An important series can be presented from the rockglacier Val Sassa (Graubünden, Switzerland; cf. Fig. 7.1). Around 1915–1920, it moved with an average velocity of about 160 cm a^{-1} (Chaix 1919, 1923). In 1940, the velocity must have already been reduced, because the average for 1921–1942 was down to 130 cm a^{-1} (Chaix 1943). In 1973, the velocity was believed to be 20 cm a^{-1} (Girsperger 1973); however, this was already doubtful because Barsch (1973) reported this rockglacier to be inactive based on the results of seismic studies (cf. Sect. 6.2.2.1). The interpretation by Girsperger was derived from inclination measurements, but the change in inclination was due to subsidence of the melting rockglacier core. This was later acknowledged by Eugster (1973). The change in velocity of this rockglacier is presented in Table 7.5 and Fig. 7.6.

Eugster (1973) summarizes some of the measurements organized by the Scientific Committee of the Swiss National Park. The data demonstrate that in around 1952/1953 the horizontal displacement 90 m above the tongue was down to 70 cm a^{-1}. For the period 1952–1963, it was only 35 cm a^{-1}. This is a

Table 7.5. Decrease in long-term movement of the Val Sassa rockglacier (Graubünden, Swiss Alps) compiled from various sources

Velocity (cm a^{-1})	Period	Reference
163[a]	1918–1979	Chaix (1923)
129[a]	1919–1921	Chaix (1923)
131[a]	1921–1942	Chaix (1943)
70[b]	1952–1954	Eugster ((1973)
51[b]	1954–1959	Eugster (1973)
35	1952–1963	Girsperger (1973)
15[c]	1942–1972	Barsch (1973)
0[d]	1972	Barsch (1973)

[a]This value represents an average of 8 points in the central part of the rockglacier above the toe.
[b]Horizontal displacement calculated for "point 25" (ca. 90 m above the front slope).
[c]Velocity calculated from the distance between the front slope and the reference point.
[d]According to sledge–hammer seismic measurements, the unfrozen layer above the toe is so thick that movement is impossible.

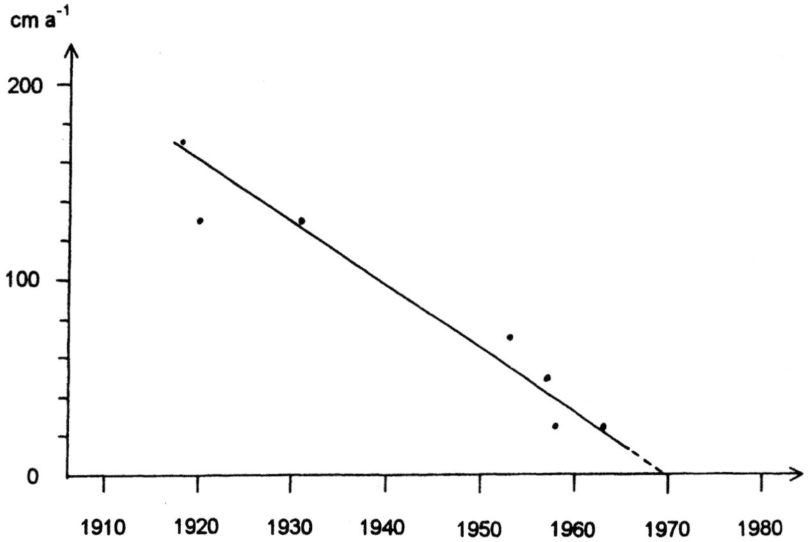

Fig. 7.6. Decrease in movement 1918–1970 of the rockglacier Val Sassa. (Compiled from different sources)

very marked decrease in velocity that, at around 1970, led to inactivity at least at the front. There are probably two reasons for this behavior:

1. The talus rockglacier Val Sassa ends at ca. 2100 m a.s.l. This is very low even if it is developed in a valley with a cool microclimate. The general warming since 1850 and especially since 1926 with the warm period 1940–1950 may have caused the climatic inactivity of this rockglacier at the lower limit of active rockglaciers.
2. Rockglacier movement is also due to the input of avalanche snow and talus/debris. It is not possible to evaluate if and how much the input, which is also controlled by climate, has been changed since the turn of the century.

As already stated by Chaix (1943), the rockglacier moves more slowly in its upper part (in this case ca. 260 m above the front). This is confirmed by Eugster (1973) for the period 1954–1959. This seems to be the rule for a number of valley rockglaciers if topography or input changes do not cause a different behavior (cf. Sect. 7.2.2).

The talus rockglacier Äußeres Hochebenkar was already on object of terrestrial photogrammetric measurements in 1936 (Pillewizer 1957). Vietoris (1972) independently started an additional program in 1951. The rockglacier moves over a break in slope at about 2570 m which creates a marked increase in velocity (maximum ca. 5 m a^{-1}). This leads to the formation of large crevasses across the rockglacier which probably indicate a future separation of the lower part from the main body (Haeberli and Patzelt 1982). This development started in around 1936. The upper edge of the rockglacier was then at 2530 m a.s.l. and crevasses were hardly visible. In 1953, the front slope was at 2490 m a.s.l. (Pillewizer 1957). Almost 20 years later, Haeberli and Patzelt (1982) reported crevasses which were more than 8 m deep, very dangerous to penetrate, and which traversed the whole rockglacier.

In table 7.6 and in Fig. 7.7, the velocities along cross-sectional profile 2 at 2625 m a.s.l. are presented according to Vietoris (1972). The measurements indicate interesting velocity changes. During the first 3 years of measurements (1951–1954), the movement was rapid with 170 cm a^{-1} in the center. In the following year (1954/1955), a marked decrease of about 35% down to 110 cm a^{-1} was measured. After a slight velocity increase (1955/1956), a lower velocity was observed during 1956–1959. After a velocity increase (1959/1960), maximum velocities were found during 1961/1962. Since that year, a trend towards lower velocities existed which was confirmed by a survey on 5 September 1972, with a maximum velocity of only 83 cm a^{-1}. The highest velocities were recorded in the years 1952/1953 and 1961/1962; the lowest (40 cm a^{-1}) in 1956/

Table 7.6. Horizontal movement (cm a^{-1}) of four boulders on a cross-sectional profile on the rockglacier Äußeres Hochebenkar (Ötztaler Alpen, Tirol, Austria) from Vietoris (1972)

	Period	Days	Pt.2	Pt.3	Pt.4	Pt.5
Profile 2	1951–1952	360	35	160	132	98
	1952–1953	357	67	210	153	72
	1953–1954	361	51	197	162	101
	1954–1955	376		107	112	49
	1955–1956	358	36	133	87	51
	1956–1957	363	25	40	80	0
	1957–1959	731	25	120	90	75
Profile 2′	1959–1960	350	69	123	61	51
	1960–1961	391	38	193	142	85
	1961–1962	370	72	222	187	128
	1962–1970	2909	45	148	124	81

Survey dates:	Profile 2	Profile 2′
	30 August 1951	14 August 1959
	24 August 1952	29 July 1960
	16 August 1953	24 August 1961
	12 August 1954	29 August 1962
	23 August 1955	16 August 1970
	15 August 1956	
	13 August 1957	
	14 August 1959	

Profiles 2 and 2′ are situated at 2630 m a.s.l. about 300 m above the snout of 1950. At that place, the rockglacier is about 310 m wide. Profile 2 was established on 14 July 1951; profile 2′ (established 14 August 1959) replaced profile 2.

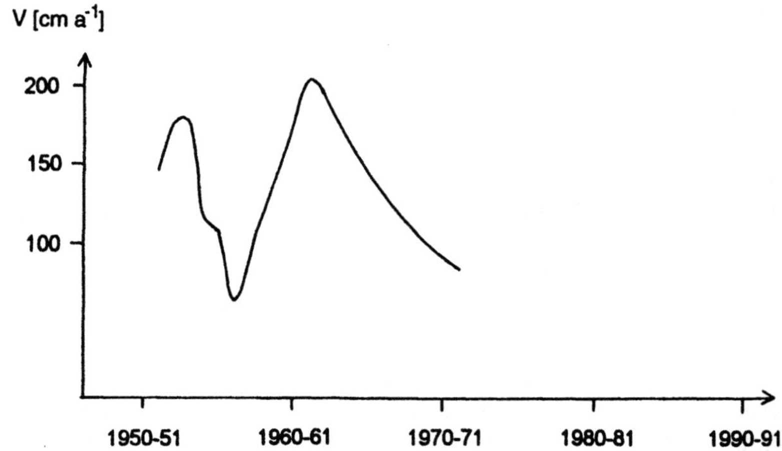

Fig. 7.7. Variations in horizontal movement of the rockgalcier Äußeres Hochebenkar. The movement rates are averages of the movement of points 3 and 4 on cross-profile 2 (cf. Fig. 7.14) in 2620–2630 m a.s.l. which was later replaced by profile 2′ due to rockglacier movement (Vietoris 1972). The marked and abrupt decrease in movement in 1955/1956 is difficult to explain. The nearly linear decrease since 1962 is perhaps related to climate

1957. An explanation of this behavior is difficult and only possible in connection with theoretical considerations.

Long-term velocity measurements (1932–1955–1971) have been published by Barsch and Hell (1975). They are based on aerial photographs of the rockglacier Murtèl I. On blowups of the three pairs of photographs, 50 to 70 well-defined points were selected and the movement vectors were determined (Table 7.7 and Figs 7.8, 7.9). The error is estimated to be ±0.7 m for horizontal, and ±0.5 m for vertical displacement. For the period 1932–1955, the vectors are much larger than the error; for 1955–1971, the displacement is in

Table 7.7. Average horizontal velocities (cm a^{-1}) on a longitudinal profile on the rockglacier Murtèl I for different periods, calculated from the data in Barsch and Hell (1975)

Period	Toe	Lower center	Center	Upper center	Rooting Zone lower		upper
1932–1955	73	37	16	1650			57
1955–1971	47	22	14	16	35		38
1971–1972	4	7	6	–		10	

The measurements 1932, 1955, 1971 were made photogrammetrically on aerial photographs.

The measurements 1971–1972 (15 September 1971-21 September 1972) are from a geodetic survey by Bruno Baur (Basel).

The data are averages measured for at least two points.

Fig. 7.8. Horizontal displacement on the rockglacier Murtèl I 1932–1955 (Barsch and Hell 1975). The photogrammetry was done at the Technical University, Karlsruhe

Fig. 7.9. Horizontal displacement on the rockglacier Murtèl I 1955–1971 (Barsch and Hell 1975). The photogrammetry was done at the Technical University, Karlsruhe

the same order of magnitude as the error. This makes interpretation difficult, but as the main result it may be noted that the movement decreased considerably from the first to the second period. Regarding the averages for both periods, the horizontal displacement goes down from 7.1 to 3.2 cm a^{-1} (± 3 cm a^{-1}). The value for 1955–1971 is confirmed independently by terrestrial measurements in 1971–1973 with 3.8 ± 1 cm a ^{1}.

The decrease in velocity in the second period is not equally distributed on the rockglacier (Fig. 7.10). It is greatest at the right toe. The velocity in 1955–1971 here amounts to only 46% of that of the first period. In the central toe, the velocity is still 64%, whereas the central part of the rockglacier displays about the same velocity in both periods. In 1955–1971 the velocity in the rooting zone had been reduced to 66%.

The data summarized in Table 7.7 demonstrate several conclusions for the rockglacier Murtèl I

1. The velocity decreases from 1932 to 1971 down to ca 45%;
2. The relative velocity distribution along the central flow line is nearly the same in all three periods;

Fig. 7.10. Horizontal surface velocity 1955–1971 compared to that of 1932–1955 on the rockglacier Murtèl I

3. The velocity decreases from the rooting zone towards the center and increases from the center to the toe.

The decrease in rockglacier velocity after 1955 has also been demonstrated, in an MSc. thesis (Pröhl 1977), on the neighboring rockglaciers Albana and Albana West (Upper Engadin, Graubünden) using the same technique on similar aerial photographs. Even if the quality of the aerial photographs of 1931 is not comparable to that of 1971, it can be demonstrated that between 1931 and 1952 both rockglaciers display a clear vector field of horizontal movement that is not recognizable for the period 1952 to 1971. The hitherto unpublished results are presented in Table 7.8. They reveal that rockglaciers close to the lower limit of mountain permafrost in Graubünden (ca. 2600 m.a.s.l. on northern exposure) have probably been close to inactivity since 1950/1970 (cf. Fig. 5.3). Thus, it seems that the decrease in movement is a regular phenomenon probably connected with the warmer atmosphere since 1926.

The movement on the rockglacier Macun has been measured by theodolite from two reference points for more than 25 years (Barsch 1969a; Barsch and Zick 1991). Originally (in 1965), 125 points were marked on the rockglacier (Fig. 7.11), and surveyed at 2-or 3-year intervals, that is in 1967, 1968, 1969, 1970, 1971, 1973, 1977, 1985, and 1988. The first results (period 1965–1967) were published in Barsch (1969a). In a high mountain environment measurements are not always easy. Thus, the quality of the data is not the same in every year. To overcome this problem a rigorous check has been carried out. The results have been published by Barsch and Zick (1991).

An initial evaluation could lead to the conclusion that the velocity, especially of the toe, decreased continuously from the first period (1965–1967). A thorough check of the data of all points reveals that only the data from the years 1967, 1971, 1977, 1985, 1988 should be used. In Table 7.9, these values are presented together with the mean for 1967–1988. It shows that the velocities for different periods starting with 1967–1971 do not differ significantly (S = 99%) from the mean 1967–1988 (= 14 cm a^{-1}). Only the velocity during 1965–1967 at the toe is much higher (24 cm a^{-1}).

There are two possible explanations. The first is related to measurement problems: it seems that in 1965 the reference points were not sufficiently

Table 7.8. Average horizontal velocity 1931–1952 and 1952–1971 of the rockglaciers Albana and Albana West, near Piz Albana, Graubünden, Swiss Alps. (Pröhl (1977)

Rockglacier	Location	Exposure	Lower limit (m a.s.l.)	Horizontal displacement ($cm\ a^{-1}$) 1932–1952	1952–1971
Albana	S of Piz Albana	S	2550	~4.8	~2.1
Albana West	W of Piz Albana	WSW	2560	~9.5	–

Table 7.9. Horizontal movement of the rockglacier Macun I (Lower Engadin, Swiss Alps) according to Barsch and Zick (1991) in cm a^{-1}

Points	(65/67)	67/71	71/77	77/85	85/88	67/88
2	25.8	17.8	13.2	13.9	14.4	14.5
5	26.6	15.2	14.7	13.3	14.1	14.2
6	25.6	14.5	14.3	12.6	13.3	13.5
8	27.1	16.0	16.4	14.3	16.2	15.5
9	24.1	14.3	14.2	14.7	14.7	14.4
10	23.6	14.5	14.3	13.6	14.7	14.1
11	22.5	13.0	12.9	14.2	14.7	13.6
14	22.6	14.0	12.5	14.6	14.0	13.8
15	24.6	14.3	12.7	14.9	14.7	14.1
16	25.5	14.6	12.7	15.4	14.7	14.4
17	22.9	14.9	13.3	15.9	14.9	14.8
18	25.8	15.4	13.4	16.0	15.5	15.1
19	24.1	14.9	12.9	15.5	14.9	15.6
20	21.4	13.5	10.4	14.0	13.6	12.8
21	23.7	14.1	13.7	12.8	14.8	13.6
22	22.1	14.0	13.2	12.6	13.5	13.2
23	19.2	12.5	11.5	11.3	12.0	11.6
24	22.7	13.4	12.8	12.0	12.8	12.6
26	21.7	14.1	12.3	12.1	13.0	12.7
27	16.2	11.0	9.8	10.3	10.4	10.2
28	21.5	13.6	13.2	12.2	12.4	12.7
30	23.3	15.1	14.2	13.5	13.7	14.0
31	25.3	15.0	15.8	14.6	13.9	14.9
32	22.4	12.4	13.3	13.0	12.1	12.8
34	28.0	16.7	16.9	15.3	15.5	16.0
35	30.7	16.9	17.0	15.8	15.1	16.2
36	31.6	15.9	18.3	15.8	15.4	16.5
37	31.6	16.8	17.2	15.7	15.2	16.2
38	29.1	16.0	17.3	14.8	15.1	15.8
39	28.5	16.8	15.7	15.3	14.5	15.5
40	24.7	15.0	14.8	14.0	13.6	14.3
41	22.6	14.1	15.1	13.7	13.0	14.1
42	21.7	13.3	15.7	13.6	12.7	14.0
43	21.5	14.7	14.8	13.9	12.8	14.1
44	20.7	14.2	14.7	14.0	13.9	14.2
45	22.7	13.3	16.2	14.5	13.5	14.6
46	16.1	10.0	11.8	11.1	10.0	10.9
47	16.0	10.1	11.0	11.1	9.8	10.7
48	14.5	9.1	11.6	10.7	9.1	10.4
49	15.5	10.1	10.3	10.5	9.7	10.2
52	14.2	7.6	10.9	10.0	10.6	9.9
53	18.1	8.3	9.8	9.8	12.5	9.9
54	13.1	6.9	10.6	8.6	11.9	9.3
55	11.1	6.2	10.1	8.1	9.7	8.5

Table 7.9. (*Contd.*)

Points	(65/67)	67/71	71/77	77/85	85/88	67/88
56	12.0	6.1	9.6	8.0	10.2	8.4
57	8.5	3.2	5.4	5.3	6.8	5.1
58	4.5	3.8	4.0	3.7	5.0	3.6
62	8.4	4.4	7.4	6.4	8.2	6.5
64	8.4	5.2	7.9	7.2	8.8	7.2
65	10.5	6.0	9.3	8.7	9.4	8.4
66	10.4	7.0	7.8	7.9	8.5	7.7
67	8.6	3.1	5.9	5.4	6.2	5.1
68	6.3	2.8	6.0	5.9	5.4	5.2
71	5.6	5.2	6.0	6.0	5.7	5.7
72	5.0	6.2	5.4	6.6	6.7	6.1
73	7.8	11.6	9.0	9.7	9.9	9.8
74	6.7	10.3	7.9	9.4	8.6	9.0
75	6.4	9.1	7.8	8.4	8.2	8.3
76	4.5	6.1	6.9	7.1	6.7	6.7
77	3.8	6.5	6.1	5.9	6.4	6.0
78	7.9	8.0	8.4	7.7	8.8	8.0
104	17.7	12.6	13.1	12.1	10.3	12.2
105	25.1	15.8	18.4	15.6	15.9	16.4
107	19.7	8.7	14.3	10.3	10.2	11.1
108	23.6	11.9	16.4	13.5	13.5	14.0
109	20.5	10.5	13.8	11.4	11.5	11.9
111	26.6	15.0	17.8	14.8	15.8	15.9
113	26.0	16.9	16.7	14.9	15.2	15.8
114	26.3	14.8	17.4	15.2	15.6	15.8
115	25.9	13.8	16.5	14.3	14.9	14.9
116	24.4	13.4	15.3	13.7	15.3	14.3

marked. Thus, after a better marking in 1967, a slight change in position of the reference points occurred. This could explain the differences. Thus, it is better to accept only the data of the years 1967–1988 as being correct.

The second explanation accepts a change in the rheological behavior of the rockglacier. This could be caused by a change in the deformation rate of the inner core of the rockglacier due to a warmer atmosphere. Because only one period (1965–1967) displays this disorientation, it seems safer to accept the first explanation. Therefore, only the periods after 1967 are marked on Fig. 7.11. It shows that the rockglacier displays the following average velocities in the course of 21 years (1967–1988):

I toe (left) 36 points, 14.2 ± 0.2 cm a^{-1}
II toe (right) 10 points, 14.2 ± 0.6 cm a^{-1}
III upper part (center) 10 points, 9.8 ± 0.3 cm a^{-1}
IV upper part (right) 5 points, 7.6 ± 0.5 cm a^{-1}
V rooting zone 10 points, 6.1 ± 0.4 cm a^{-1}

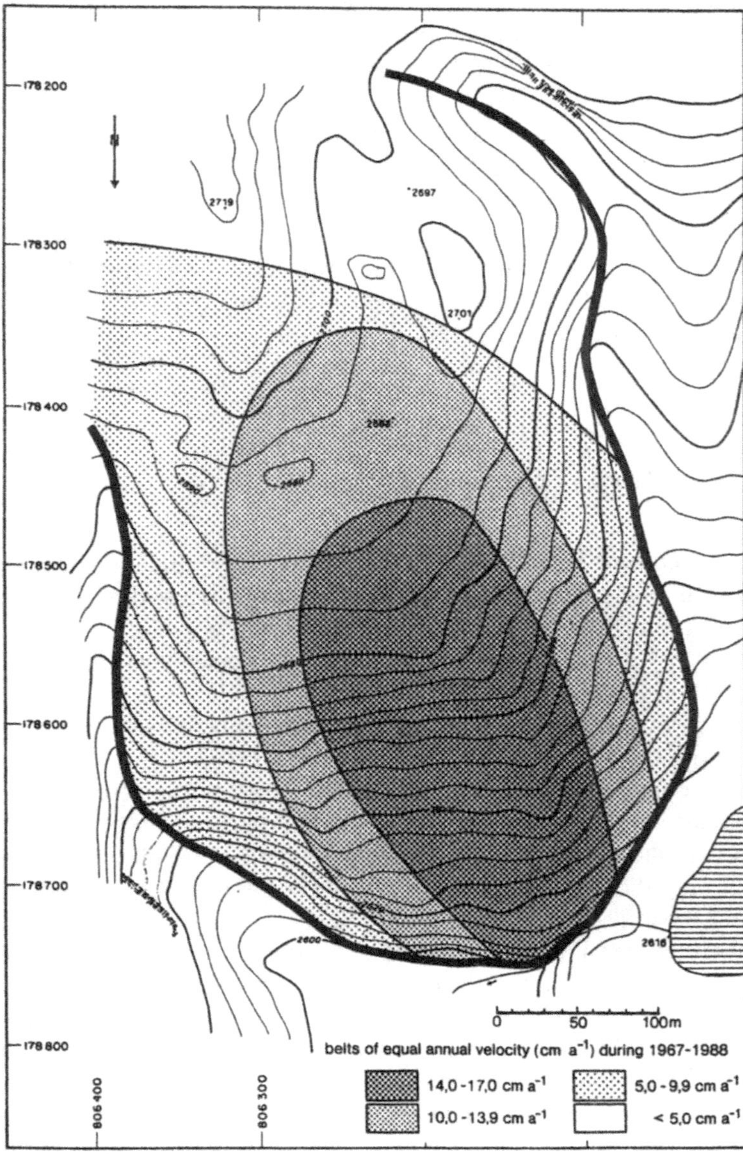

Fig. 7.11. Horizontal displacement 1967–1988 on the rockglacier Macun I. (Barsch and Zick 1991)

Summarizing, there are very important indications which demonstrate that rockglacier movement changes relatively quickly over time. Excluding local, individual conditions and causes, it has to be stated that velocity fluctuations exist. The statement by Haeberli (1985a) that rockglaciers "advance at a slow, but almost constant rate, barely influenced by post-glacial type climatic changes", has thus to be corrected. In most cases, the velocity changes are probably due to a change from activity towards inactivity (Val Sassa, Murtèl I). In other cases, individual causes can perhaps not totally be excluded (Äußeres Hochebenkar). On the other hand, the rockglacier Macun displays a constant velocity over more than two decades, which can be assumed to be an indication of a steady-state creep at an altitude highly favorable for rockglacier permafrost development (cf. Sect. 10.6).

7.1.6 Monthly and Seasonal Measurements

Monthly measurements have been reported from only three rockglaciers (Barsch and Hell 1975; Haeberli 1985a). The northerly exposed active rockglaciers Murtèl I and Muragl (Upper Engadin, Swiss Alps) which have been checked during 1971, 1972, and 1973, and the rockglacier Gruben (Wallis, Swiss Alps).

On the first two rockglaciers, horizontal accuracy is better than 1 cm a^{-1}. Rockglacier movement is small during this period (Murtèl I, 3.8 cm a^{-1}; Muragl "north lobe", 20.8 cm a^{-1}). Table 7.10 presents the results as arithmetic averages for all measured points (50 points on Murtèl I, 14 on Muragel) because the data set is not large enough to be discussed in greater detail. Even if only the average values are accepted, the results are surprising. The horizontal (as well as the vertical) movements are larger during summer than during winter, even though marked contrasts exist between the two rockglaciers. In particular, during the summer 1971 the horizontal movement of the rockglacier Murtèl I is by an order of magnitude higher than that during the following winter. This is in general also true for the following year, but the differences are much smaller. On the rockglacier Muragl, the difference between summer and winter movement is detectable but not very marked (summer 130% of winter movement) Whether or not this was caused by the warm summer of 1971, during which the 0 °C isotherm was situated ca. 260 m higher than during summer 1972 (Barsch and Hell 1975, 138), cannot be conclusively demonstrated even though the relatively high vertical displacement (cf. Sect. 7.1.6) seems to confirm this conclusion.

From the Gruben rockglacier, Haeberli (1985) reports similar measurements (Table 7.11). He ensured that the targets (aluminium poles) were fixed with concrete into drillholes in large boulders. The boulders were large enough to be frozen into the rockglacier permafrost. An additional lichen check was

Table 7.10. Seasonal movements on the rockglaciers Murtèl I and Muragl I, Upper Engadin, Swiss Alps according to Barsch and Hell (1975)

	Horizontal movement				Vertical movement							
	Murtèl I		Muragl I		Murtèl I				Muragl I			
	Field data (mm/ month)	(%)	Field data (mm/ month)	(%)	Field data (mm/ month)	(%)	Due to slope (mm/ month)	(%)	Field data (mm/ month)	(%)	Due to slope (mm/ month)	(%)
Summer 1971 (Sept./Oct.)	22.3	59	–		–5.4	245	–3.5	65	–		–	
Winter 1971/ 1972 (Oct./Aug.)	1.6	4	–		–0.8	36	–0.3	38	–		–	
Summer 1972 (Aug./Sept. and Sept./Oct.)	8.9	23	21.4	10	–2.0	90	–1.4	70	–16.8	175	–5.3	32
Winter 1972/ 1973 (Oct./July)	1.3	3	16.0	8	–2.5	113	–0.2	8	–5.5	58	–3.4	61
Annual average 1971/1973 (mm/year)	38.0		208.0		–2.2		–6.0		–	–9.6	–36.7	–

The velocities given for summer or winter are arithmetic means of all data (50 points on Murtèl I, 14 points on Muragl I). They have been calculated for a period of 30 days = 1 month.

Measuring dates are: *1. Murtèl I*
1971: 15 September/15 October
1972: 9 August/21 September/17 October
1973: 17 July
2. Muragl I
1972: 7 August/20 September/18 October
1973: 18 July

The percentage given is related to the average movement per year.

The vertical displacement can be caused by the slope (normal surface slope), by losses due to melting or by gains due to additional inflows from above. In the table the normal variation due to the surface slope (Murtèl ca. 9°, Muragl 14° is indicated). The percentage indicates how much the slope explains of the total vertical displacement.

done to preclude individual erratic movements of the boulders. The results (Fig. 7.12) are summarized as follows:

1. The horizontal displacement shows simultaneous velocity changes in the lower part of the rockglacier. Peak velocities were measured in autumn 1981 and in early summer 1982; the rockglacier moved more slowly in winter and early spring. In contrast, the points on the upper part of the rockglacier displayed faster movements during early spring 1982 and minima in winter 1981/1982 as well as in summer 1982. Thus, the rockglacier shows a harmonica-like movement, which is not really understood.
2. On the rockglacier, no remarkable thickness changes are recorded. The surface lowering corresponds to measured horizontal displacement and the surface gradient of about 10 °. Minor variations are explained by the extreme roughness of the rockglacier surface and by the unknown critical distance over which the deforming rockglacier core averages the surface slope. Thus, flow seems to be in general parallel to the surface.
3. The ratio

$$\frac{\left(\overline{V}_s - V_{s,min}\right)}{\overline{V}_s}, \tag{7.1}$$

where $\overline{V}_s$ = mean annual surface velocity, $V_{s,min}$ = minimum observed seasonal surface velocity is assumed (as on glaciers) as a good approximation of the relative amount of sliding (cf. Haeberli 1985a). On average, on the Gruben rockglacier this ratio amounts to 34%. That 34% of the horizontal displacement of the rockglacier Gruben is really caused by sliding at the base of the

Table 7.11. Vertical and horizontal displacement of the rockglacier Gruben 1979–1982 according to Haeberli (1985)

Pole (cm)	VD (cm)	HD (cm)	Loss to gradient (cm)	Loss/gain	
150	–8	84	–15.7	+7.7	$\phi = +5.8$
151	–11	66	–11.6	+0.6	
152	–10	108	–19.0	+9.0	
153	–71	254	–44.8	–26.2	$\phi = -24.7$
154	–72	213	–37.6	–34.4	
155	–19	236	–42.1	+23.1	
156	–68	260	–45.8	–22.2	
157	–109	255	–45.0	–64.0	

Surface slope is about 10°.
The points 150–152 are situated in the upper right part; the points 153–157 are close together on the lower part (tongue) of the rockglacier.
Survey dates: 16 August 1979, 8 August 1982.

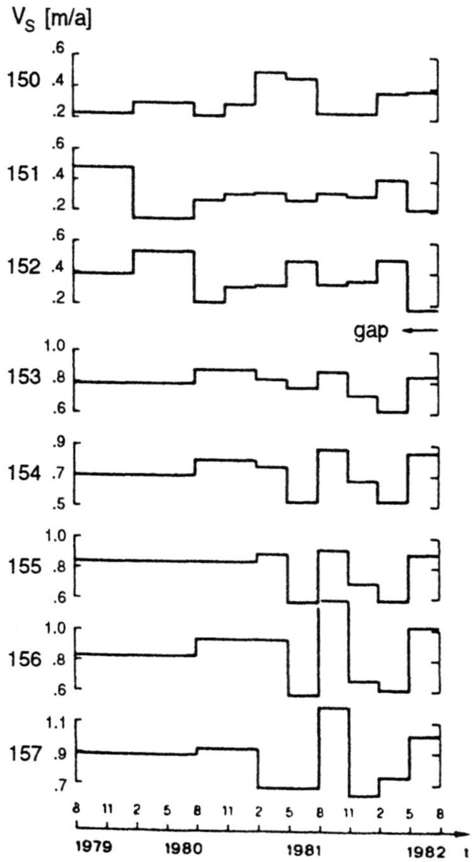

Fig. 7.12. Seasonal velocity variations on the rockglacier Gruben (Haeberli 1985a). Measurement error is assumed to be within ± 0.01 m a^{-1}. The points *153–157* display simultaneous velocity changes (faster movements in autumn 1981 and in early summer 1982; slower movements in winter and early spring). The points *150–152*, which are about 150 m higher up on the rockglacier, show a different velocity rhythm

rockglacier seems unlikely in the light of our present knowledge. This unanswered question will be discussed later (cf. Sect. 7.5).

Summarizing, it seems that the horizontal movement of active rockglaciers during summer is greater than during winter (at least in its lower part), although published data are still not sufficient to allow a clear conclusion of this kind. In any case, more measurements (on other rockglaciers and in other climates) are needed to define the pattern of summer and winter movement in active rockglaciers. Perhaps, the high-precision data collecting by Haeberli and Schmid (1988) will help to find a satisfying answer. The rheological consequences of a possible greater summer movement of rockglaciers will be discussed in Section 7.7.

7.1.7 Vertical Displacement

The vertical movement of a rockglacier surface is not so easily summarized:

1. The displacement is normally much smaller than the horizontal movement.
2. To record the vertical changes of a relative rough surface, sophisticated techniques have to be applied, i.e., the professional help of a geodetic engineer is required.
3. Vertical displacement has to be explained by three different factors(cf. Fig. 7.13);
   - settling due to interior melting
   - mass gain by freezing of meltwater at the permafrost table
   - change in elevation according to the surface gradient of the rockglacier.

Therefore, only a limited number of measurements can be reported. On the rockglacier Jaeger (Calkin et al. 1987), the subsidence was between 6 and 122 cm (±5 cm) for 5 years (1978–1983) with an average of 11 cm a^{-1}. The Pika rockglacier displays 6–94 cm subsidence (± 5 cm) for the same period, with an average of 10 cm a^{-1}. During 1982 on two lobes of the Jaeger rockglacier, a rise of 30–60 cm was recorded, which is difficult to explain. If may be that is either due to compressive flow or to an increased internal accumulation of ice.

On the rockglacier Äußeres Hochebenkar (Figs 7.14, 7.15), the vertical displacement 1936–1953 (17 cm a^{-1} has been documented by Pillewizer (1957). He describes the following gains and losses:

2700–2600 m a.s.l. an irregular loss of ca. 5 m (ca. -29 cm a^{-1});
2580–2530 m a.s.l. a vertical loss of nearly 10 m (ca. -59 cm a^{-1});
2450–2500 m a.s.l. a vertical gain of 10–15 m (ca. + 59–88 cm a^{-1}).

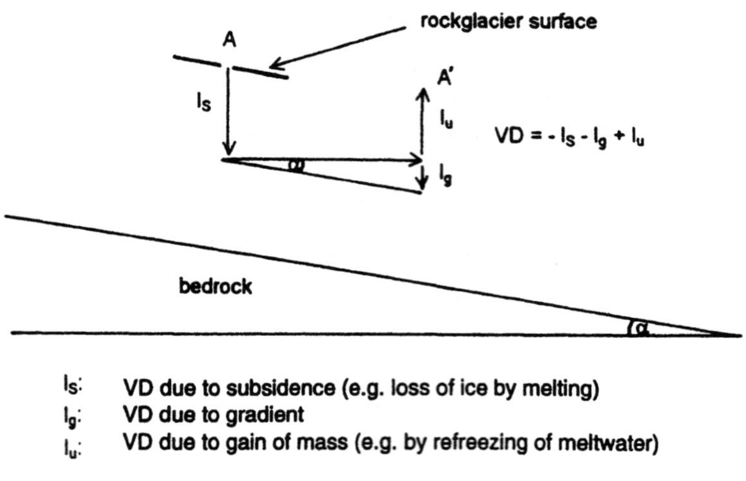

Fig. 7.13. Components of the vertical displacement of a rockglacier surface

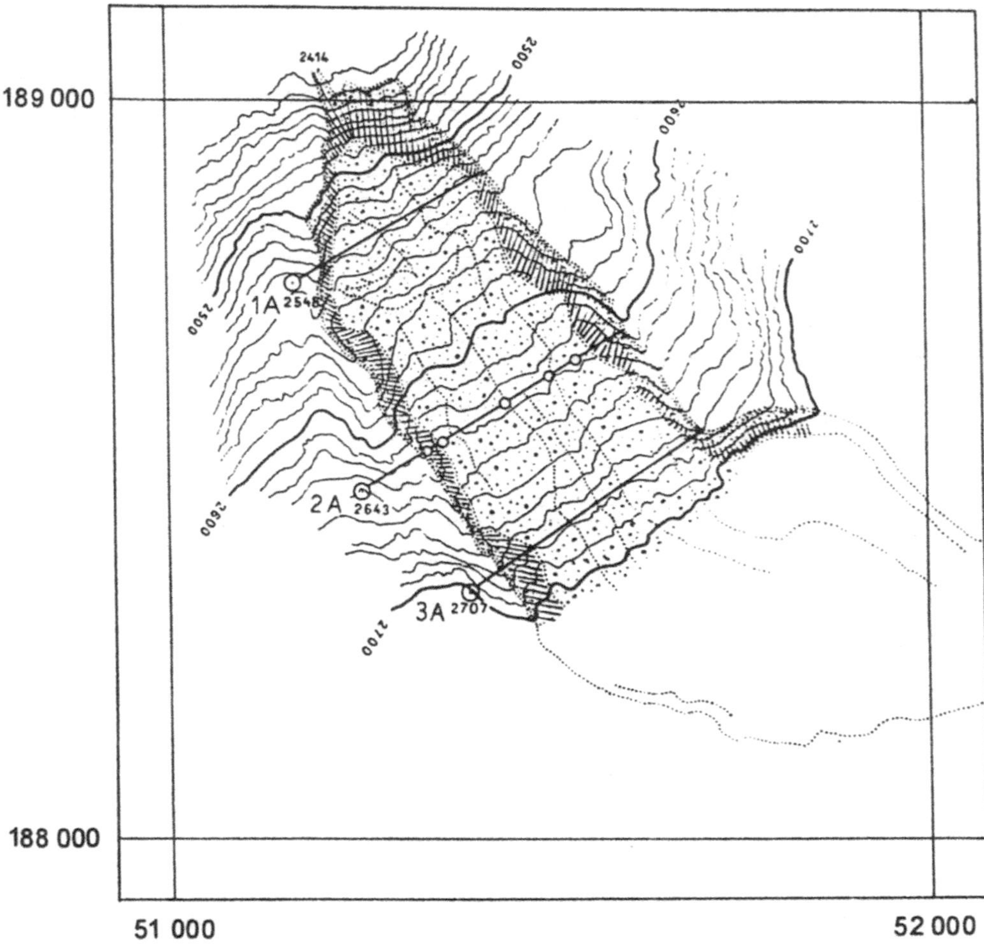

Fig. 7.14. Map of the rockglacier Äußeres Hochebenkar (Pillewizer 1957). The positions of the three cross-sectional profiles discussed in the text are marked on the rockglacier

Vietoris (1972,p.186) presents additional measurements on the three cross-sectional profiles discussed by Pillewizer. Compared with the 1953 profiles, the highest profile (ca 2680 m a.s.l.) is several meters thicker 19 years later; profile 2 (2625–2635 m a.s.l.) presents slightly higher (at the edges) or the same values (central part) between 1960 (29 July) and 1972 (5 September). Profile 3 displays a loss (several meters in its central part). Thus, the situation 1953–1972 seems to be the reverse of that of 1936–1953. The question is, is this behavior is due to pulses in the rockglacier movement, to quite different inputs of ice (e.g., re-frozen meltwater at the permafrost table), or to different talus inputs? Again, this can only be decided on theoretical grounds or by using larger data sets.

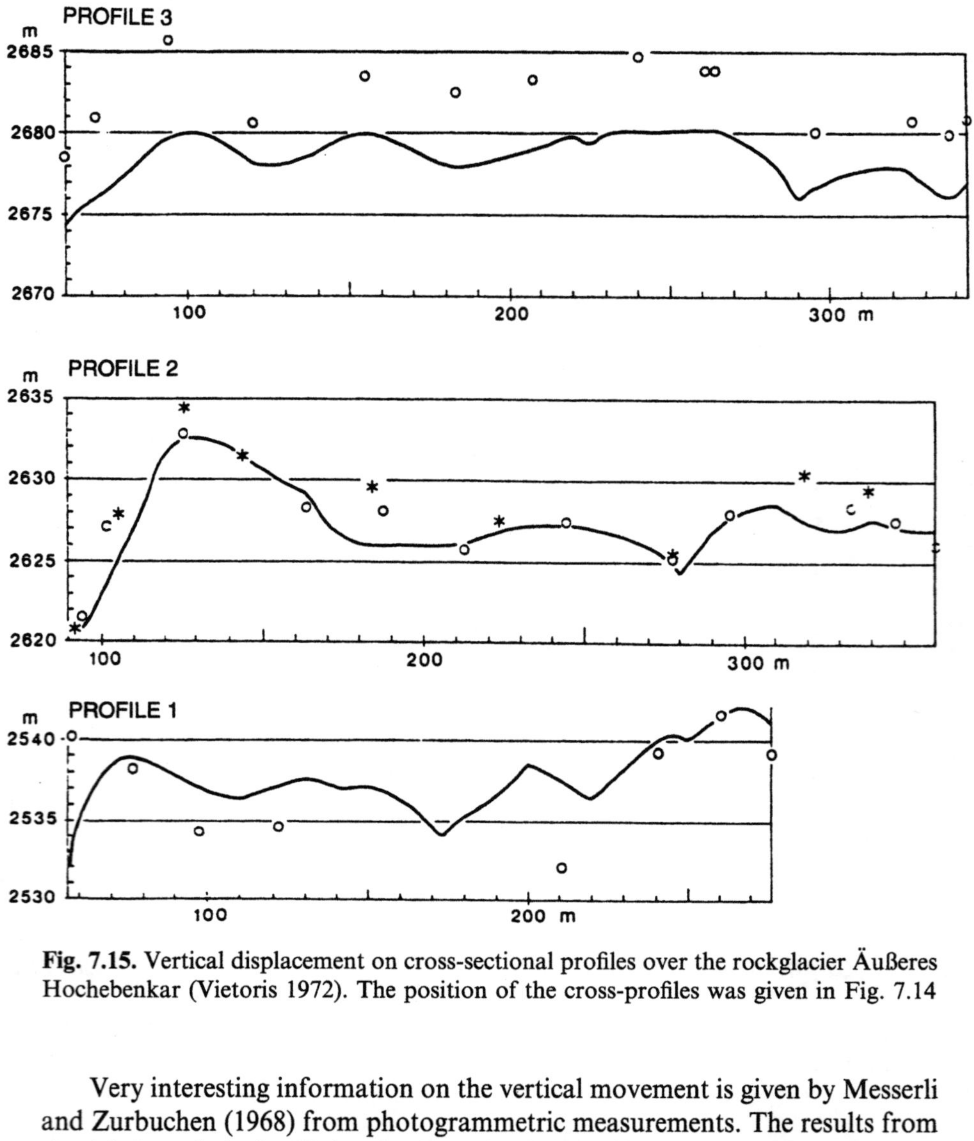

Fig. 7.15. Vertical displacement on cross-sectional profiles over the rockglacier Äußeres Hochebenkar (Vietoris 1972). The position of the cross-profiles was given in Fig. 7.14

Very interesting information on the vertical movement is given by Messerli and Zurbuchen (1968) from photogrammetric measurements. The results from the debris rockglacier Weissmies (Laggintal, S Wallis) and the talus rockglacier Grosses Gufer (Aletsch, N Wallis) are presented in Table 7.12. Despite marked horizontal velocities, the rockglacier Weissmies lost about 50 000 m^3 in 6 years (1958–1964), the rockglacier Grosses Gufer presents no marked gains or losses. Thus, the tongue of the Weissmies displays an average lowering of its surface of more than 30 cm a^{-1}. Subsidence decreases by one-third towards the rooting zone. Despite this general loss, zones with marked volumetric gains exist (Figs. 7.16, 7.17). They are found at the talus (upper right) and at the right lateral slope; they are also very marked at the right-and left-hand sides of the tongue. According to the authors, the position of the front seems to have been stable

Table 7.12. Annual vertical displacement (VD) of the rockglaciers Weissmies (Wallis) and Grosses Gufer (Aletsch, Wallis) calculated from the data in Messerli and Zurbuchen (1968)

	Weissmies[a]		Grosses Gufer[b]	
	Points	VD (cm a^{-1})	Points	VD (cm a^{-1})
Talus[c]	1–3	-11.1	46–48	-1.4
Rooting[d]	8–11	-22.1	12, 14, 41, 42	-10.2
Center right	21–24	-22.9		
Center mid	65, 67, 68	-26.6	19, 20, 101–105	-10.1
Center left	40, 43, 48, 49	-17.1		
Front left	45, 50, 51	-32.2	5, 32	-15.0
Front middle	–	–	54, 55	-27.1
Front right	31, 47, 52, 53	-32.5	36, 52	-0.9

[a]Weissmies 1958–1964 (23 September 1958 to 10 September 1964)
[b]Grosses Gufer 1950–1962 (September 1950 to 14 September 1962)
[c]At the debris rockglacier Weissmies, talus describes the connection to the talus slopes of the Tälligrat in the upper right of the rockglacier. At the Grosses Gufer, the points are on the talus slopes of the upper right.
[d]Rooting = rooting zone of the rockglacier.

during the 6-years period, despite the clear-horizontal displacement above the front of ca. 60 cm a^{-1} (center). Therefore, the increase in volume at the right and left front is probably best explained by compressive flow.

On the Grosses Gufer, the measured vertical lowering is small at the front (between 0 and 8 cm a^{-1}). Due to gradient alone, which is about 24° on the lower part of the rockglacier, the horizontal movement of ca. 50 cm a^{-1} explains a lowering of ca. 20 cm a^{-1}. This means that the rockglacier Grosses Gufer shows a volume increase at the front of 12–20 cm a^{-1}.

The different behavior of the two rockglaciers is astonishing. The regional situation in the Wallis seems more or less the same for both rockglaciers; this is also true for the local conditions. Exposure is northern (Grosses Gufer), and eastern for Weissmies. They end in 2310 m a.s.l. (Grosses Gufer) and 2470 m (Weissmies). Material is added to the Grosses Gufer from the talus slopes between the Eggishorn and the Bettmerhorn, to the Weissmies rockglacier by the glacier during glacier advances, if the glacier overtops its lateral moraine. The lower limit of active rockglaciers in northern exposure in the N Wallis is around 2520 m a.s.l., in southern Wallis ca. 2600 m (Barsch 1980). Taking the difference between easterly and northerly exposures into account and without discussing here the reaction time of active rockglaciers to a warmer climate, the rockglacier Weissmies seems to have been in around 1950/1960 at the lower limit of activity. In contrast, no signs of inactivity are recognizable on the rockglacier Grosses Gufer at this time.

Terrestrial measurements of the vertical displacements on the rockglacier Murtèl I were done in 1971–1973 (Table 7.13). The maximum subsidence was

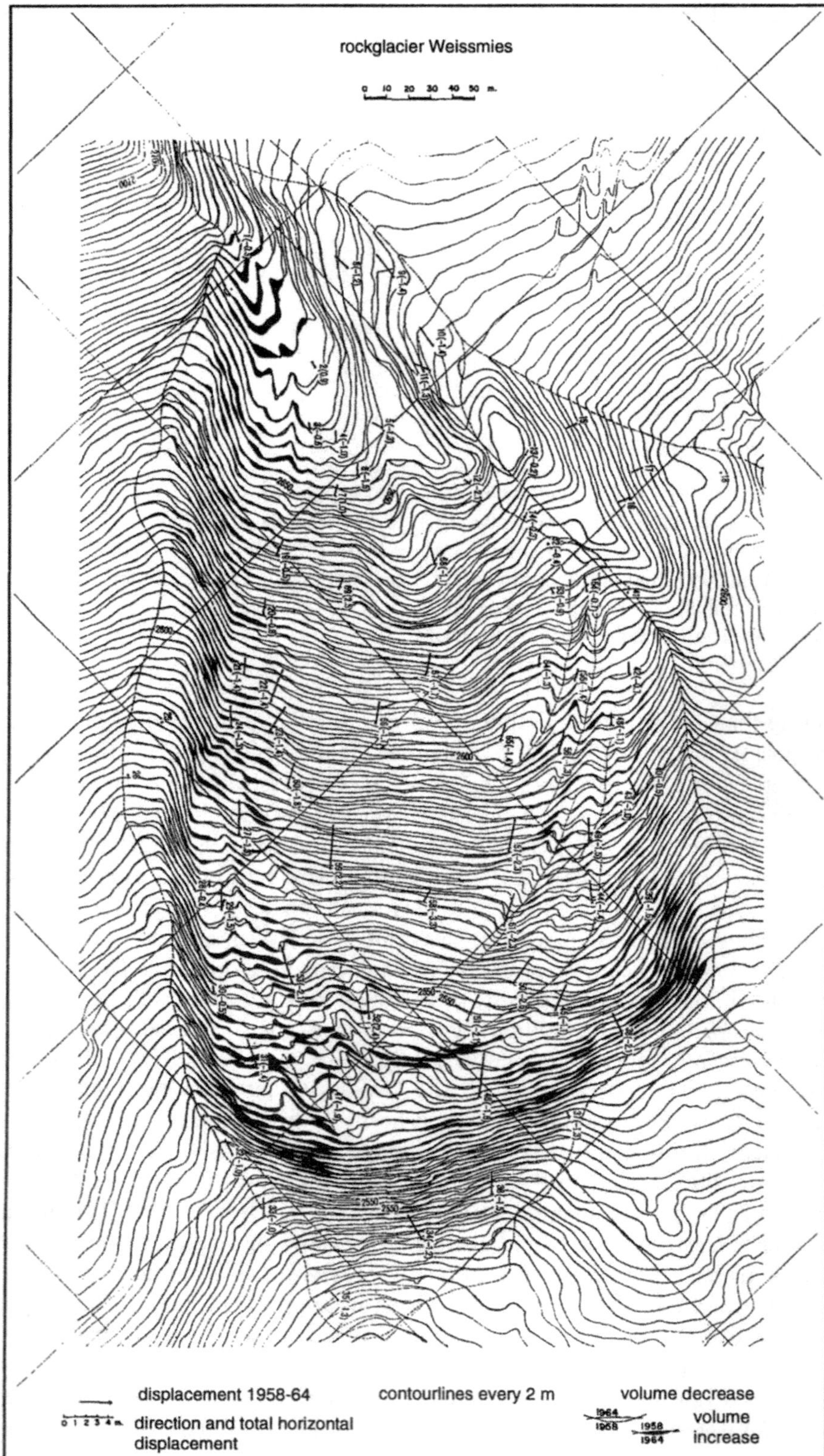
rockglacier Weissmies
0 10 20 30 40 50 m.
2650
2600
2550
displacement 1958-64
0 1 2 3 4 m.
direction and total horizontal displacement
contourlines every 2 m
volume decrease
1964
1958
volume increase
1958
1964

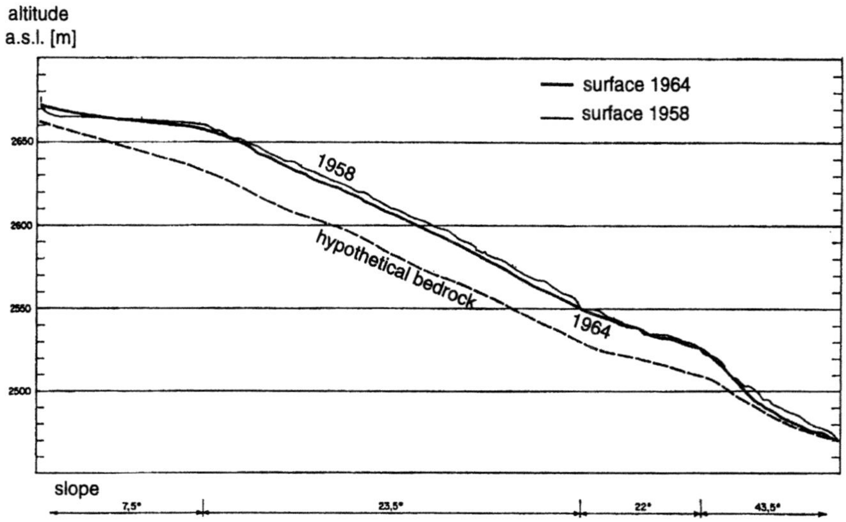

Fig. 7.17. Changes in the surface position on a longitudinal profile over the Weissmies rockglacier between 1958 and 1964 (Messerli and Zurbuchen 1968)

encountered in late summer 1971 (together with the maximum horizontal movement). This relatively high vertical displacement may be due to increased melting during summer 1971. Because the vertical downward movement is greater than would have been expected from the surface slope of the rockglacier, it may indeed indicate an effect of melting, though subsidence is always larger in summer than during winter (except summer 1972 to winter 1972/1973 on Murtèl I).

Some measurements of vertical displacement on the Val Sassa rockglacier are given by Eugster (1973, p. 374). These values are especially interesting because the rockglacier was in the final stages of becoming inactive. The data in Table 7.14 indicate that all points during all periods experienced a lowering of the surface (except point 26 in 1953/1954). In most cases this was a loss, because the lowering is higher than expected from the surface gradient, although the absolute differences are not very big. Despite a marked trend towards inactivity (cf. Sect. 7.1.3), collapse due to melting does not seem to be very intense during the short period of the measurements.

Haeberli (1985a; cf. Table 7.11) reports his measurements on the vertical displacement of the Gruben rockglacier. The overall trend of the 8 points along a flow line is negative, but there are marked differences. Assuming an average surface slope of 10°, the vertical displacement due to the gradient can be calculated for each horizontal displacement. It shows that in the upper part the surface lowering is smaller than it should be according to the gradient. This is

Fig. 7.16. Vertical displacement on the Weissmies rockglacier 1958–1964 according to Messerli and Zurbuchen (1968)

Table 7.13. Long-term horizontal and vertical displacement on the rockglacier Murtèl I according to Barsch and Hell (1975)

	Point	HD (cm a^{-1})	VD (cm a^{-1}) 1932–1955	LG (cm a^{-1})	Gain/loss (cm a^{-1})	HD (cm a^{-1}) 1955–1971	VD (cm a^{-1})	LG (cm a^{-1})	Gain/loss (cm a^{-1})
Front	64	10.9	−3.9	−1.5	−2.4	6.3	+3.8	−0.9	+4.7
	65	10.4	−5.2	−1.5	−3.7	6.9	+0.6	−1.0	+1.6
	66	11.7	−8.2	−1.6	−6.6	4.4	+2.5	−0.6	+3.1
					ϕ −4.2				ϕ +3.1
Center	47	9.6	−6.5	−1.4	−5.1	5.0	0.0	−0.7	+0.7
	48	7.8	+0.4	−1.1	+1.5	7.5	−1.9	−1.1	−0.8
	49	11.7	−3.0	−1.6	−1.4	3.1	+3.8	−0.4	+4.2
	50	8.3	−0.9	−1.2	+0.3	5.0	+2.5	−0.7	+3.2
					ϕ −1.2				ϕ +1.8
Upper part	22	15.7	−7.4	−2.2	−5.2	–	–	–	–
	28	8.3	−6.1	−1.2	−4.9	3.8	+1.3	−0.5	+1.8
	29	6.5	−11.3	−0.9	−10.4	3.8	−2.5	−0.5	−2.0
					ϕ −6.8				ϕ −0.1
	ϕ	7.1	−6.0	−1.0	−5.0	3.2	+2.0	−0.5	+2.5

Measurement error of the original data is ±0.7 m for HD and ±0.5 m for VD for each period (1933–1935, 1955–1971).

HD: Horizontal displacement
VD: Vertical displacement
LG: Lowering due to gradient

Table 7.14. Horizontal and vertical displacement 1952–1959 on the rockglacier Val Sassa (Graubünden, Swiss Alps) calculated from data published by Eugster (1973)

Point	Distance to snout (m)	Horizontal (cm a^{-1})	Vertical (cm a^{-1})	Loss to gradient (cm a^{-1})	Gain/loss (cm a^{-1})	Period
25	ca. 90	70	–15	–18.8	+3.8	1952–1954
25		51	–18	–13.7	–4.3	1954–1959
17	ca. 240	50	–16	–13.4	–2.6	1954–1959
22	ca. 650	6	–2	–1.6	–0.4	1954–1959
26	ca. 785	60	–30	–16.1	–13.9	1952–1953
26		30	+10	–8.0	+18.0	1953–1954
26		50	–16	–13.4	–2.6	1954–1959

The average surface slope is ca. 15°.
All points are situated on a profile along a center line.
Point 26 is on the central part of the next lobe.

only possible (local topographic variations excluded) if there is a mass gain in this part which can be calculated to be 5.8 cm (1979–1982). On the tongue, the average loss (probably due to melting) is 24.7 cm for the same 3-year period. Even if the data set is very small, it seems that the rockglacier loses mass at its tongue. It is not possible to decide whether this is a general trend.

The vertical displacement on the rockglacier Murtèl I was studied photogrammetrically on aerial photographs for a longer period (Barsch and Hell 1975; cf. Figs. 7.18, 7.19, Table 7.13). For the whole period (1932–1971), a general average loss of 2.8 ± 1.0 cm a^{-1} has been calculated. The average for all measured points is a lowering of 6.0 +/- 2.2 cm a^{-1} (1932–1955) and a rising of 2.0 ± 3.1 cm a^{-1} (1955–1971). Assuming an average surface slope of 8°, the surface lowering during the first period should be 1 cm a^{-1}. This means that during 1932–1955 a general loss (melting) of 5 cm a^{-1} took place. During 1955–1971, the surface lowering due to the gradient should be 0.5 cm a^{-1}. Together with the rising surface, a total gain of ca. 2.5 cm a^{-1} can be calculated.

Despite the error for the direct measurement of the vertical displacement (± 50 cm), the general picture is confirmed in detail (Table 7.13). During 1932–1955, the surface was lowered markedly at the front and in the upper part, it was less pronounced in the center. During the second period, a slight gain was recorded at all parts of the rockglacier.

It is highly remarkable that the maximum lowering of the surface occurs in times of higher horizontal velocities, even if the vertical changes are corrected according to the mean surface slope. This could be interpreted as a consequence of extending flow. Unfortunately, the data are not detailed enough to allow a final answer; correlations with warmer years as well as with winter cover, etc. are also not possible.

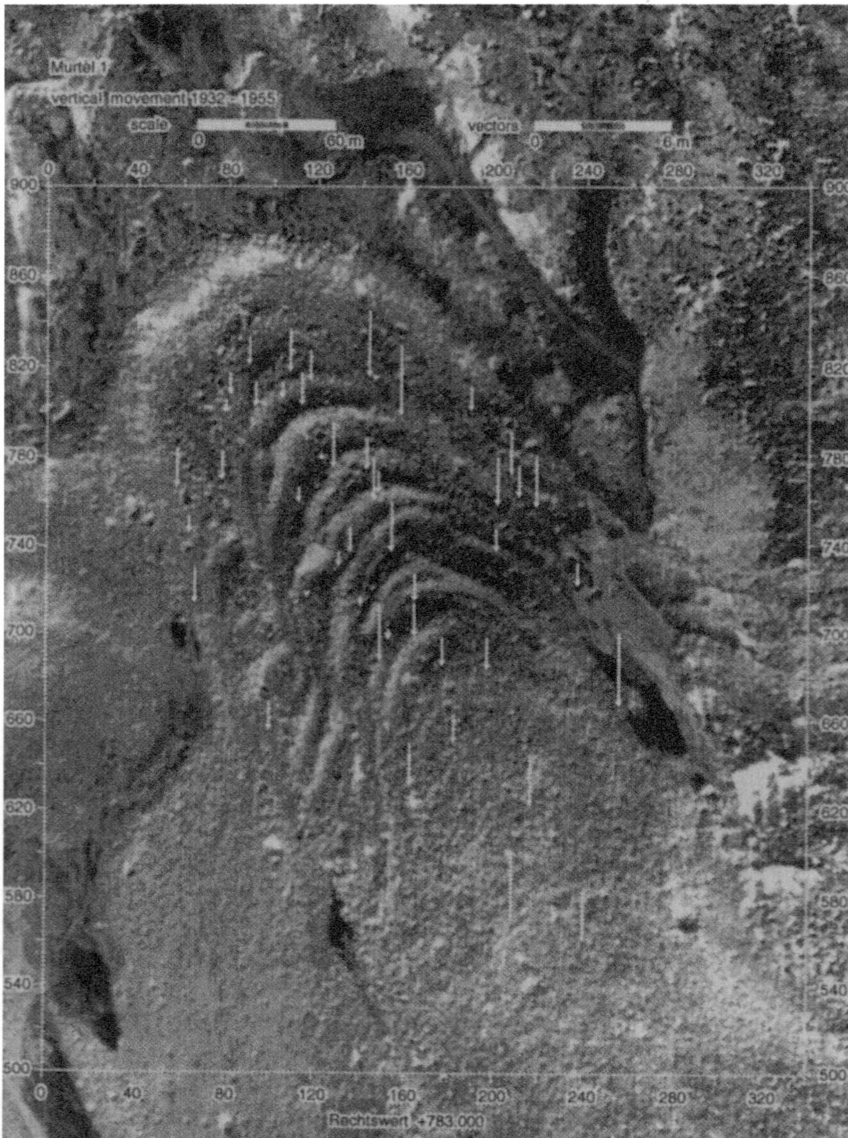

Fig. 7.18. Photogrammetrically determined vertical displacement 1932–1955 on the rockglacier Murtèl I. (Barsch and Hell 1975)

Fig. 7.19. Photogrammetrically determined vertical displacement 1955–1971 on the rockglacier Murtèl I. (Barsch and Hell 1975)

In general, more sophisticated measurements on the vertical displacement on active rockglaciers are desirable to evaluate their actual behavior, to forecast their future development, and to monitor the effects of global warming and other climatic effects on these systems.

7.1.8 Conclusion

Active rockglaciers display a range of annual horizontal displacements between 1 or 2 and 100 or 200 cm a^{-1}. Higher values are possible but these are related to special topographic (rockglacier Äusseres Hochebenkar) or special geomorphic situations (rockglacier Gruben). Horizontal velocity is influenced by gradient and mass balance, by extending and compressive flow.

Velocities are often constant over decades. A very good example is the rockglacier Macun I (Fig. 7.20). Over a period of 21 years (1967-1988), the toe displayed a movement of 14.2 cm a^{-1}. The error on the left-hand side is ±0.3 (36 points), on the right-hand side ±0.6 cm a^{-1}. In the rooting zone, velocity amounts to 6.1 ± 0.4 cm a^{-1} (10 points). A similar constancy is displayed by the advance of the rockglacier Val da l'Acqua (Fig. 7.21) between 1920 and 1980 according to Jackli (1978).

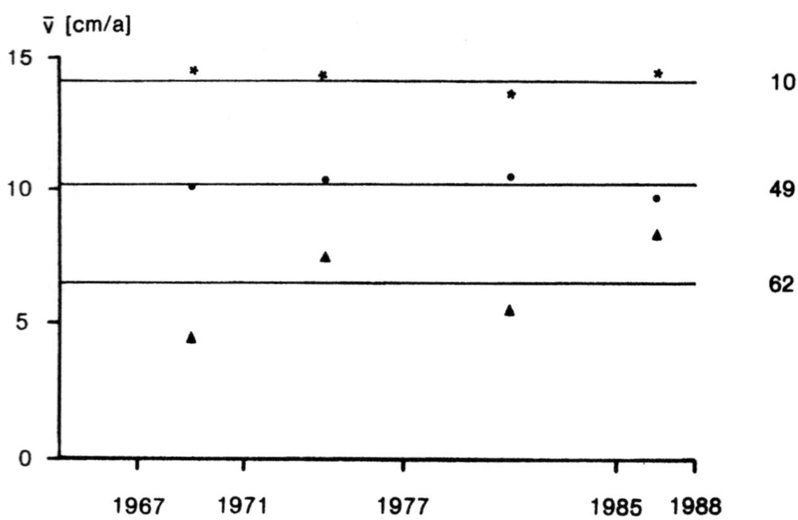

Fig. 7.20. Horizontal velocity 1967–1988 of three points on the rockglacier Macun I (Engadin). Though at *point 62* a slight trend to an increase in velocity and at *point 49* a decrease seems possible, the interpretation favors that of a constant velocity based on possible measurement errors. The *solid lines* display the average velocity of the points discussed

It is not always easy to decide whether reported velocity variations are real or just a consequence of inaccuracy of the measurements, but there is no doubt that velocity variations in time occur on rockglaciers. Real velocity differences may be due:

1. To long-term changes in activity (e.g., from active to inactive status; cf. the Val Sassa rockglacier, Table 7.5);
2. To differential movements along the stromstrich or on cross-sectional profiles;
3. To differential movements caused by input changes;
4. To probable changes between extending and compressive flow (cf. the Arapaho rockglacier);
5. To seasonal changes (cf. the examples of the rockglaciers Murtèl I or Gruben); or
6. To pulses in rockglacier movement which have been detected on the rockglaciers Gruben and Äusseres Hochebenkar and which are not really understood.

It can be said that we are well informed today about the annual horizontal movement rates of many rockglaciers. Still lacking are more differentiated data

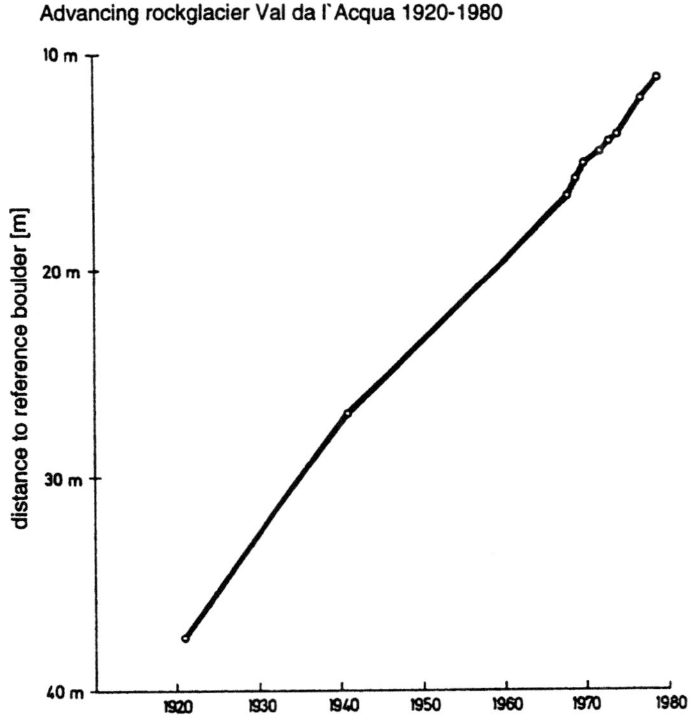

Fig. 7.21. Horizontal displacement at the front slope of the rockglacier Val da l'Acqua (Jäckli 1978). The average advancement over nearly 60 years is 46 cm a^{-1}

on seasonal movement rates and on the detailed differentiation of movement on single rockglaciers over time.

Data on the vertical displacement of a rockglacier surface are very limited. More measurements are needed because this information is decisive for the evaluation of the future development of a rockglacier (or of rockglacier permafrost).

7.2 Geometry of Movement

The foregoing section presents the available data on the amount of rockglacier creep. These data describe a pattern which can be summarized as the geometry of rockglacier movement. It is the vector field that represents the velocity at the rockglacier surface. There are three points to be discussed:

1. The surface flow pattern in general.
2. The velocities on long and cross-sectional profiles.
3. The relationship between surface flow and inner deformation.

This will be done on the following pages.

7.2.1 The General Flow Patterns of Active Rockglaciers

The geometry of the flow of rockglaciers can be shown by several methods if enough data are available. The following examples have been chosen:

1. The construction of flow lines as a possible way of presenting a graphical expression of the movement.
2. The definition of zones of equal movement or the construction of lines of equal velocity as a representation of the movement pattern.
3. The relationship between (differential) movements and surface relief, especially to show the formation of ridges and furrows by compressive flow.

1. Flow lines are constructed by using flow vectors on a rockglacier. Haeberli (1985a) developed flow lines for some of the better known rockglaciers of the Alps (Macun I, Murtè l I, Gruben, Grosses Gufer, Weissmies). In Figs. 7.22 and 7.23 flow lines are presented for the debris rockglacier Macun (based on 21 years of measurements; cf. Sect. 7.1.3), and the talus rockglacier Grosses Gufer. They reveal a clear stream-like pattern through the central parts of the rockglaciers and they connect source areas (talus slopes, morainic deposits) over the rockglacier rooting zone with the rockglacier front. Their form is controlled mainly by topography and geomorphic history.

2. The presentation (Fig. 7.24) of zones of equal velocity of the rockglacier Macun displays similar information to the flow lines. Maximum movement is

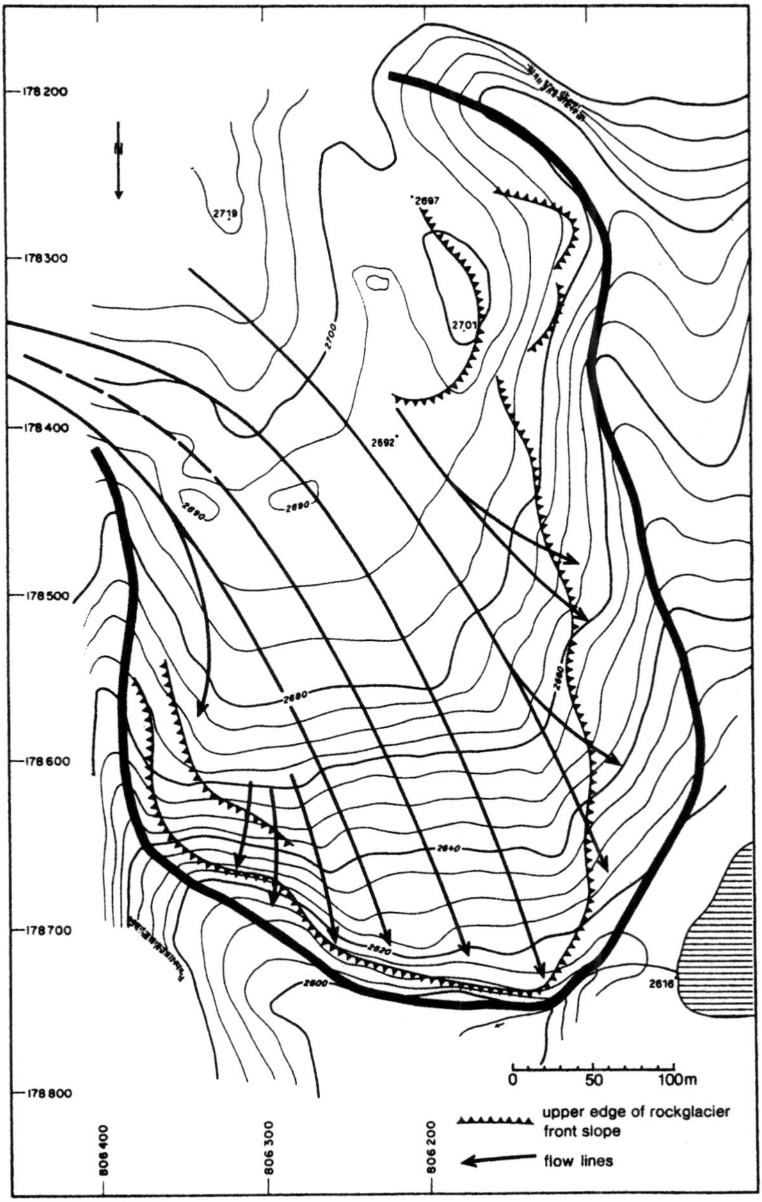

Fig. 7.22. Flow lines 1967–1988 on the rockglacier Macun I

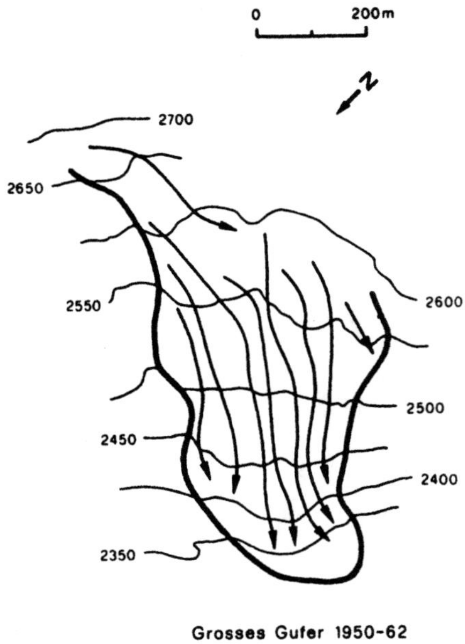

Fig. 7.23. Flow lines on the rockglacier Grosses Gufer (Haeberli 1985) according to data from Messerli and Zurbuchen (1968)

found around the center above the toe. This may be due to the fact that rockglaciers have here a greater thickness which causes this zone of high velocity. If they creep onto higher ground they get thinner, and thus they have to allow a greater flow through a smaller cross-sectional area. This seems logical, even if the rheologic explanation regarding the constancy of mass is difficult. In the example of Macun, thickness decreases from 90 to 60 m, and velocity increases from (9.8 to 14.2 cm a^{-1}). This is a marked velocity increase, which can be explained by the decrease in thickness.

Topographic and climatic influences may cause lower velocities immediately above the front slope. Generally, rockglacier movement seems to be less at or below the rooting zone. It is not clear yet if this is caused by changing inputs or by topographic effects (slope gradients). In any case, relationships between movement velocity and mass balance in the rockglacier rooting zone are not well known.

3. Haeberli et al. (1979) did a very careful lichenometric study of the Gruben rockglacier (Wallis). They found a clear boundary between the lichen-covered and the lichen-free areas. The latter were covered by the Gruben glacier during its 1850 advance. On all other parts of the rockglacier surface the lichen diameter varied in areas of the same geoecological potential between 2 and 4 cm. This is probably due to the high altitude (2800–2900 m). Fresh (lichen–free) boulders have not been detected outside the areas covered by either the glacier in 1850 or by almost perennial snowfields. Thus, it can be concluded that there are no places with different shearing movements, where fresh boulders are pushed up to the surface. The surface of the rockglacier is obviously passively riding on the creeping frozen core.

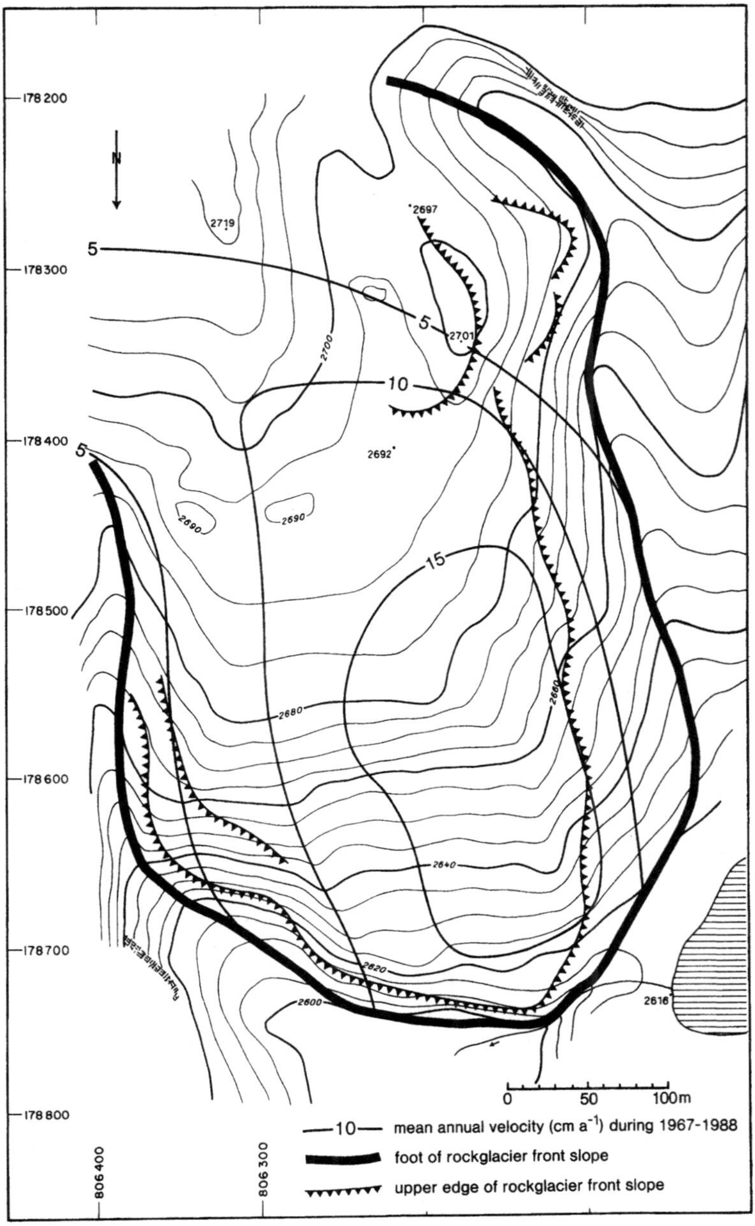

Fig. 7.24. Lines of equal surface velocities on the rockglacier Macun I. (Barsch and Zick 1991)

On the other hand, it seems to be accepted that the intensity of the furrow and ridge topography on the surface is correlated with compressive flow. If the rockglacier mantle is today only passively riding on the deforming inner core, then the formation of the furrows and ridges had to be done during an early stage of the rockglacier development. This does not preclude the present-day increase of ridges, for example.

One expression of extremely different movements of the rockglacier surface is the presence of inner front scarps or inner front slopes. They indicate that an upper lobe is overriding the lower part of the rockglacier. This is only possible if marked differences in activity exist which allow the development of shearing planes along which the overriding takes place. Regarding taxonomy, these are the multi-unit rockglaciers (cf. Chap. 4).

These points will be discussed by the following three examples.

7.2.1.1 Gruben Rockglacier

Figure 7.25 has been constructed after Haeberli et al. (1979). It displays the very interesting movement pattern of the Gruben rockglacier. Its western part shows velocities between 2 and 82 cm a^{-1} which is in the normal range of rockglacier velocities. The southeastern part displays flow lines almost perpendicular to those at the toe and to the longitudinal ridges in the northeastern part: in addition velocities exceeding 300 cm a^{-1} are connected with vertical displacements of around 1 m a^{-1}. This is the area in which the Gruben glacier pushed against and moved onto the rockglacier during the big general advance in the Alps of 1856. The glacier probably caused, then, a compression of the rockglacier that reacted with the development of ridges parallel to the surface slope and the flow lines during this situation. These longitudinal ridges may be taken as the surface expression of the stress field at this time.

It is probable, according to Haeberli et al. (1979), that the present-day flow lines started more to the east at the talus slopes and not at the rock spur. After the melting of the glacier a new stress field developed and an adjustment began. Haeberli et al. (1979) explain this movement as a rotational slip of massive ice with a shear plane at depths of 12 to 15 m. The ice bodies are either buried snowbank ice (deposited by avalanches) or buried glacier ice. This massive ice is melting because it is not in equilibrium with the present thermal conditions. It was deposited on frozen material, as is indicated by the presence of water at the bottom of the crevasses. The flow is extending and is directed into the basin, which was occupied by the Gruben glacier during its Little Ice Age advances between 1600 and 1850.

There are some points of interest:

1. The flow lines at the lower part of the rockglacier start at the foot of the talus slopes to the north of the rockglacier. They demonstrate a direct transport from the source area to the snout.

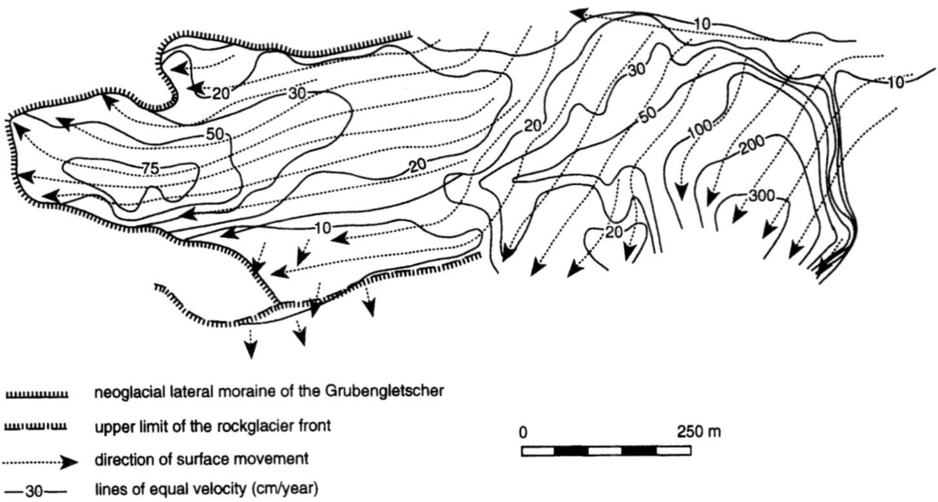

Fig. 7.25. Flow lines and velocity zones at the Gruben rockglacier. (Haeberli et al. 1979; Haeberli 1985a)

2. The flow in this part is extending. It follows the slope of the rockglacier surface. The lack of transverse ridges can be explained by this fact.
3. Surface relief and flow lines are discordant in the uppermost parts, where the flow is presently directed into the now ice-free valley. This valley was filled by ice during the glacier advances between 1600 and 1850. The surface relief still represents the general flow direction of the past.

The disturbance of the rockglacier by the last-century glacier advance does not seem to be extreme. It added, apparently, only a limited amount of material (till and ice) to the rockglacier, but the melting of the glacier led to an adjustment of the oversteepened slopes by sliding and creeping. This part of the rooting zone of the Gruben rockglacier was probably influenced more by glacial than by periglacial processes throughout the Holocene. The true rockglacier developed to the west of this zone (cf. Fig. 7.25).

7.2.1.2 Macun Rockglacier

The rockglacier Macun I (Barsch 1969a; Barsch and Zick 1991; cf. Figs. 6.7, 6.8, 7.11, 7.20) is the largest rockglacier in the cirque of Macun (South of Lavin, Lower Engadin, Graubunden). It is a north facing debris rockglacier that has been studied since 1965. According to seismic soundings, it has a maximum thickness of between 80 and 100 m. Horizontal surficial movement 1967–1988 is characterized by a zone of maximum velocity along the *stromstrich* (central flow line).

Velocity decreases towards the sides and towards the rooting zone. In the upper left part, a huge mass of debris displays no measurable movement. It is probably glacial debris which has not yet been incorporated into the rockglacier movement. Its age is difficult to assess, but an assignment to glacier advances during the last centuries (Little Ice Age) is possible.

It cannot be ruled out that the rockglacier will need to change its movement pattern in the near future as it pushes against a rock spur at its left toe. As a consequence, it will probably move more towards the right in the years to come. Up until 1988, a change in surface velocity had not been detected, i.e., the rockglacier has shown a nearly constant velocity for more than two decades. Therefore, it can be used as a good example of the secondary creep of rockglacier permafrost.

7.2.1.3 Arapaho Rockglacier

This rockglacier has been intensely studied. S.E. White (1987) recorded the absolute distances between 11 marked boulders on the top of three transverse ridges that run more or less parallel. The measurement site, at 3570 m a.s.l, has the best developed furrow and ridge relief on the rockglacier. Measurements were made using a steel tape; accuracy is thought to be ±0.5cm. The results are presented in Table 7.15 (cf. also Fig. 7.26). Two points should be especially noted:

1. The transverse ridges are spreading in a direction perpendicular to the flow.
2. The distance between the ridges, i.e., parallel to the flow, is generally decreasing.

Thus, in the direction of movement, the rockglacier displays a longitudinal compression. This should show up in the development of the height of the ridges, which can only be monitored using precise leveling. On the other hand, part of the compression is compensated for by lateral extension which is especially marked at the left-hand side (1966/1967, 1970/1971). Checking the change in the triangle defined by the points 5,7, and 11 between 1966 and 1975 indicates that it has been slightly enlarged (from 363.57m^2 to 366.63 m^2). This demonstrates that extension at least at the right-hand side of the test field is dominant.

This may be explained by the direction of flow. The rockglacier moves towards the left, the test field is on its right-hand side. Thus, it may be that the transverse ridges indicate, on average, a compressive flow; however, on the right-hand side perpendicular to the direction of flow an opposite tendency exists. Without data on vertical changes it is difficult to evaluate the situation correctly.

In any case, an alternation of extending and compressive flow is difficult to understand. It is also surprising that the values for the horizontal desplacement

Table 7.15. Extending and compressive flow at the Arapaho rockglacier (Colorado) according to S.E. White (1987; Table 10.1; cf. Fig. 7.21)

1. Extending flow on ridges normal to the direction of movement (cm)

	Upper ridge		Central ridge		Lower ridge	
	Pt. 1–4	Pt. 4–6	Pt. 7–8	Pt. 8–10	Pt. 3–5	Pt. 5–11
1966–67	–	–	7.0	0.0	–	5.0
1967–1968	–	–	–4.0	1.5	–	9.0
1968–1970	–	–	4.0	0.5	–	7.0
1970–1971	2.0	4.0	0.0	0.5	13.0	4.0
1971–1972	4.0	1.0	4.5	–0.5	–9.5	5.0
1972–1975	6.0	2.5	1.5	0.0	5.5	12.0
Sum:	(12.0)	(7.5)	13.0	(2.0)	(9.5)	42.0

2. Compressive flow parallel to the direction of movement (cm)

	Left side		Center left		Center right		Right side	
	A	B	A	B	A	B	A	B
1966–1967	27.0	0.5	2.0	0.0	–2.5	–	–2.5	–
1967–1968	–12.0	–2.0	–1.5	0.0	–0.5	–	–3.0	–
1968–1970	–17.0	–2.5	–2.5	–1.5	0.0	–	–6.0	–
1970–1971	11.0	1.0	–0.5	–0.5	0.0	–1.0	–2.0	–2.0
1971–1971	–13.0	–1.5	–0.5	–2.5	3.0	–3.0	–3.5	–5.0
1972–1975	2.0	–6.0	–6.0	–4.5	–1.0	–6.0	–8.0	–7.5
Sum:	–2.0	–10.5	–9.0	–9.0	–1.0	(–10.0)	–25.0	(–14.5)

( ), Values referring only to part of the time; pt., point; A, change in distance between upper and central ridge; B, change in distance between central and lower ridge.

		A	B
Left side:	pt.	1–2	2–3
Center left:	pt.	6–7	7–5
Right:	pt.	6–8	8–11
Right side:	pt.	9–10	10–11

Measurements: mid-summer 1966; 10 July 1967; 24 August 1968; 28 August 1970; 23 July 1971; 12 August 1972; 6 August 1975

are so different from year to year. The only possible explanation seems to be that there are pulses in rockglacier movement which are not yet really understood. this has already been indicated by the movement of the rockglaciers "Äusseres Hochebenkar" (cf. Sect.7.1.5) and Gruben (cf. Sect.7.1.6).

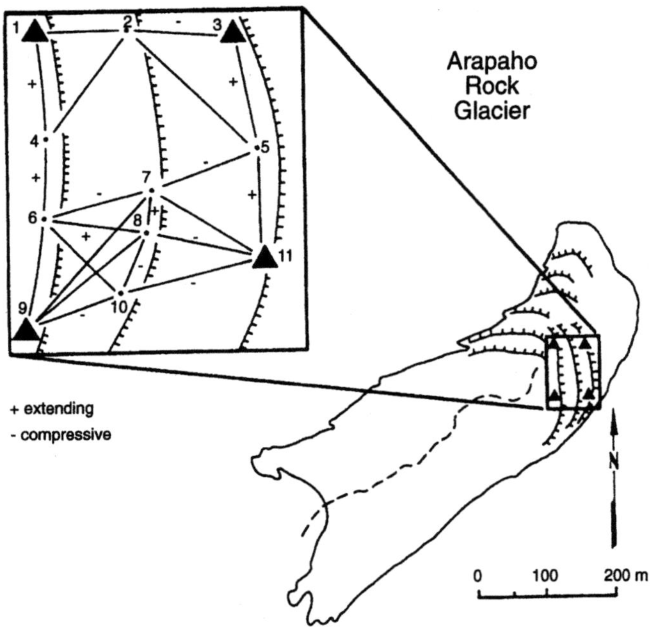

Fig. 7.26. Compressive (-) and extending (+) flow on the Arapahoe rockglacier. (S.E. White 1987)

7.2.2 Horizontal Velocity on Longitudinal and Cross-Sectional Profiles

Horizontal displacements (discussed above) on active rockglaciers show marked differences between the stromstrich (central flow line) and along cross-sectional profiles. The latter normal to the stromstrich, is easily described. As already pointed out, movement across an active rockglacier is greatest at the center and least at its outer edges. In general, it is assumed that the velocity cross-sectional profile forms a parabola, i.e., a second-order polynomial of the following form (or an equation of the Gauss type):

$$V = a + bx + cx^2. \qquad (7.2)$$

Some examples are presented in Table 7.16 (cf. Fig. 7.27) for rockglaciers with different velocities. To allow a comparison, the rockglacier width has been rescaled to ±100% of the stromstrich. In general, the velocity cross-sectional profiles are typical for large moving debris masses.

Based on a still small data set, the movement along the stromstrich (longitudinal profile) can be provisionally divided into three types (Fig. 7.28):

1. Velocity increases from the rooting zone towards the lower center and decreases from there towards the toe (Macun type).

Table 7.16. Examples for velocity representation in cross-sectional profiles on active rockglaciers

Murtèl I	y	=	5.157	+	6.580x	−	$0.0005x^2$
Macun I	y	=	14.233	−	0.014x	−	$0.0014x^2$
Gruben	y	=	48.609	−	0.010x	−	$0.0057x^2$
Galena Creek A	y	=	43.018	−	0.073x	−	$0.0045x^2$
Galena Creek B	y	=	76.160	−	0.164x	−	$0.00007x^2$
Äuss. Hochebenkar	y	=	180.310	+	0.014x	−	$0.0195x^2$

The polynomials fit the velocity distribution in rockglacier cross-sectional profiles. The width of the rockglacier has been set to the right and to the left of the stromstrich (center) to ± 100% to allow a direct comparison
Data sources: Barsch and Hell (1975) for Murtèl I,
Barsch and Zick (1991) for Macun I,
Haeberli (1985a) for Gruben,
Potter (1968) for Galena Creek,
Vietoris (1972) for Äusseres Hochebenkar.

2. Velocity decreases from the rooting zone towards the lower center and increases form there towards the toe (Murtèl type).
3. Velocity increases generally from the rooting zone towards the toe (Gufer type).

The three types are probably topographically controlled. Thus, the first type is caused by compressive flow at the toe as a consequence of a topographic threshold. The second type is probably due to the fact that the rockglacier has overcome a topographic threshold, i.e., the front slope has reached the rock sill and a steady general surface slope from the center towards the toe has been developed again. It is probable that the Murtèl type can merge into type three. The third apparently represents a steady state with an increasing thickness and an increasing velocity. As examples for the three types, the following rockglaciers can be used.

The Macun type (rockglacier Macun I) is taken as the normal case. It displays the increase in velocity from the rooting zone to the toe over a small central peak, i.e., from about 7.6 cm a^{-1} (±0.5cm) to 9.8 cm a^{-1} (±0.3cm) and over 16 cm a^{-1} to 14.2 cm a^{-1} (±0.4 cm) over a distance of ca. 450 m. The little peak is probably also due to the increasing thickness of the rockglacier at this place (cf. Fig. 6.8). It seems that the rockglacier Val Sassa during its time of activity belonged to this type.

A number of other examples can be named, in which movement first increases and then decreases from the lower center downwards towards the toe. A good example is the Tungsten rockglacier. After an initial increase (210 to 250 cm a^{-1}), velocity decreases to 40 cm a^{-1}. This is probably due to compressive flow at the toe (or the onset of inactivity?). The same behavior is displayed by the rockglacier in the Combe de Laurichard (Francou and Reynaud 1992, cf. Fig. 7.5).

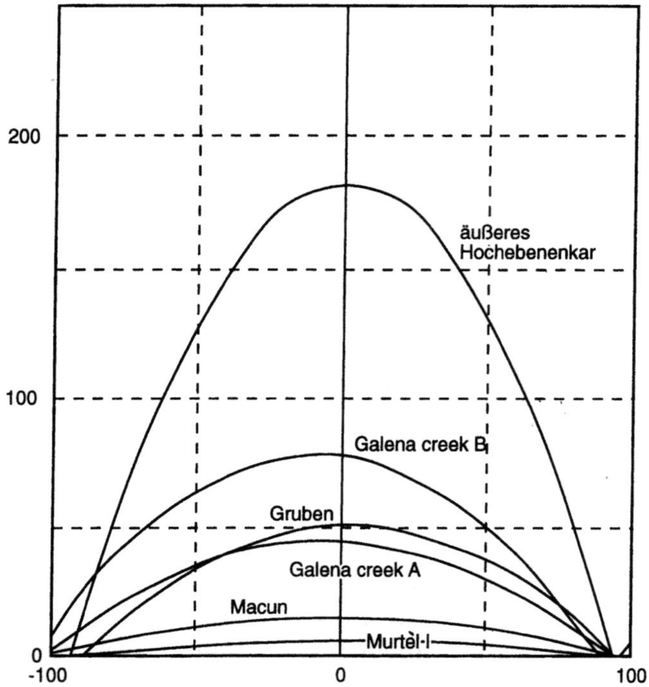

Fig. 7.27. Idealized velocity cross-sectional profiles on active rockglaciers. For better comparison, width is set right and left of the stromstrich as 100. Velocity is given as (cm a^{-1})

The rockglacier Murtèl I displays the second type. It is characterized by a decrease of velocity from the rooting zone towards the center and by a velocity increase from the center towards the toe. This is probably due to highly compressive flow in the center, which in this case is demonstrated by the very well developed surface relief of this rockglacier.

The talus rockglacier Grosses Gufer (Messerli and Zurbuchen 1968) represents the third type. Velocity increases from the rooting zone towards the toe. It is also a rockglacier with almost no surface relief. This is probably due to a uniform debris input or to generally extending flow.

The three types are schematically summarized in Fig. 7.28. A mathematical description of the velocity distribution along the stromstrich of active rockglaciers is not intended. The data set is still too small.

7.2.3 Surface and Subsurface Velocity

The velocity of a rockglacier is greatest at its surface (Johnson and Nickling 1979; Wagner 1992). Towards the interior it decreases. The form of the velocity

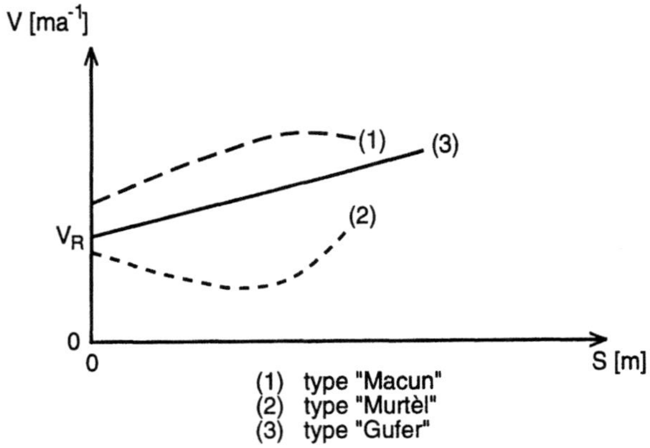

Fig. 7.28. Velocity V versus distance S from the rooting zone along the stromstrich (center flow line) on active rockglaciers. The different types are discussed in the text

gradient with depth has often been discussed. The classical analysis is given by Wahrhaftig and Cox (1959 Fig. 7.29). It is concentrated on the form of the front slope, and it discusses different assumptions for a possible velocity gradient with depth. Wahrhaftig and Cox have already pointed out that the apparently stable form of the front slope and the material that is transported to the front are two different phenomena. The lower parts of the rockglacier move more slowly than the upper layers, therefore, material has to be added from above (S.E. White 1971a). This material forms the well known talus apron around the rockglacier front slope. The height of the talus apron is an indicator for the steepness of the velocity gradient ("the higher, the steeper"). It seems that on most rockglacier frontslopes a steady state has been developed, which is demonstrated by the similarity of rockglacier frontslopes over years.

With reference to Fig. 7.30, Wahrhaftig and Cox (1959) propose that the types b, c, or d are possible descriptions for rockglacier movement. Johnson and Nickling suggest possibility b, which seems only justified if the inclination measurements in the borehole of RG II (cf. Fig. 6.17a) represent the whole gradient curve. Considering an average talus height of 30% at the front (S.E. White 1976; Evin 1983b), Haeberli prefers possibility c (and in some cases a combination with sliding).

According to the inclination measurements in the borehole through the rockglacier Murtèl I (Wagner 1992), type d with a high velocity near the base

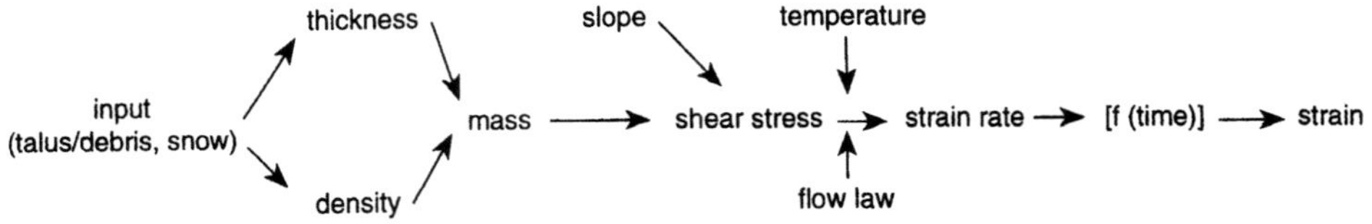

Fig. 7.29. Conceptual model of the parameters affecting the deformation of ice-supersaturated mountain permafrost bodies

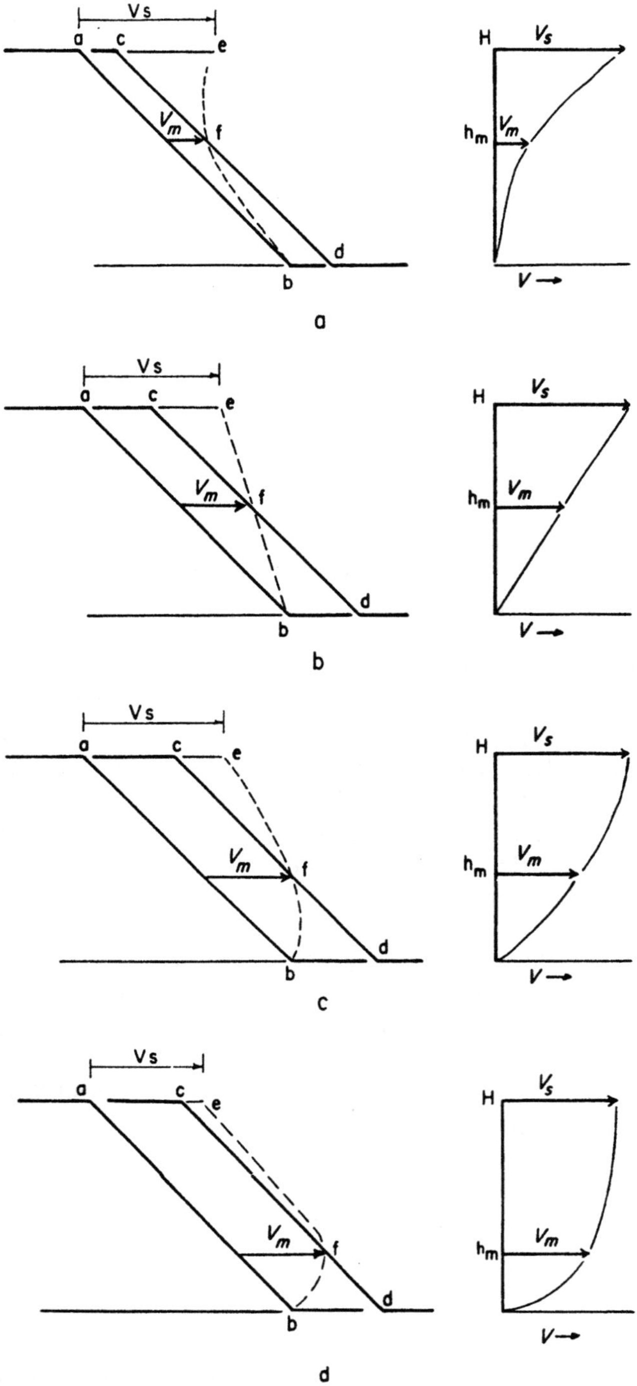

Fig. 7.30a-d. Profiles of the front slopes of active rockglaciers and the possible velocity gradients (Wahrhaftig and Cox 1959). It is assumed that the height (h_m) represents the position of the average velocity of the front. Material from above is added to stabilize the form of the front slope (see text)

seems to be the most appropriate description. On the rockglacier Murtèl I, about 75% of the total movement occurs near the base in a thin layer at 28–32 m depth. If this can be assumed for the whole rockglacier, a number of consequences need to be discussed. The first is related to the height of the talus at the front: it has to be not more than 30%. The second refers to some calculations. the velocity gradient in the Murtèl I rockglacier can roughly be described by the following equation (cf. Fig. 6.18a):

$$Y = 0.17 \cdot 2.3^{x} \; , \tag{7.3}$$

where Y = height above the base (m),

X = velocity ($cm\ a^{-1}$) .

This equation describes a strong velocity increase near the bed of the rockglacier. Thus, the assumption seems justified that the height (h_m; f in Fig. 7.30) of the mean velocity of the rockglacier front (V_m) occurs at about 30% of the total height.

Following the equation Haeberli (1985a), the average velocity (V_m) of the rockglacier can be calculated:

$$V_m = (1 - h_m) \cdot \overline{V}_s \tag{7.4}$$

where $\overline{V}_s$ = mean surface velocity parallel to slope, h_m = ratio of talus height to total front height.

For the rockglacier Murtèl I (with $V_s = 6\ cm\ a^{-1}$; $h_m = 0.3$):

$$V_m = 4 cm\ a^{-1} \; . \tag{7.5}$$

In this case, the difference between surface velocity and mean velocity, i.e., the movement of the total rockglacier, is not very big.

It has often been observed that despite a measurable surface movement, on some rockglaciers the front seems to stay at the same place for (Messerli and Zurbuchen 1968; Jackson and MacDonald 1980, p. 847). This can probably be explained by a greater difference between V_m and V_s. As a consequence, the height (h_m) of the talus apron around the rockglacier front should be relatively high.

7.3 Rheologic Considerations

As discussed in Chapter 5, the inner core of rockglaciers consists of frozen, ice-supersaturated talus or debris. Thus, ice plays a very important part in this material. On the one hand it gives unconsolidated material a greater strength, on the other hand it is the component of the material which is most easily deformed. Physically, the deformation of such a solid is a function (Fig. 7.31) of:

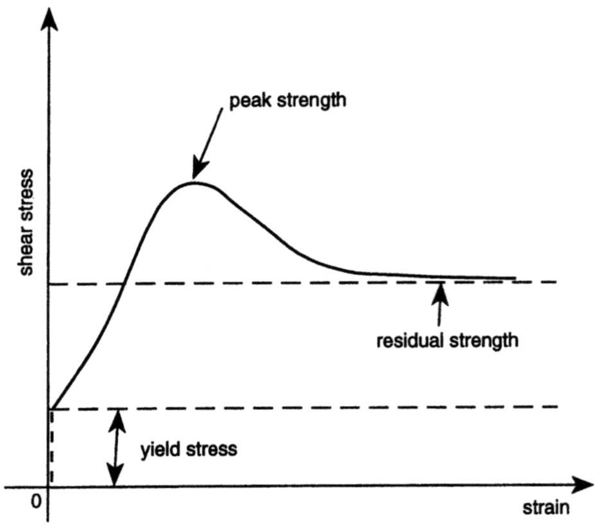

Fig. 7.31. General stress/strain diagram of material with plastic behavior

- The shear stress applied
- The density (as an indicator of the volume ratio between ice and mineral components)
- The thickness (h_p, as an indicator of the normal stress)
- The temperature T
- The grain sizes and the grain forms (influencing the internal friction, etc.)
- The form, type, and size of ice crystals
- The density of the ice (content of air bubbles, etc.).

As already indicated , a solid body like this has a strength which has to be overcome in order to cause a deformation. In the case of ice and snow, a small shear stress normally causes a deformation which after a short time is permanent. Therefore, ice and snow can be considered rheologically as fluids with a high viscosity. On the other hand, the viscosity of ice and snow is not constant but controlled by shear stress, temperature, and the positions of grain boundaries. Therefore, the behavior of both materials can be discussed as pseudo-plastic or, with a certain simplification, as ideal plastic.

According to Paterson (1981), the assumption that polycrystalline ice behaves as a solid rather than a fluid was recognized in about 1948. This led to studies on the relationship between the applied stresses and the related strains. If ice is considered to be a perfect plastic, its yield stress (i.e., the minimum stress required for continuous deformation) is often thought to be around 100 kPa ($=10^5$ N $m^{-2}=1$ bar).

The general behavior of a plastic material under stress is described in Fig. 7.31. If a stress larger than the yield stress is applied, deformation begins. Initially, strain rate (and deformation) grows with shear stress; after a maximum shear stress (ϕ_p, at peak strength) strain rate decreases and reaches a

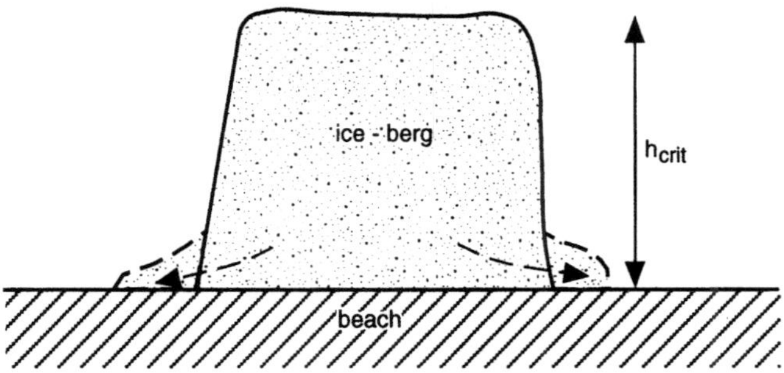

Fig. 7.32. Model of stranded iceberg. If the thickness of the stranded iceberg exceeds a critical value (normally 35–50 m), the ice deforms plastically at the base and creeps as indicated

constant value (residual strength). Often the difference between the peak and residual strain rates is small.

The classic relationship between the final strain rate and the shear stress for ice is given by Glen's flow law (Glen 1952; Paterson 1981; Haeberli 1985a):

$$\dot{\varepsilon}_{xy} = A \cdot (\tau_{xy})^n \ , \qquad (7.6)$$

where $\dot{\varepsilon}$ = final strain rate (a^{-1}), τ = Shear stress between 50 to 200 kPa, A = constant (temperature-dependent), n = constant.

Glen (1952) gives the following values for ice:

$$A = 0.148 \ a^{-1} bar^{-4.2} \ , \qquad (7.7)$$

$$n = 4.2 (\text{today, n is assumed to be between 1.9 and 4.2}),$$

$$\dot{\varepsilon}_{xy} = 0.148 \cdot (\tau_{xy})^{4.2} \ . \qquad (7.8)$$

The plastic deformation or creep of ice is illustrated by the example of the stranded iceberg, which slowly deforms at its base (Fig. 7.32). If the (temperature dependent) thickness exceeds a critical threshold, the creeping process starts. The plastic deformation or the creep of ice is divided into three forms (Paterson 1981; Haeberli 1985a); The primary, the secondary, and the tertiary creep (Fig. 7.33). After an initial phase of elastic deformation the nonreversible deformation begins:

1. The primary ("damped") creep is characterized by decreasing strain rates.
2. The period of secondary creep ("steady-state creep") occurs if the strain rates become more or less constant.
3. The tertiary creep ("undamped creep") is reached if strain rate increases after a period of time. It leads to failure.

As McRoberts (1975) following Vialov (1965) puts it, this differentiation is more or less academic; it is nothing more than a convenient model. Normally, glaciers are believed to be in a permanent state of secondary creep under low to

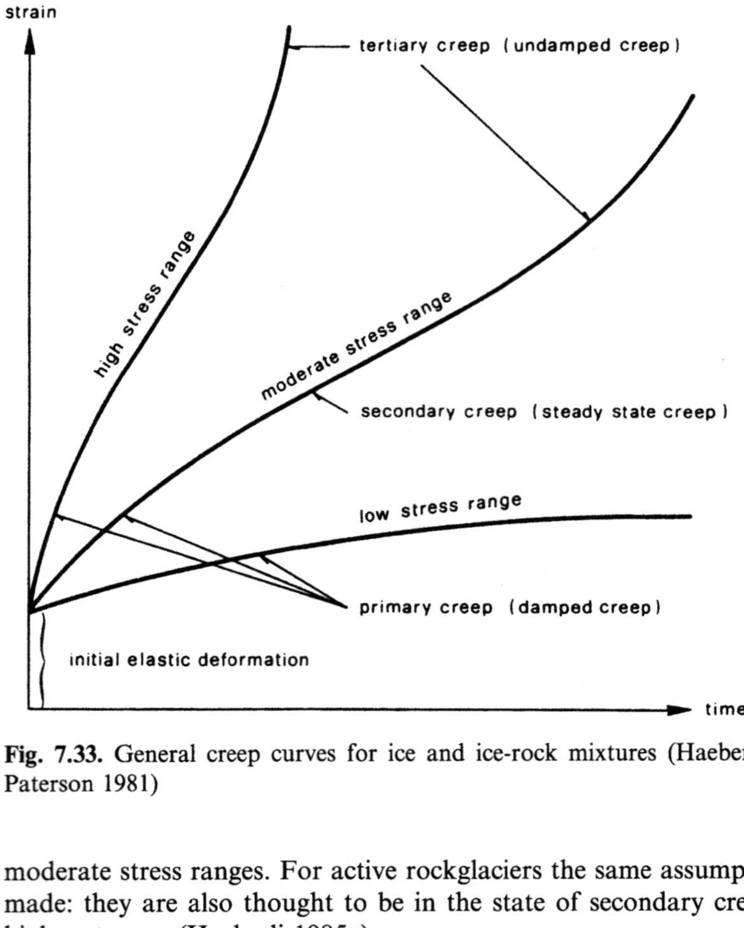

Fig. 7.33. General creep curves for ice and ice-rock mixtures (Haeberli 1985a; after Paterson 1981)

moderate stress ranges. For active rockglaciers the same assumption has been made: they are also thought to be in the state of secondary creep but under higher stresses (Haeberli 1985a).

If ice or ice-rich frozen soils are considered as a material with an ideal plastic behavior, all possible elastic responses are normally neglected. Considering the uncertainties in the rheological approaches to supersaturated mountain permafrost under natural conditions, this is an acceptable simplification. Unfortunately, the movement of rockglaciers in terms of the creep of supersaturated permafrost has not caught the interest of too many physicists, who work mostly with glaciers, or engineers, who work in construction. Therefore, information on the rheology of this system comes from laboratory tests on frozen soils over relative short time scales that have often been connected with research in ground freexing to stabilize unconsolidated materials during construction (Jessberger 1965, 1980). Important review papers which are helpful in understanding the physics of these systems include those of Vialov (1965), Goughnor and Andersland (1968), Hooke et al. (1972), McRoberts (1975), and Roggensack and Morgenstern (1978).

7.3.1 Shear Stress and Strain Rates in Active Rockglaciers

Active rockglaciers are cohesive bodies of frozen talus. In such a body the normal stress is traditionally calculated by

$$\sigma = \delta \cdot z \cdot g \cdot \cos\alpha \ , \tag{7.9}$$

where δ = density, z = depth (measured normal to surface), g = acceleration due to gravity, α = surface slope, and the basal or tangential stress is accordingly

$$\tau = \delta\, z \cdot g \cdot \sin\alpha \ . \tag{7.10}$$

Using this formula, the basal shear stresses for some of the better known rockglaciers can be computed. This has already been done several times, initially by Wahrhaftig and Cox (1959). They estimated the basal shear stresses in acxtive rockglaciers to be between 100 and 200 kPa (originally 1–2 bar). Haeberli (1985a) calculated basal shear stresses for the well-known rockglaciers in Table 7.17 to be between 110 and > 350 kPa (average 250 kPa). He thought this represented a marked difference between ice glaciers and rockglaciers. Table 7.17 shows values which are 10–30% smaller than those of Haeberli (1985a) because lower values for the average density (1.25 Mg m^{-3}) and, in the case of Murtèl I, lower values for h_p have been used here. Wagner (1992, p. 161) assumes the shear stress in the rockglacier Murtèl I to be 20–40 kPa (2–4 10^4 N m^{-2}).According to Eq. (7.9), the basal shear stress in the rockglacier Murtèl I is about 60–70 kPa. Thus, basal shear stresses seem to be smaller than hitherto assumed.

Under stress, frozen debris deforms plastically, e.g., Fig. 7.34 (Jessberger 1965). Additional information is given by Goughnor and Andersland (1968) who demonstrate that the ratio of clasts to ice is quite important for the deformation rate. Working with an ice–sand system, they concluded that above a threshold of 42 vol.% of sand (Figs. 7.35; 7.36) deformation rate decreases rapidly. The stress–strain relationship for the samples on which the definition of this threshold is based is given in Fig. 7.37, which shows that the strength of the sample increases with sand content. Unfortunately, the low temperature (–12.03 °C) during the experiment does not allow a direct comparison with most of the active rockglaciers.

Strain rates for active rockglaciers have been calculated only on few occasions. They are normally about $10^{-4}a^{-1}$. For the Fruben rockglacier the longitudinal strain rate has been calculated by Haeberli et al. (1979) as $10^{-3} - 10^{-4}a^{-1}$(frozen debris), $10^{-2} - 10^{-3}a^{-1}$(ice).

Wagner (1992) calculated the strain rates for the rockglacier Murtèl I based on Glen's flow law:

layer 12–27m: ca. $15 \times 10^{-4}a^{-1}$ with $A = 1.5 \times 10^{11}Pa^{-0.9}s^{-1}; n = 0.1$
layer 12–29m: ca. $10 - 150 \times 10^{-4}a^{-1}$ with $A = 1.0 \times 10^{-23}Pa^{-2.9}s^{-1}$; n = 2.9.

Table 7.17. Basal shear stresses and creep rates of active rockglaciers according to Haeberli (1985a)

		Kintole	Prafleuri	Macun I	Murtèl I	Gruber
l_r	[m]	1575	855	600	275	1000
ΔH	[m]	600	255	145	75	150
$\overline{\alpha}$	[°]	21	15	14	10	9
$\overline{h}_p$	[m]	> 50	> 30	80[a]	28–32[a]	100
$\overline{z}_b$	[m]	200–300	~150	80–100[a]	~50	100–200
τ	[k Pa]	> 250	> 110	> 350	200–250	250
$\overline{v}_s$	[cm a^{-1}]	–	–	~10–15[a]	~6[a]	~30
τ^a	[k Pa]	~220[a]	~95[a]	~240[a]	~65[a]	~200[a]

If new information was available, these were used in this table (cf. Footnote a).
The information on the Hurricane Basin rockglacier have been omitted because in the original table the data were based on the wrong impression that Brown (1925) described a longitudinal profile (cf. Sect. 6.3.1).
The values for the basal shear stress have been calculated with new data according to the equation (Paterson 1981; Haeberli 1985):

$$\tau = f\ \delta\ g\ h_p \sin\alpha \ [\text{k Pa}],$$

Where $f = 1$ (form factor),
$\delta = 1.25\ [\text{Mg m}^{-3}]$,
$g = 9.81\ [\text{ms}^{-2}]$.

l_r, length of rockglacier; ΔH, altitudinal difference between head and toe of rockglacier; $\overline{\alpha}$, surface slope; $\overline{h}_p$, thickness of deforming permafrost body; $\overline{z}_b$, bedrock depth; τ, basal shear stress; $\overline{v}_s$, mean surface velocity.
[a]Revised values in relation to the original table published by Haeberli.

Sources	Kintole and Prafleuri	:	Fisch et al. (1978)
	Macun I	:	Barsch (1969a, 1971b), Barsch and Zick (1991)
	Murtèl I	:	Barsch and Hell (1975), Wagner (1992)
	Gruben	:	Haeberli et al. (1979), Haeberli (1985)

7.3.2 The Rheological Description of Active Rockglaciers

It has been assumed that rockglacier movement follows the creep law for pure ice formulated by Glen (1952; cf. Wahrhaftig and Cox 1959; Anderson and Morgenstern 1973; Barsch et al. (1979; Haeberli 1979). If neither the thickness of permafrost changes nor sliding occurs, the surface velocity of the rockglacier should be directly proportional to $\sin^3$ (average surface slope). According to Haeberli (1985a), the average surface slope over a reference distance of 500 m is best correlated with rockglacier surface velocity, but this is only acceptable for the discussion of whole rockglaciers.

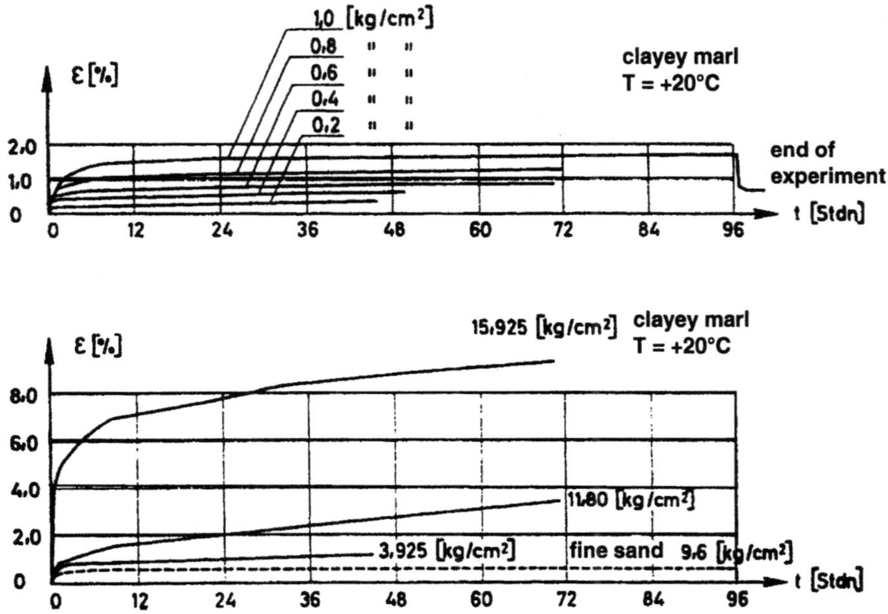

Fig. 7.34. Creep curves (strain against time in hours) for frozen and unfrozen materials. (Jessberger 1965)

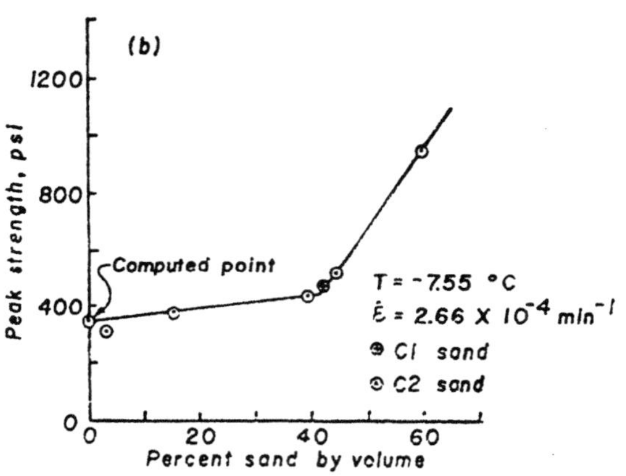

Fig. 7.35. Peak strength and mineral content in ice-sand mixtures. (Goughnor and Andersland 1968) 1 PSI equals 57. 8525 g/cm^2/or ca567 N m^{-2}

On the other hand, rockglaciers are, even with a high ice content, an ice–rock mixture. The deformation of ice-supersaturated permafrost is discussed in general terms above. In a rheological context, strain, temperature, and time are especially important. Based on the data from the drilling through the rockglacier Murtèl I, Kroh (1991) discussed the deformation of rockglacier permafrost at temperatures around –1 to –3 °C

The discussion is based on the general relationship between stress and strain rates: either

$$\dot{\varepsilon} = a \cdot \sigma^{b}, \quad \text{or} \tag{7.11}$$

$$\dot{\varepsilon} = c \cdot e^{\sigma} \tag{7.12}$$

with a,b, and c as constants. Strain rate is comprises an elastic and a visco-plastic part.

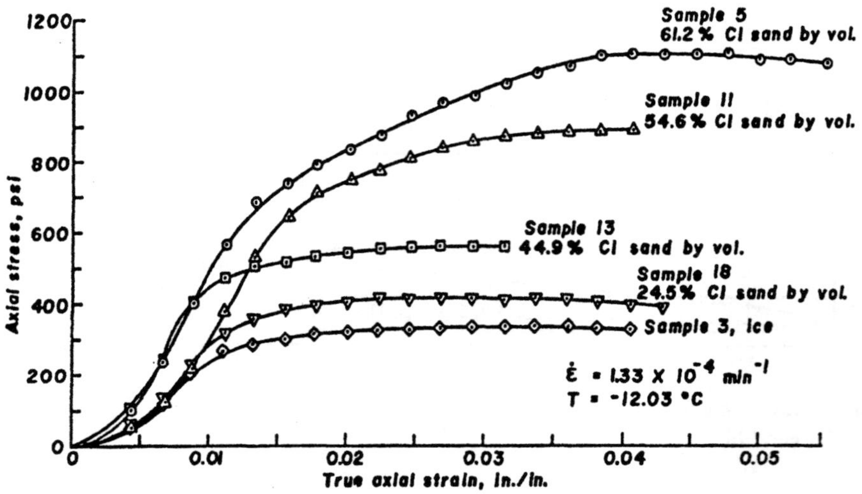

Fig. 7.36. Examples of strain-stress curves for frozen material at –12.03 °C. (Goughnor and Andersland 1968)

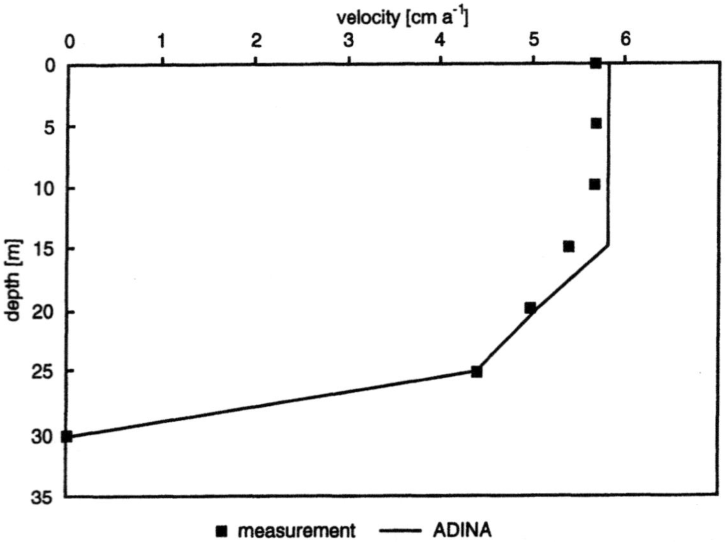

Fig. 7.37. Measured and calculated deformation in the rockglacier Murtèl I. (After Kroh 1991)

Unfortunately most creep laws describe only the primary and secondary creep. The equations of Goughnor and Andersland (1968) or Ting (1983) are exceptions and have the form

$$\dot{\varepsilon} = K_1 \cdot e^{(n_1 \cdot \varepsilon)} + K_2 \cdot e^{(n_2 \cdot \int \sigma \cdot d\varepsilon)} \text{ and}$$

$$\dot{\varepsilon} = A.\sigma^B \cdot t^c, \quad \text{respectively,} \tag{7.13}$$

with A,B, and C determined by experiments.

Following a proposition by Meisner (1985; cf. Kroh 1991), the creep law has at time t_{eff} the form

$$\varepsilon_{eff} = A \cdot \sigma_{eff}^{B} \cdot t_{eff}^{c}[h^{-1}] \tag{7.14}$$

with length in meters (m), t in hours(h) and stress in kN.

The parameters A,B, and C of Eq.(7.14) are determined as follows:

1. C = 1, because the creep velocity of rockglacier permafrost is assumed to be constant (linear) over time (example Macun I).
2. B = -1.96 + 0.6 log σ. This is taken from a relationship found by Herzog and Hofer (1981).
3. The parameter A has been found by model runs using a program used for creep (Adina) for stresses and strains measured in the borehole Murtèl 2/1987. The parameters are presented in Table 7.18.

A comparison of measured and calculated velocities is displayed in Fig. 7.37. The comparison demonstrates that both values are close together. Equation (7.13) describes the creep of the rockglacier Murtèl I fairly well. The lack of data precludes a discussion of other rockglaciers and, thus, a more general and more complete treatment of the creep of rockglacier permafrost.

Table 7.18. The parameters A, B, and C for different layers in the rockglacier Murtèl I. (Kroh 1991)

Layer (m)	Material	Density (m^{-3})	A	B	C
15–20	Frozen sand	1.2	1.885×10^{-7}	–0.55	1.0
20–25	Frozen sand	1.2	1.455×10^{-7}	–0.48	1.0
25–30	Frozen sand	1.5	4.393×10^{-6}	–0.43	1.0

The layer 25–30 m comprises more or less the high viscosity layer in the rockglacier.

$$A = \left[\left(\frac{m^2}{kN}\right)^B * h^c\right]$$

7.4 Rockglacier Movement and Climate

Active rockglaciers display average annual temperatures between –1 and –3 °C. Because ground temperatures in high mountain systems are normally higher than mean air temperatures, MAAT is also below zero, often below –2 °C.This demonstrates that rockglaciers can occur down to the lower limit of the cryogenic belt in high mountain systems. This is the area where Holocene climate fluctuations have probably the greatest effects.

The summer temperature (June–September) for Bever (Upper Engadin, Graubünden) is presented in Fig. 7.38. Following the cold summers in around 1910, a marked increase can be observed from 1915 to 1955. This trend has been strengthened since 1980. Qualitatively, a marked influence on the temperature of rockglacier permafrost cannot be ruled out. Even if mountain permafrost temperature is also influenced by the height of the winter snow cover (c.f. Sect. 6.2.2.5), it seems that the observed warming is responsible for the inactivity which has been observed since 1950/1970 at many rockglaciers in Graubünden. Allowing for a lag of about 20 years, this seems to be a very direct influence. It seems that during the Holocene, periods of activity and inactivity changed many times. Information obtained from tree-ring analysis confirms this conclusion. Figure 7.39 shows that in the Alps "cold" and "warm" periods of more than 50 years followed one another. At least at the lower edge of the rockglacier belt, the warmer periods should lead to increased melting, the colder perhaps to increased activity.

Unfortunately, at the moment the lack of data prevents a better correlation between climate and rockglacier movement.

7.5 Discussion of Rockglacier Movement

Rockglacier movement has inspired many geomorphologists to present rheological models. The first comprehensive, physics–based study of rockglacier movement was done by Wahrhaftig and Cox (1959) who analyzed the relationships between the physical properties of the constituent materials of an active rockglacier, its morphology, and recorded movement. Their results confirm that rockglacier movement is associated with the creep of interstitial ice, which seems to be in accord with viscous or visco-plastic flow laws.

Thus the often asked question, whether the whole body of an active rockglacier moves as a unit (i.e., if the bouldery layer is passively riding on the deforming frozen talus) or whether the single ridges display individual movements was answered by Wahrhaftig and Cox. If, for instance, as Vere and Matthews (1985, p.402) state: "most of the rockglacier surface was sufficient stable to allow undisturbed lichen growth by the mid-nineteenth century", this displays that they still believe in surface instabilities due to horizontal move-

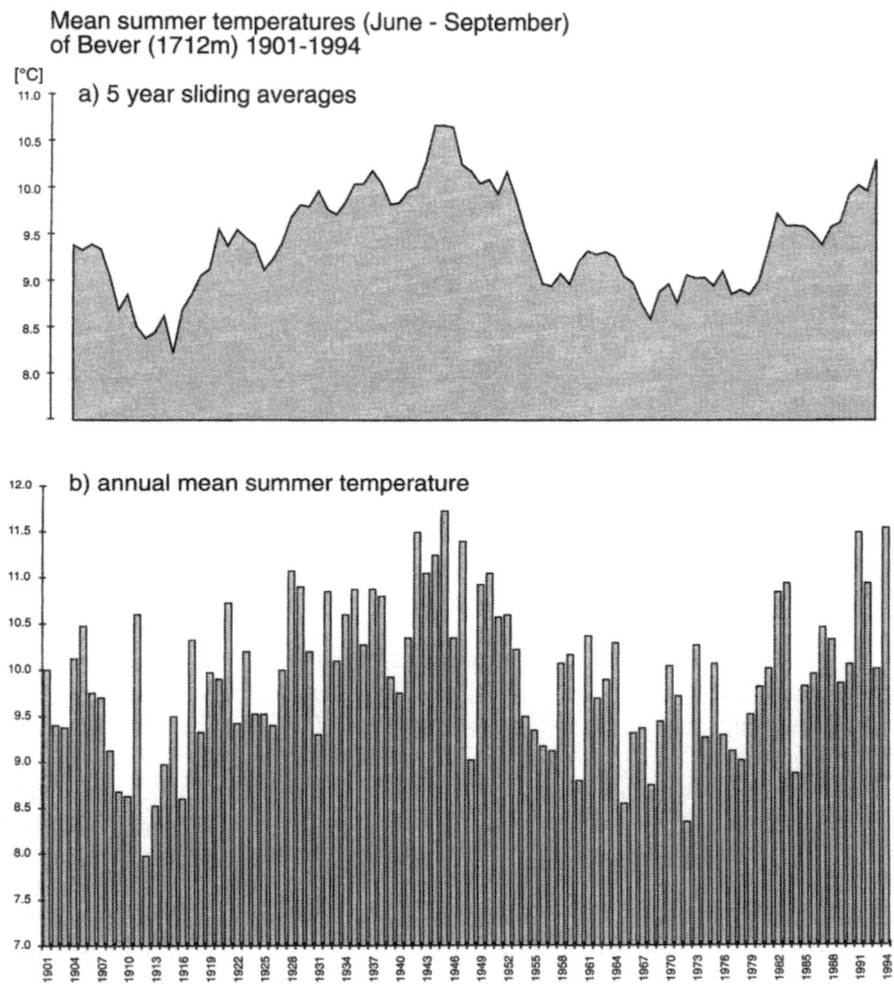

Fig. 7.38. Mean summer temperatures in Bever and Buffalora (Graubünden) as averages of the mean monthly temperatures of June to September. The generalized curve presents 5-year sliding averages (Barsch and Hell 1975). The station Bever was closed in 1983. Thus, the data series up to 1994 was based on the measurements of the station Samedan

ment. At the moment, the rockglacier mantle has to be considered as stable (cf. the lichenometic studies of Haeberli et al. 1979), although vertical displacements have been recorded. It is not yet clearly understood if and how the deformation of the core affects the surface relief of the mantle.

In a more general approach, Johnson (1987), lists eight theoretical possibilities for the movement of rockglaciers:

1. Granular flow
2. Sliding (on a water film like a temperate glacier)

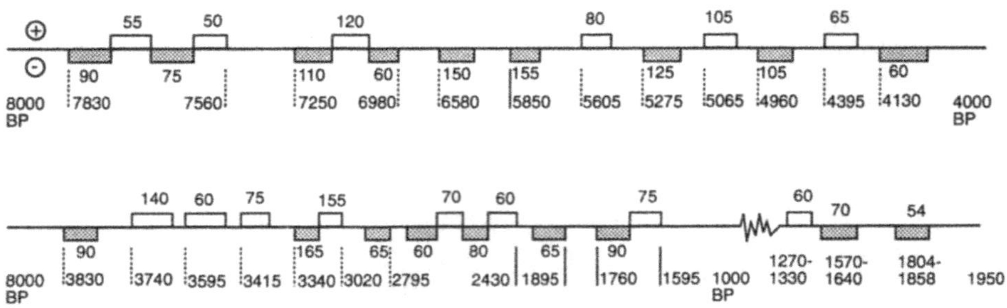

Fig. 7.39. "Warm" and "cold" periods of more than 50 years since 8000 B.P. in the central Alps, which seem to be long enough to influence rockglacier behaviour. The information is based on tree-ring analysis of subalpine forests presented by Bircher (1982) and Renner (1982)

3. Air-cushion mechanism
4–6. Slab mechanisms
7. Hydrostatic flow(mudflow mechanism)
8. Creep

According to the information obtained, of these only creep of the rockglacier permafrost can be accepted as an explanation for rockglacier movement. The other theoretical possibilities are of interest only for other forms of mass movements.

The interesting hypothesis for "rockglacier (?)" movement by Giardino (1979, 1980, 1983; cf. also Shroder and Giardino 1987) also belongs to these general mass-movement theories: the bouldery features of Mount Mestas (Colorado) are the example which illustrates the theory (1980). According to Wahrhaftig (1987), they are not rockglaciers (cf. Chap.8). Giardino believes that water is trapped between the frozen surface and the permafrost layer underneath, and that this causes the movement. This is a process of sliding (cf. Chap. 8) and not in accordance with the information collected about rockglaciers.

It may be, therefore, summarized: rockglacier movement is the creep of ice-supersaturated mountain permafrost. According to the present state of knowledge sliding can be neglected. The creep is often constant over decades, but not independent of climatic changes. Warmer periods apparently lead to inactivity.

Even if the general movement of rockglaciers seems to be understood, quite a number of questions are not yet answered. In some cases pulses in rockglacier movement have been recognized which are not explained. Also, the existence of seasonal changes in the surface velocity of rockglaciers is difficult to understand.There are long-term (and often sudden) changes in rockglacier velocity which oppose the often assumed constant long-term deformation. They are probably caused by climatic changes; however, the exact mechanism of the influences is still not sufficiently discussed. In a number of cases the relationship between horizontal and vertical displacement is surprising, because sometimes an apparent increase in thickness goes together with periods of high velocity. Also, seasonal changes in vertical displacement have not been explained yet. It may be that many of these effects are connected with annual or seasonal balances (regarding water and ice content) near the permafrost table, but no data are available from there either.

Therefore, rockglacier movement is generally understood but many questions regarding the details are still unanswered.

8 Rockglacier Genesis and the Relation to Similar-Looking Landforms

In the discussion of rockglacier genesis, three terms,

Chaos – Continuum – Order

are of greatest importance. In the literature, a large number of contradicting hypotheses have been proposed: this is chaos. A number of proposals have been made to clarify this situation. The logical first hypothesis would include rockglaciers in a group of broadly similar-looking landforms (ranging from glacial and landslide deposits to real rockglaciers under other names) as a continuum (Johnson 1974, 1983; Corte 1987a; Giardino and Vitek 1988). This approach is not really helpful. It is correct (Giardino and Vitek) that relief forms a continuum which can be considered at different scales and with different geometric or geomorphic intentions, but it seems unwise to use the continuum aspect to mask clear differences. The only way to create order as a basis for a better understanding is a process-oriented definition, "because form is a response to a process operating over time" as Giardino and Vitek (1988) state. Based on the definition in Chapter 1, the genesis and the phenotypes of active (Sect. 8.1.1), inactive (Sect. 8.1.2) and relict (Sect. 8.1.3) rockglaciers will be discussed first. In comparison to this, all physiognomic similar forms will be discussed in this chapter (Sect. 8.2)

As stated in Chapter 1 (cf. also Chap. 7), rockglaciers are the visible expression of the steady-state creep of ice-supersaturated mountain permafrost. Active rockglaciers are a creep phenomenon, and they are therefore often very similar to features created by other types of flow. In rockglacier formation, the creep process is influenced by a number of environmental parameters related to relief, geomorphic history, climate, rock type, weathering intensity etc. Despite the great variety of possibilities, a number of standard situations can be described for active rockglaciers.

8.1 Rockglacier Genesis

Based on the material presented earlier in this volume on the internal structure and composition of rockglaciers, their movement, and their geometry, their formation will be discussed here. This includes not only the different rockglacier types (cf. Chaps. 1, 3, 4), but also the different activity states (active, inactive, relict).

8.1.1 The Formation of Active Rockglaciers

Three conditions have to be met to allow the development of an active rockglacier:

1. The prospective area for rockglacier development must be part of the discontinuous mountain permafrost belt (temperature control).
2. Unconsolidated, ice-supersaturated debris must be available in sufficient thickness (quantities) in talus slopes, as moraines, or as other deposits (material control).
3. Relief must allow the development of a rockglacier; that is a gradient must exist which facilitates the instability of the deposit and allows, thus, the creep of the permafrost body towards the flatter ground below (relief control).

If these conditions apply, then creep caused by the plastic deformation of the ice in the permafrost body begins. The first ridges appear and the rockglacier develops out of the unstable frozen deposit. Depending on topography and the inputs of rock material (talus and debris) and of snow and ice (as refrozen meltwater), different types of rockglacier with respect to shape (cf. Chaps. 3, 4), surface relief, velocity, etc. develop. In the following some standard situations are discussed (cf. Figs. 1.4, 1.5).

8.1.1.1 Talus Rockglaciers

These are developed beneath talus slopes in discontinuous mountain permafrost belts of all major mountain systems. Normally they display a lobate form, but as a function of talus input and topography, tongue-shaped forms are also possible (cf. Figs. 1.2, 1.4, 1.5).

Their formation can be summarized as follows (Fig. 8.1). The production of talus in the rock walls above the talus slope increases the mass of the ice-supersaturated debris. Thus, the pressure (shear stress) on the lower and on the basal parts of the system increases. These basal parts of the ice-supersaturated talus slope become unstable if the shear stress, which is a function of ice content, thickness of the talus deposit, the gradient in front of the talus slope, exceeds a critical threshold. By plastic deformation part of the base of the talus slope creeps out: the first ridge of a rockglacier forms. Depending on talus production and the incorporation of snow (avalanches) and refrozen meltwater, more ridges will develop. The shape and velocity of the new talus rockglacier will be strongly controlled by topography. The velocity is depen-

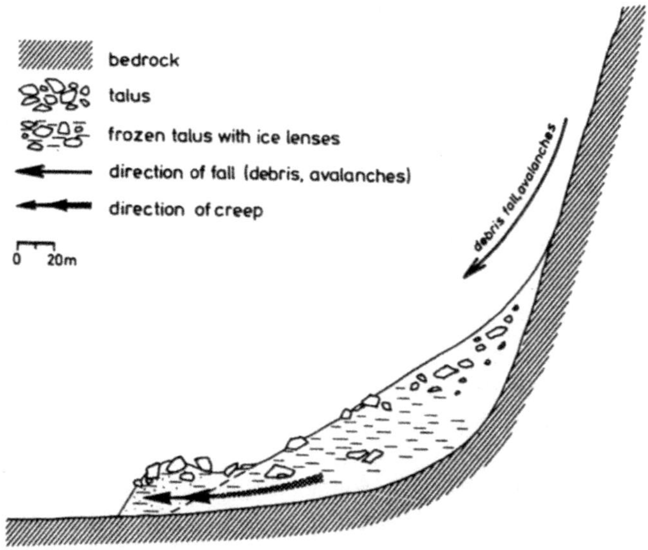

Fig. 8.1. Model of the development of talus rockglaciers (below mountain talus slopes) in discontinuous mountain permafrost environments

dent upon the gradient of the slope over which the rockglacier moves and can be too high in relation to the annual input from the talus slope. As a consequence, a depression between the talus slope and the talus rockglacier develops. In extreme cases and if the already creeping body is thick enough, the rockglacier may even lose the connection to its source.

8.1.1.2 Debris Rockglaciers

As defined in Chapter 1, debris rockglaciers transport material of glacial origin, i.e., till which has been piled up by a glacier (e.g., as an end moraine). If the three controls (temperature, material, and relief) are overcome, the moraine can develop into an ice-supersaturated permafrost body (incorporation of refrozen meltwater), which starts to creep (Fig. 8.2). In these cases, material can only be added to the rockglacier if the glacier pushes it against the inner part or top of the moraine. In the Alps, this happened during the advances of 1820 or 1856, which lasted about 50 years (1804–1858). Similar glacier advances occurred about ten times during the Holocene (Patzelt 1977b), but according to tree-ring chronologies (Bircher 1982; Renner 1982), about 17 cold periods of more than 50 years have already been during the last 8000 years (cf. Fig. 7.39).

A good example of a debris rockglacier is the rockglacier Macun I (cf. Figs. 6.8, 7.19) where during glacial advances, material was pushed into the rooting zone (head) of the rockglacier and incorporated into it. Movement studies (cf. Sects. 7.1.3, 7.2) show that the debris, which has probably been

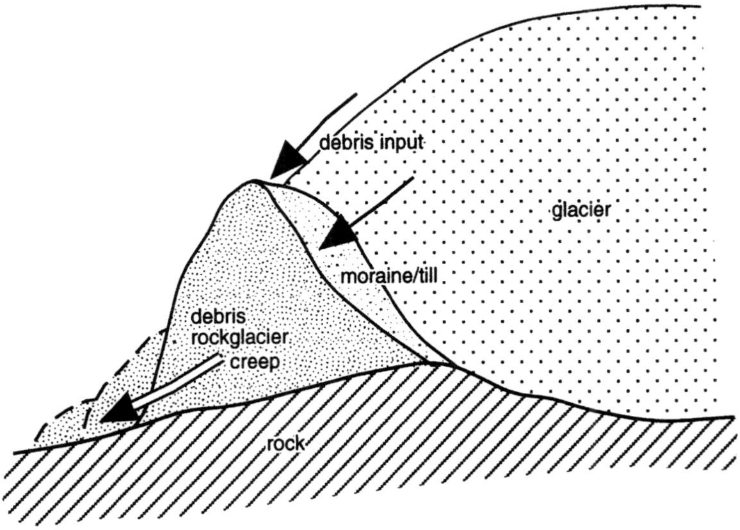

Fig. 8.2. Model of the development of a debris rockglacier in mountain permafrost environments

added during the last century, is still only partly incorporated into the rockglacier movement.

In the case of small debris rockglaciers, i.e., where the debris input even during glacier advances is small, it is easy to observe which parts are already moving and which are still stable. The rockglacier Macun V (Barsch 1969a) displays in its moving parts the normal two-layered front and side slopes and the normal bouldery mantle on the surface. In contrast, the stable parts show surface materials with the character of moraine with a mixture of various grain sizes (sand, silt, till, boulders).

It is often assumed that in the rooting zone of a debris rockglacier masses of glacier ice are incorporated into the rockglacier. A study by Barsch and King (1989) revealed, for the rockglaciers in the Morenas Coloradas (Argentian Andes near Mendoza), that this is not the case. Under a climate with an intense radiation, glaciers tend to become debris covered if they are stationary. Thus, it becomes difficult to define the end of the debris-covered glacier and the start of the rockglacier. Geoelectric soundings reveal that the thermokarst area on the glacier can be clearly differentiated from the rockglacier and that no large bodies of "glacier-derived ice" exist in the rooting zone of the rockglacier.

This may be different in cases where the glacier advanced onto a rockglacier. One example (Fig. 8.3) is the rockglacier La Tscheppa, Upper Engadin, Swiss Alps (cf. Barsch 1977a). Here, the glacier moved just onto the rockglacier. During mapping, glacier ice was not found although the surface relief had been subdued by the advancing glacier. In other cases, the glacier deposited some ice onto the upper part of the rockglacier. An early example is given by Fisch et al. (1978) from the Chessi rockglacier in a tributary of the

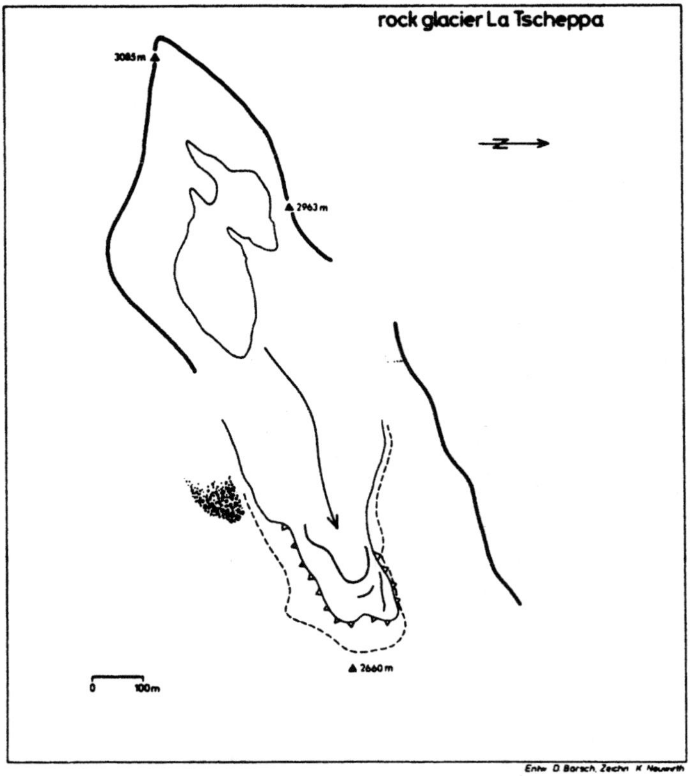

Fig. 8.3. Model rockglacier La Tscheppa (Upper Engadin), (Barsch 1977a)

Mattertal (Wallis). Here (cf. Sect. 6.3.3, Fig. 6.18) during geoelectric soundings, a thin layer of glacier ice was found on the upper part of the rockglacier. A similar example is reported by Haeberli (1985a; cf. also King et al. 1987) from the Gruben rockglacier. In this case, thermokarst lakes are found on the rockglacier in the uppermost part where glacier-derived ice was deposited during the advances of the last century. The glacier ice does not seem to be in equilibrium with the present climate despite the fact that it is supported by the rockglacier permafrost underneath. These examples demonstrate that glacier ice may be added on top of existing rockglaciers, but it seems that this is of no importance for the development of the rockglacier itself. Thus, active debris rockglaciers form a separate system which is more areal than genetically connected with the glacial system.

8.1.1.3 Special Rockglaciers

Talus and debris rockglaciers are differentiated by the genesis of the material incorporated. The two main sources are talus and glacial debris. It is also possible to derive material from other sources. In the literature, examples are given for rockglaciers developing out of mining dumps (Gorbunov 1983) or out of debris-flow deposits (Ono and Watanabe 1986; cf. Chap. 4). These rockglaciers obey the same laws as the other two types; to avoid problems with the given classification, a special class has been introduced.

8.1.1.4 Problematic Cases

It is to be expected that a classification with only two or three types of rockglaciers would present problems if different sources of the rock material in them are involved. Therefore, some problematic and interesting cases will be discussed on the following pages.

The debris rockglacier Weissmies (Messerli and Zurbuchen 1968; Fig. 8.4; cf. also Figs. 1.1, 7.17 and Sect. 7.1.2) develops out of the right lateral moraine of the Weissmies glacier just below the place where the moraine starts after the Tälligrat (arette). The rock wall below the arette is heavily weathered. The material arrives on the glacier as rock fall; it is then transported by the glacier to the first place where it can be deposited outside the glacier as lateral moraine. The deposition and the incorporation of meltwater was apparently high enough to overcome the stability threshold. Thus, rockglacier development started. In addition to the debris deposited by the glacier, some of the talus slopes from the Tälligrat grade into the upper right-hand part of the rockglacier. Following the melting of the glacier since the beginning of this century, the rockglacier is now cut off from the glacial-debris input. According to the photogrammetric measurements by Messerli and Zurbuchen (1968), it was still moving during the period 1958–1964.

In a number of cases, the development of one rockglacier out of morainic deposits of two different glaciers has been observed. In Fig. 8.5 an example from the Argentinian Andes west of Mendoza is shown.

8.1.2 Inactive Rockglaciers

According to the definition (cf. Chap. 1), an inactive rockglacier (Fig. 8.6) does not move (at the moment). Its velocity at the toe is below a threshold of 1 cm a^{-1}, which is taken from the usual accuracy of yearly measurements. This means that creep is practically zero, which may be due to heavy melting of interstitial ice. In the lower parts of the rockglacier in particular, the unfrozen mantle may be more than 10 m thick (cf. Sect. 6.3.6, Barsch 1973), thus, forming an obstacle to further movement. In addition, the frozen core may become so thin that the possible shear stress is too weak to provoke any creep. This is called climatic inactivity. In Fig. 8.7A the situation of a climatic inactive

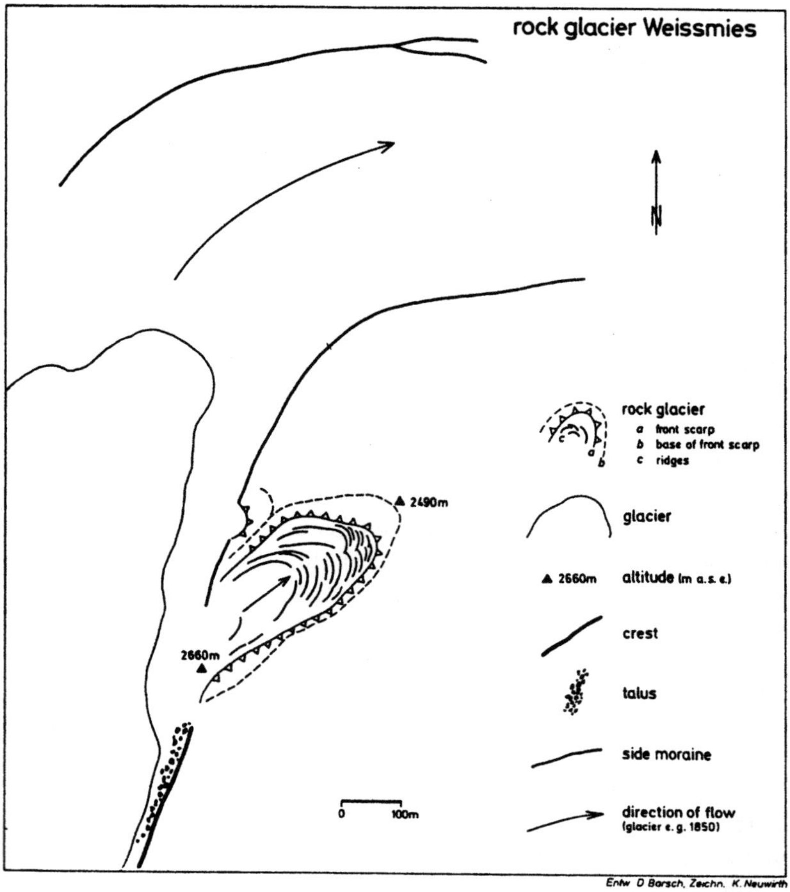

Fig. 8.4. Model of Weissmies rockglacier (southern Wallis). (Barsch 1977a)

rockglacier is presented. It is controlled at the lower parts by heat input large enough to enlarge the summer unfrozen mantle, which in turn prevents any further movement. The only exception may be in the uppermost parts of the rockglacier, where shearing may occur and new lobes may be created. These new lobes form a new rockglacier generation and so the multiunit rockglacier develops.

Dynamic inactivity also occurs (Chap. 1). In this case, creep is or approaches zero due to decreasing rockglacier thickness in relation to gradient. The summer time unfrozen mantle (as active layer) at the toe is of the same thickness as on active rockglaciers in the neighborhood. The dynamically inactive rockglaciers are mainly found in areas with continuous permafrost, where the temperature requirement is met but the input of snow, refrozen meltwater, talus debris etc. may be too low to allow further movement. In Fig. 8.7B, a model to describe this situation has been constructed.

Fig. 8.5. Rockglacier developing two extremely different lobes (Argentinian Andes, W of Mendoza)

Examples of dynamically inactive rockglaciers are found on Spitzbergen (cf. Fig 1.8) or in the mountains of Alaska. In the Brooks Range in the continuous permafrost zone, both active and inactive rockglaciers have been mapped (Ellis and Calkin 1979, 1982; Ellis 1982; Calkin et al. 1987). This data set has been discussed in a master's thesis (Wehier 1994) together with that from the Alaska Range (Wahrhaftig and Cox 1959) which is situated generally in the discontinuous permafrost zone, however, it is possible that the rockglacier range (1280–1890 m a.s.l) already belongs to the continuous mountain permafrost belt.

From the map of Ellis (1982), an area with about 115 rockglaciers (16 inactive ones) was selected and incorporated into a GIS. Inactivity does not

Fig. 8.6. The inactive rockglacier Albana (Upper Engadin) in August 1972. Indicators for inactivity are the boulder apron, the debris slides, and the vegetation patches on the front slope, which is about 70 m high

seem to be related to radiation or climate. Ellis (1982) has already mentioned "that no correlation was found between the incoming energy and rock glacier's state of activity". In addition in eastern, southern and southwestern exposures, active rockglaciers are mapped below inactive ones. This demonstrates that both are found at the same altitude. The possible explanation for inactivity is given by the following results:

1. The active rockglaciers are on average smaller than the inactive ones.
2. Four of the 16 inactive rockglaciers are creeping onto very flat surfaces ($<5°$). In the Alaska Range, 9 out of 53 are found in similar positions.

Therefore, besides a change in input, it may be that extending flow on flat surfaces led to dynamic inactivity. The suggestion that higher energy inputs are responsible for the observed inactivity can be excluded.

It is often difficult to document inactivity in the field if no recent geodetic measurements are available; however, there are indicators (cf. Fig. 8.6) such as:

- the existence of a well-developed boulder apron around the foot of the front slope,
- the occurrence of vegetation patches on the fine material at the front (and side) slopes,
- traces of small debris slides at the front slope,
- a progressively more gentle front slope.

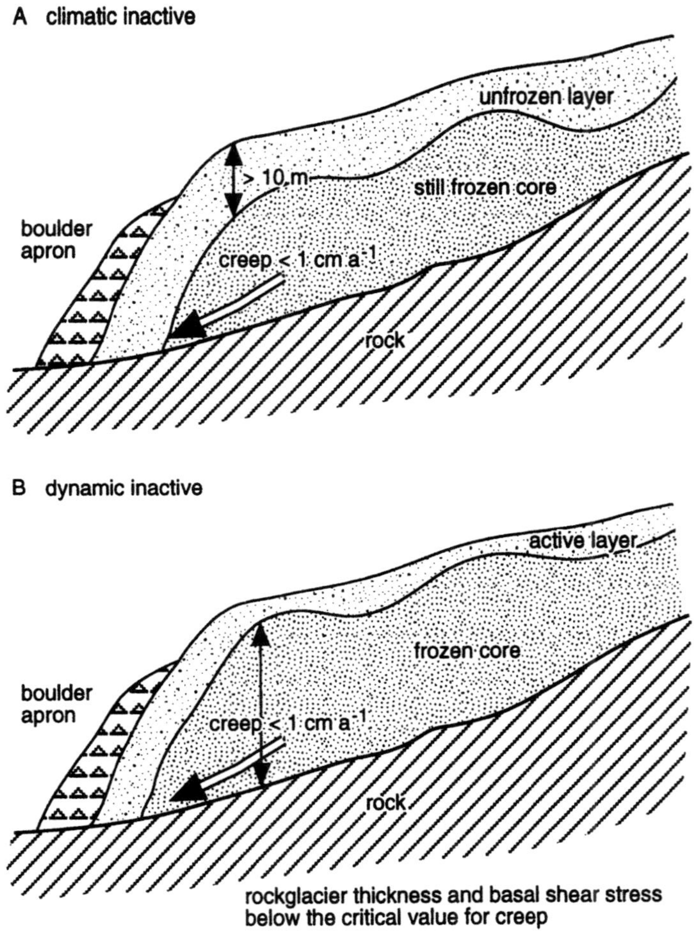

Fig. 8.7A,B. Model of inactive rockglaciers. **A** Model of a climatic inactive rockglacier; **B** model of a dynamic inactive rockglacier

The latter is already an indication of longer lasting inactivity. The development of inactivity over a period of years has been observed in a number of cases. For instance, Barsch (1973, 1983a) visited rockglaciers described by Domaradzki (1951): 25 years after the original description they showed signs of inactivity that had been reported before. As a famous example indicating this situation, the rockglacier Val Sassa (Swiss National Park, Graubünden/Grison) can be named. Between 1920 and 1970, it changed from well-documented activity (up to 1.6 m a^{-1}) to inactivity (cf. Sect. 7.1.3). The same development is indicated by the photogrammetric analysis of old (1932) aerial photographs. A marked decrease in activity can be demonstrated between the periods 1932/1955 and 1955/1971 (cf. Sect. 7.1.3). The rockglacier Albana (cf. Fig. 8.6 and Table 7.8) was still active during the first period and inactive around 1972. This probably indicates that rockglaciers at the lower limit of discontinuous mountain permafrost that have been active during colder periods of the Holocene, e.g., during the "Little Ice Age", are inactive today. This does not mean that these rockglaciers were formed during the Little Ice Age, but that they were active during this or similar cold periods of the Holocene. Warmer periods lead to an enhanced melting of the interstitial ice and to inactivity. Climate has, as already stated, a marked effect on the movement of rockglaciers. The altitude of inactive rockglaciers outside the polar permafrost zones is, therefore, considered to be a valuable indicator for the amplitude of Holocene climatic variations.

8.1.3 Relict (Fossil) Rockglaciers

Relict (fossil) rockglaciers (cf. Figs. 1.9, 1.10) contain no ice at present. Therefore, they do not display any horizontal or vertical movement. In the field they are rocognized by the "normal" set of furrows and ridges, which are subdued as a result of the melting of interstitial ice. However, the former more prominent relief is still visible. At the front and side slopes, the two layered system (finer-grained core–coarser mantle) can still be observed. In many cases, only the parts made out of fine-grained material bear a vegetation cover; the coarse-boulder ridges are often only partly or not at all covered by trees or shrubs.

Unfortunately, a variety of similar landforms resulting from bergsturzes or even from landslides exists (cf. Sect. 8.2). Thus, an intense check has to be made before a deposit can be identified as a relict rockglacier. Besides deciding whether the feature is a talus or a debris rockglacier, the following characteristics should be checked for:

1. A (naturally subdued) surface relief of ridges and furrows is still visible.
2. At the front and side slopes (perhaps also at inner ridges) not only the two-layered stratigraphy of the former rockglacier is detectable, but the outer ridges are normally the highest ones. In addition, all grain sizes occur at the surface.
3. The deposit is at the front ridge is still more than 5–10 m thick.
4. The area of the deposit (in case of the talus rockglacier) comprises one-fourth to one-third of the area of the rock shed. In any case, there must be a source area for the material incorporated in the relict rockglacier.

In addition, rockglaciers are scarcely found alone. If, therefore, doubts exist in interpreting a debris deposits as a relict rockglacier, the surrounding area should be checked for other equivalent forms.

Relict rockglaciers have been described in most mountains where active or inactive rockglaciers exist today. From a paleoclimatic point of view, it is interesting to discuss relict rockglaciers in areas which do not display a recent subnival belt today. An example might be upland Britain and Ireland. Here, relict rockglaciers (Ballantyne and Kirkbride 1986) have been described in

- Jura (Dawson 1977);
- Western Ross (Sissons 1975; cf. also the reply to Whalley's suggestions: Whalley 1976a, Sissons 1976a,b);
- Cairngorms (Sissons 1979; Chattopadhyay 1984);
- Lake District (Sissons 1980);
- Erigal Mountains, Donegal (Wilson 1990).

The relict rockglaciers in the southwestern United States deserve similar interest. Pioneer work has been done by Blagbrough (1967). In addition, the so-called protalus ramparts (Blagbrough and Breed 1967) should be considered as relict rockglaciers (cf. Sect. 8.2). Barsch and Updike (1971a,b; cf. Fig. 8.7) provided additional information on Kendrick Peak in Arizona (relict rockglaciers are also found in the Escudillo Mts. in eastern Arizona and in the Navajo Mts.; Fig. 8.9). It can be demonstrated that during the Wisconsin period the average altitude of the lower limit of rockglaciers was at 2585 m on Kendrick Peak, whereas the snowline (as ELA = equilibrium line altitude) was at 3300 m on nearby San Francisco Peaks. This shows that the difference between snowline and lower limit of discontinuous mountain permafrost is about 800 m, demonstrating an arid climate during part of the last glaciation. In arid to semiarid mountains, this difference is often as much as 1200 m or more, as demonstrated by Barsch and Happoldt (1984) for active rockglaciers in the Andes of Mendoza.

The features presented in Fig. 8.8 and 8.9 demonstrate the former existence of discontinuous permafrost down to the altitudes around their toes (cf. Sect. 10.1). As it is assumed that they were formed during the main Wisconsin, they display information on the climate during this time (cf. Sect. 10.42), i.e., in the higher mountains of Arizona a well-developed periglacial frost climate existed.

As already mentioned in Chapter 5, the mountains of Africa are practically devoid of rockglaciers. Clear features have been described from the Rif (Mensching 1955), from the Atlas (Raynal 1960) and from Mt. Kenya (Mahaney 1980). From the Hoggar and from the Mouskorbé (Tibesti) problematic features have been described by Rognon (1967) and Messerli (1972) which may be relict rockglaciers. According to Messerli, the one from the Mouskorbé starts at an altitude of about 3000 m with a southwesterly aspect, at 2600 m it bends towards the south, and it ends at 2300–2100 m a.s.l. It has a length of 2 km, a thickness of more than 40 m, and a volume of more than 15×10^6 m^3. Messerli favors an explanation as a rockglacier (excluding bergsturz or debris flow), and he dates it tentatively as a deposit from the penultimate glaciation. In the Hoggar, similar features exist. Despite all the microclimatic differences between the two mountain systems, the main feature ends at nearly the same altitude (2200 m). It probably belongs also to the penultimate glaciation.

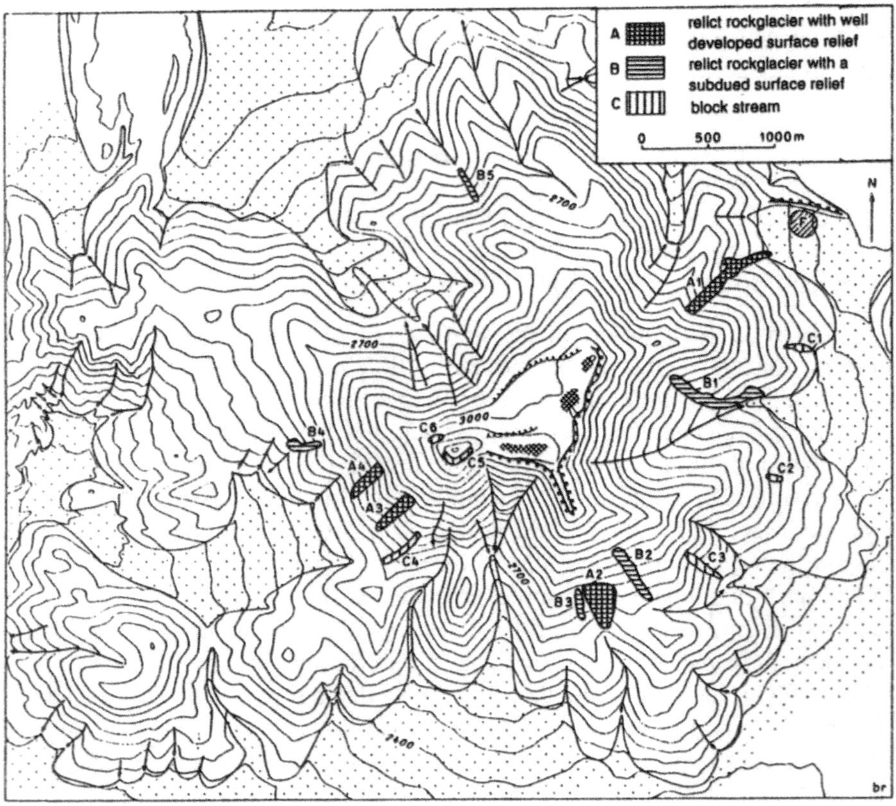

Fig. 8.8. Relict rockglaciers on Kendrick Peak (N Arizona). (Barsch and Updike 1971b)

8.2 Published Hypotheses of Rockglacier Formation

The discussion about internal structure in the previous chapter demonstrates that the available evidence favors the conclusion that the inner core of an active rockglacier is frozen, ice-supersaturated talus or debris. The supersaturation in ice often amounts to more than 60 vol%, sometimes to 90 vol%, which occurs as ice lenses or as bodies of interstitial ice. The frequently discussed internal "huge glacier ice bodies" are apparently nonexistent or a wrong interpretation. This fact rules out every hypothesis which is in favor of a pure glacial formation of active rockglaciers (see Sect. 8.2.2). Thus, terms implying a glacial model should be avoided. To underline this point, a listing and discussion of most rockglacier development theories presented so far will be given in the following sections.

The discussion of these theories which have been put forward to explain rockglacier formation, is not only of historical interest but of importance for

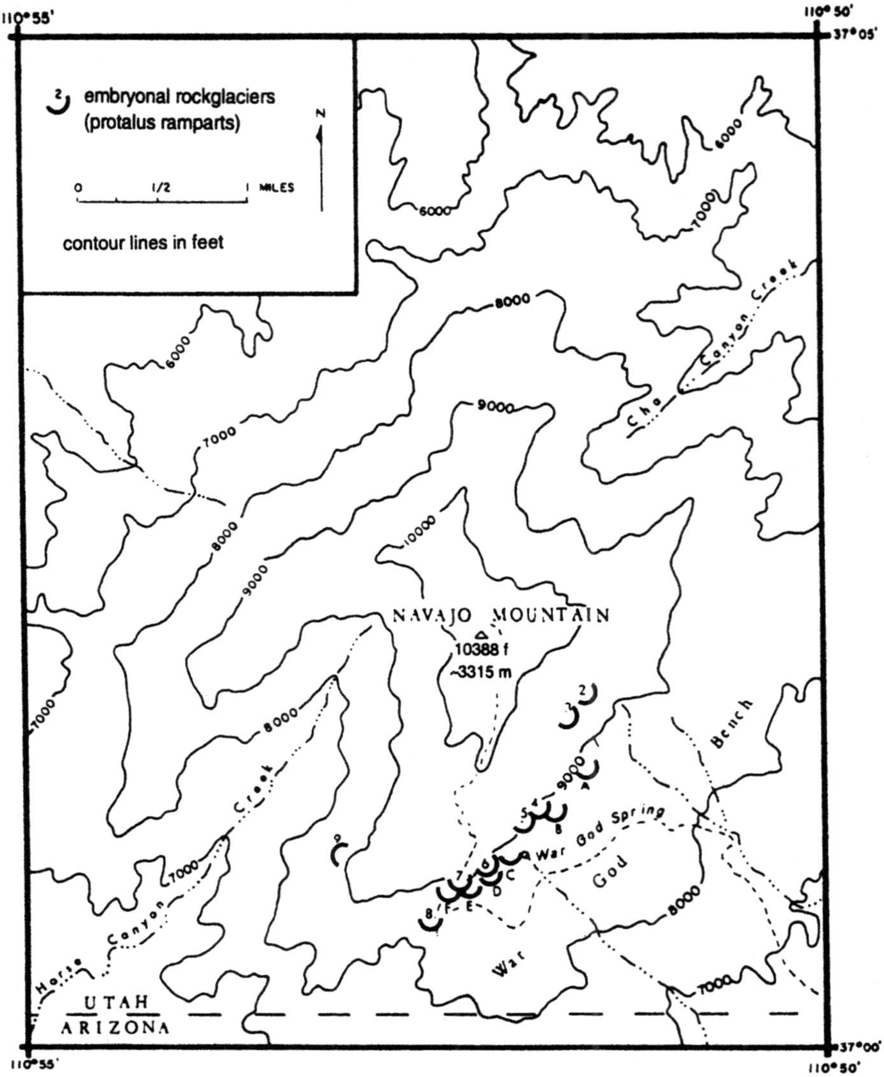

Fig. 8.9. Relict rockglacier from Navajo Mountain (N Arizona/Utah). (Modified from Blagbrough and Breed 1967)

our present-day research. In a number of cases, parts of former hypotheses are used again, mostly without knowing that they have already been published. In other cases, models are used that Haeberli (1989a) refers to as concepts "which can obviously not be 'understood' but must be believed". The problem is that in many cases no geophysical measurements or only a very limited amount of field evidence are presented. In addition, good field observations have been interpreted according to a priori hypothesis without discussing other possibilities. In other cases, more measurements are unrealistically suggested because one borehole through rockglacier permafrost gives "only" information at one point! (Whalley and Azizi 1994). Thus, the list of proposed concepts is long and only the more prominent ones are listed below:

- Bergsturz (Cross and Howe 1905; Howe 1909)
- Debris-covered glacier (Liboutry 1965; Whalley 1974)
- Ice-cored rockglacier (Potter 1972)
- Ablation complexes (Johnson 1974)
- Special moraines (Gerhold 1969)
- Push moraines (Haeberli 1979)
- Ice-cored moraines (Östrem 1964a)
- Ablation-complex rockglacier (Evin 1987b)
- Rockslide debris stream (Shroder 1973b)

These hypotheses can be divided into three main groups

1. Special mass-movement theories
2. Glacial theories
3. Cryogenic/periglacial theories

They will be discussed in this order.

8.2.1 Mass-Movement Hypotheses

In the following section, distinction is drawn between bergsturz and landslide. A bergsturz is a large slope failure which moves an enormous quantity of rock material (of all grain sizes) during a very short time (several seconds up to a few minutes) over a couple of kilometers. At least partly, a bergsturz moves in free fall. As a general rule, a minimum volume of 10^6 m^3 is required. Events of smaller magnitude (volume $< 10^6$ m^3) are refered to as *felssturz*. Sometimes a *sturzstrom* develops out of a bergsturz (Heim 1882, 1932) with lateral ridges, which may be confused with a relict rockglacier deposit (see Sect. 8.1.1).

In contrast, a landslide is a (slow to rapid) mass movement of limited extent in fine-grained (clayey or silty) material moving on well-defined shear planes. Frequently, landslides form stream-like features or lobes, which move slowly downslope, e.g., as earth streams, earth flows or talus streams (*Schuttstrom*). These stream-like features are not necessarily triggered by landslides; they are noncatastrophic (i.e., slowly moving) landforms caused by slow mass wasting.

8.2.1.1 The Bergsturz Hypothesis in General

One of the first of all the explanations of rockglacier formation is the one forwarded by Cross and Howe (1905; cf. also HOWE 1909). They presented extremely instructive photographs of the famous rockglaciers in the San Juan Mtns, Colorado. The transport of these rock masses was intensely discussed. Both authors assumed the hypothesis that the debris was deposited by a bergsturz onto small glaciers which transported it to its present position. This seemed to be a plausible explanation, but the authors soon found it difficult to apply this hypothesis to all rockglaciers in the San Juan Mountains (Howe 1909, p. 50). Howe especially names the rockglaciers of the Imogene basin and from Slide Rock Ridge as examples. Under the influence of the reports from the Elm bergsturz (8 September 1881, Switzerland) and the Frank bergsturz (28 April 1903, Alberta), Howe proposed his bergsturz hypothesis as a general explanation for the genesis of rockglaciers. This is quite understandable considering the illustration by Heim (1882, 1932), which presents an astonishing physiognomic similarity between the *sturzstrom* of the Elm bergturz and a rockglacier (Fig.8.10).

The Example of the Elm Bergsturz

The bergsturz Elm was a man-triggered event. Since 1868, a quarry (*Schiefersteinbruch*) had existed on the north-facing slope above the village of Elm. The schists dip at about 30° towards the south into the mountain. The quarry was about 300 m above the valley bottom and 450 m below the local mountain summit (1747 m). It was organized as an open pit about 180 m wide along the slope and undercutting the mountain by ca. 65 m.

At 1736 hours on 8 September 1881, the main bulk of the bergsturz came down very fast, hitting the valley and rushing partly up the opposite valley side to an altitude about 115 m above the valley bottom. The greater part of the main bulk of debris was diverted by the Düniberg (today Tüniberg on the new topographic map "Elm") and directed down the valley in a stream-like form. Witnesses described the flow as a fast moving black train emerging suddenly from a cloud of dust. With an apparently abrupt stop the whole mass came to a halt. The whole deposition was very sharply limited and it showed marked flow structures (transverse ridges) as well as side ridges.

Albert Heim (1932) mentions four phases of the bergsturz:

1. The *breakoff*. The mountain undercut by the quarry became unstable, slid towards the valley, and fell down onto the terrace of the quarry.
2. The *jump*. The whole mass jumped out into the valley and hits the ground near the foot of the slope.
3. The *shooting uphill*. Part of the mass flowed uphill on the opposite side of the valley (Düniberg).
4. The *sturzstrom (bergsturz debris stream)*. The greater part of the material was diverted by the Düniberg and flowed about 1.5 km down the valley until it came to a sudden stop.

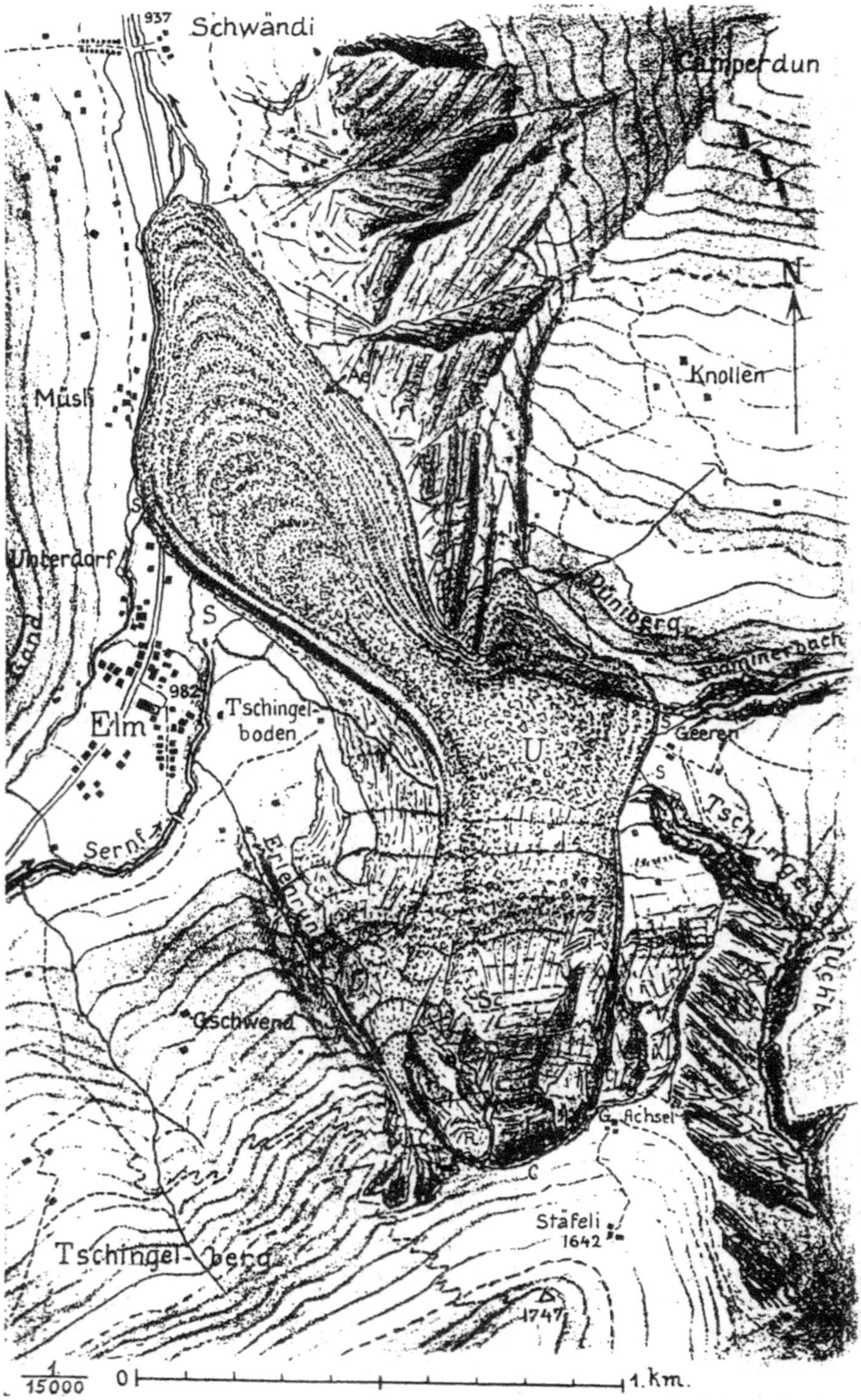

Fig. 8.10. Sturzstrom Elm (Switzerland). (Heim 1932)

Conclusion

The *sturzstrom* is shown in Fig. 8.10. This figure is the map published by Albert Heim (1932). A similar map was published by Heim (1882) in his first reports of this bergsturz. This map was probably known to Howe, who used the physiognomic similarity between the *sturzstrom* Elm and some of his rockglaciers as a possible explanation of their origin. Unfortunately, the original landform of the *sturzstrom* at Elm is no longer visible. The area has been reclaimed and only relicts of the former deposit are found today. Figure 8.10 is, therefore, a historical document.

The longitudinal and transverse ridges on the *sturzstrom* are due to a very rapid flow that had to overcome an enormous friction at its base and probably at its sides. This friction slowed down the movements at the base and at the sides and caused the formation of the ridges. These are the prominent parts of the surface relief of the *sturzstrom*. The movement of the bergsturz debris stream is a catastrophic one of high speed and high energy, whereas the real rockglacier movement is a noncatastrophic movement displaying slow creep as a consequence of the deformation of ice. Genetically, a rockglacier and a *sturzstrom* are two totally different systems, even if their appearance is occasionally very similar. The first is mainly climatically controlled with respect to talus production, talus or debris accumulation, and permafrost; the second is caused by slope instability and has no relation to special mountain belts. If both are mixed up, only physiognomic criteria are decisive. The system controlling the landform is not considered. This, however, is not scientifically sound. The theory put forward by Howe (1909) that rockglaciers are formed by catastrophic landslides or bergsturzes is understandable regarding the time when it was proposed and regarding the apparent physiognomic similarities (cf. Fig. 8.11). Today, however, it is no longer acceptable.

It is often not easy to differentiate in the field on first view between a *sturzstrom* and a fossil rockglacier. Thus, in Verkilsdalen (Rondane Mountains, Southern Norway), Dawson et al. (1986) believe a feature to be a *sturzstrom* which has been described by Barsch and Treter (1976) as a probable inactive talus rockglacier. The area itself has, at 1200 m a.s.l., a calculated MAAT of −3.0 °C (Dawson et al. 1986). Barsch and Treter (1976) report, according to sledge-hammer seismic measurements, frozen talus with an active layer of 5 m at 1360 m a.s.l. ca. 120 m upstream of the front slope. Thus, rockglacier movement cannot be ruled out.

This example demonstrates that a final decision on whether a deposit is a relict rockglacier or a bergsturz is not always easy in the field. It needs intense fieldwork because both features belong to quite different systems which have nothing in common but a physiognomic similarity under special circumstances. Both should be, therefore, strictly differentiated even if cases exist where bergsturz material has been incorporated into rockglaciers. The apparent similarity between relict rockglaciers and *sturzstrom* deposits is not a genetic relationship. Unfortunately, the berg-

Fig. 8.11. Bergsturz lobe S Lac de Tseuzier (near Anzere, Wallis, Swiss Alps). On air photographs the deposit seems to resemble a well-developed fossil rockglacier. Terrain inspection reveals the true nature of the deposit. Note the irregular surface and the unsorted grain sizes at the surface

sturz hypothesis to explain rockglaciers has recently been taken up again (Johnson 1984b).

8.2.1.2 Landslide Influences

In contrast to a bergsturz, a landslide is a rotational or translational slip with little or no free fall. On unstable slopes it involves large quantities of unconsolidated material which often move (after the first failure) downslope in large streams. A certain physiognomic similarity of these stream-like features with rockglaciers led to the suggestion that rockglaciers are partly or totally moving by internal sliding without the influence of mountain permafrost. A number of different cases have to be considered; these will be discussed on the following pages.

Talus Streams

In mountain terrains, slowly moving stream-like features exist which are built up by local talus. They are situated in the forest belt; the existence of mountain permafrost in their inner part is in the most cases unlikely. According to their physiognomic appearance, they are named talus streams or (in German)

Fig. 8.12. The talus stream (Schuttstrom) Campo/Rovana (Valle Maggia/Tessin, Switzerland). (Heim 1932)

"*Schuttstrom*". Genetically, they are topographically constrained landslide masses or landslide lobes as the following examples may demonstrate.

A talus stream or *schuttstrom* (Fig. 8.12) endangered the village Campo (Tessin, Switzerland), which is situated on a terrace which was originally 20 to 30 m above the Rovana river. Heim visited the area in 1897 (Heim 1898; cf. heim 1932, pp 51ff): the talus steam was moving onto the terrace and the village. It is about 900 m wide at its lowermost part and about 350–400 m at its upper part. The average slope is 22°, its length is 2300 m, its volume is ca. 120×10^6 m^3. The altitude of the crest above the rock stream is around 2100 m, the village is at 1250 m a.s.l., and the valley bottom (Rovana river) at 1030 m. The forest line is at 1850 m; the whole talus stream is situated in the forest belt; no influences of periglacial or permafrost activity exist.

The surface relief of the talus stream is characterized by well-developed ridges and furrows perpendicular to the direction of flow. Trees (*Larix* sp.) are tilted downvalley several times. Small ponds exist between the ridges. In the uppermost part, a scarp has developed which forms the upper end of the slide: in this part, the rock material is still more or less intact. The material is fractured down to boulder size 300 m below. The whole talus stream is a slow-moving landform that moves with changing velocity downvalley. The flow increased during the last century because of man-induced erosion of the Rovana river; however, the development of the talus stream itself is much older. Regarding the geologic situation (Heim 1898), the initial sliding started in the uppermost part on cleavage planes in the schists (*Schieferungaflächen*). Its activity is a function of the water available in the *schuttstrom.*

Following the work of Chaix, Mercanton (1923) published movement measurements of a huge talus stream, the "pierrier de Saleinaz" (Val Ferret, Wallis). This talus stream has developed between 2000 and 1400 m; this is far below the recent lower limit of discontinuous alpine permafrost in this area. It is about 1 km wide and apparently consists of several parallel units. In parts, trees grow on this feature and between its different units in the upper part. According to measurements carried out between 22 November 1920 and 4 November 1923, the lower section, at an altitude of 1550 m (ca. 30 m wide), displayed an average velocity of 2.4 cm a^{-1} (average 7 cm in 1077 days for 8 points). This is not much, but it demonstrates the movement in these talus streams.

There are more similar examples. The two described in detail may demonstrate sufficiently that these nonperiglacial talus streams are caused by "land"-sliding processes. In this case, even if their surface relief displays similarities to that of rockglaciers, they are still landslides and not rockglaciers. Their movement is due to internal wetting of fine-grained material and not to the creep of mountain permafrost.

In a number of papers, Shroder (1972, 1973a,b, 1975, 1977, 1978, 1987) and since 1980 Giardino discuss features around Mt. Mestas (Colorado) and from Table Cliffs Plateau (Utah), which they call rockglaciers (Fig. 8.13). The general problem has already been pointed out by Wahrhaftig (1987) when he

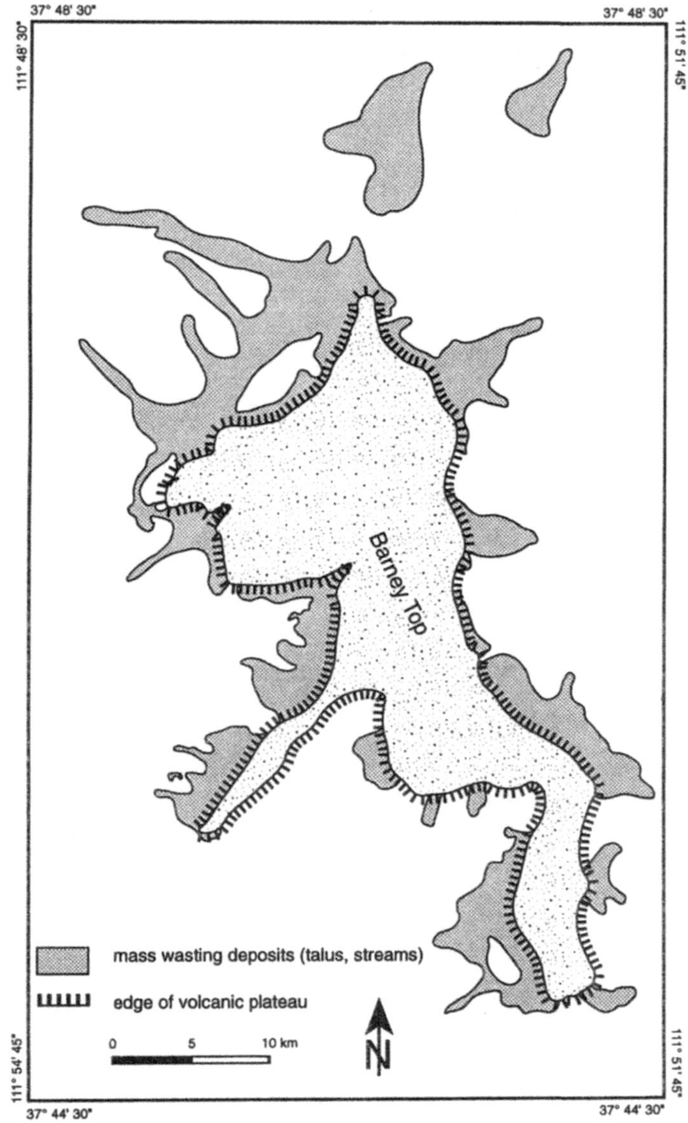

Fig. 8.13. Mass wasting deposits (talus streams) around Barney Top (Table Cliff Plateau, Utah) (Shroder 1987). The deposits are regularly distributed around the volcanic plateau with an altitude of ca 3000 m

writes: "The summit of Mt. Mestas, 3520 m, is 76 m below the two lowest rockglaciers (at 3596 m) observed on the 4377 m Blanca Massif, and these two rock glaciers are inactive". In addition, Wahrhaftig points out that the features around Mt. Mestas do not display a preferred orientation, whereas the active rockglaciers in the Blanca Massif are only located below the radiation-protected north-and east-facing slopes. Also, heights of front and side slopes "do not appear sufficient for the necessary shear stress for ice deformation". Thus, Wahrhaftig assumes that these features, although perennial ice was detected at some places, are probably not rockglaciers.

The features at Mt. Mestas and at Table Cliffs Plateau seem to be similar to the talus streams discussed from the Alps. They are huge, but special, landslide lobes. This is indicated by the following facts (Shroder and Giardino 1987):

1. The features are below timberline in the southern Rocky Mountains, and, with a high probability, are not connected with discontinuous mountain permafrost: they are located well below the belt of discontinuous mountain permafrost.
2. Their movement is positively correlated with precipitation.
3. According to tree-ring analysis, 14 peak periods of event responses have been detected since 1750.

Regarding the first point, it has to be noted that Johnson (1967) reported interstitial ice which was observed 10 m below the surface at the toe of one of these features near La Veta pass during road construction in 1963. It is located at an altitude of ca. 2700 m (9200 ft.) at a latitude of 38° N on a west-facing slope. According to the general geoecologic setting, this looks like an azonal occurrence of permafrost (sporadic alpine permafrost according to Barsch 1977c) if the ice does not melt during summer. According to information presented by Pewe (1983), this "permafrost occurrence" is more than 100 m lower than the mountain permafrost in the surrounding areas. In addition, interstitial ice was hitherto only found at a very few places. If these features are to be accepted as active rockglaciers, sufficient ice has to be demonstrated in a number of them. Furthermore, an evaluation of the distribution of mountain permafrost has to be carried out in this area.

On the other hand, the ice lies so deeply buried that the rockglacier could be climatic inactive (cf. Chaps. 1, 6). If movement occurs it can only be due to sliding. This assumption is backed by the observation that movement occurs in relation to wetter years. It has happened about 14 times in the last 230 years. This does not correspond to the steady-state creep found for decades in well-developed active rockglaciers (cf. Sect. 7.1.3); it is also not related to periods of higher or lower temperatures which, as one would expect, would cause changes of the internal ice content.

Perhaps, it should be noted that Patton (1910) gave a very good description of the talus streams of La Veta. His considerations are still worth reading. They back the landslide explanation. In addition, thse tongue-shaped debris slides or talus streams have already been accepted as "tongue-like complex landslides" by earlier workers (Flint and Denny 1958). Therefore, the argument that it "is impossible to distinguish between different types of rockglaciers and other boulder deposits everywhere because of their great similarity in size, thickness and microrelief" (Shroder 1987, p.202) is not acceptable. According to the definition given in Chapter 1 these features cannot be accepted as rockglaciers. These forms are related to sliding following an excess humidity in certain layers. It is a theory which was originally put forward by Boesch (1951) and Domaradzki (1951) as "*Durchtränkungsfliesßen*". For decades this hypothesis has no longer been accepted because it does not explain the restriction of active rockglaciers to the discontinuous mountain permafrost belt and to all collected geophysical results on active rockglaciers.

The same is true for the talus streams (*Schuttströme*) of the eastern Karawanken (van Husen 1976), which can be up to 1.5 km long and probably 30 to 50 m thick. The talus streams have a convex cross-sectional profile and follow the main valleys. The material consists of only up to ca. 65% of material > 2 mm and is derived from the local clay and silt stones. A well-developed surface relief, the bouldery mantle, and the stratigraphic differentiation into two layers, which are all found on rockglaciers, are missing. The author has already pointed out these features are not rockglaciers. According to van Husen, the material of these talus streams collected during the Würm glaciation at the head areas of the main valley. During late Pleistocene and early Würm, permafrost started to melt and these huge masses moved downvalley. Due to consolidation, these talus streams are inactive today, i.e., they are not moving at present.

Earth Streams

In all mountain terrains, tongue-shaped bodies of (normally) fine-grained debris slowly move downslope. They are part of the overall denudation and they too belong to the class of mass-wasting forms. They are, for instance, described in detail by Fischer (1967). They can be defined as stream- or tongue-shaped landslides with low downslope movement.

The reference to sliding in the above definition implies a strong correlation of this mass-movement form to the subsurface material (e.g., marls, clays (often with thinner beds of limestone), or sandstones). It precludes a simple correlation with cryogenic/periglacial conditions (e.g., solifluction/gelifluction). Similar, but normally much smaller, forms exist as solifluction lobes or solifluction streams in the periglacial environment (Furrer 1954; Washburn 1979; Gamper 1985). A periglacial environment, however, is not necessary for the development of earth streams.

Good examples are given by Fischer (1967) from the South Tirolean Dolomites. The largest of his earth streams has a surface area of 9.1 km^2(length 3.9 km, slope ca. 6°). Shear cracks develop in their upper parts into which water penetrates and causes a volume increase of the clay particles; in the lower parts compressive flow occurs which causes small transverse ridges on the surface of the earth stream. Flow velocity is often quite variable. Slower ones move at 1–5 cm a^{-1}. Fischer (1967) reports maximum velocities of 80 cm a^{-1} over a period of 17 years (1916–1933), and of 60 and 20 cm a^{-1} at closely located points on the same earth stream but at different times. This is due to the change between wet and dry years. The thickness of these earth streams is estimated to be 10 to 30 m. Their age is difficult to determine, but it is probable that they developed throughout the Holocene. Fischer (1967) reports that part of the material of a late glacial bergsturz has been transported with an average speed of 40 to 50 cm a^{-1} a distance of over 4000 m by the Premaju earth stream. This transport would only have been possible if the earth stream had been moving throughout the Holocene.

These earth streams are quite different from active rockglaciers in their environmental context, even though the sizes and velocities of the two features are similar. In any case, their movement is not due to interstitial ice. Thus, according to the definition in Chapter 1, they are not rockglaciers.

Special Landslide Influences

Landslide deposits may increase instability in their accumulation areas. This is especially the case if overloading occurs. Vick (1981) presents an interesting example from the Clear Grit and Davis rockglaciers (Iowa Gulch, Mosquito Range, Colorado). Rockglacier formation has been enforced by the deposition of smaller landslides onto a frozen moraine bench; this increased creep by increasing the strain. According to the information given by the author, the study area is part of the discontinuous alpine permafrost belt.

8.2.2 The Glacial Hypothesis

The glacial hypothesis was first suggested by Cross and Howe (1905) but later questioned by Howe (1909) himself; it still exists today. The origin of the hypothesis can be traced to two sources: one is the word "glacier" in the term rockglacier (cf. Chap. 2), and the other one is based on the fact that rockglaciers, debris-covered glaciers, and (ice) glaciers are found in the same neighborhood. This areal, not genetic, connection may be demonstrated for an area with well-developed rockglaciers and debris-covered glaciers such as the Andes of Mendoza (cf. Fig. 8.14). Both systems are clearly separated (Fig. 8.15), even if decaying blocks of dead ice from the last glacier advance still exist near the rooting zone (Fig. 8.16). Geoelectric soundings revealed mostly mountain permafrost, in the upper parts of the rockglacier (cf. Sect. 8.1.1).

The glacial hypothesis has many supporters (Lliboutry 1965, 1986; Klaer 1974; Whalley 1974a). Even Kinzl (1932) wrote in his famous study on the glacier advances of the seventeenth, eighteenth, and nineteenth century in the Swiss Alps that rockglaciers are nothing other than the preserved glaciers of the last century. Under the influence of the Wahrhaftig and Cox (1959) paper, this hypothesis was restricted to the debris rockglacier (S.E. White 1971b; Potter 1972).

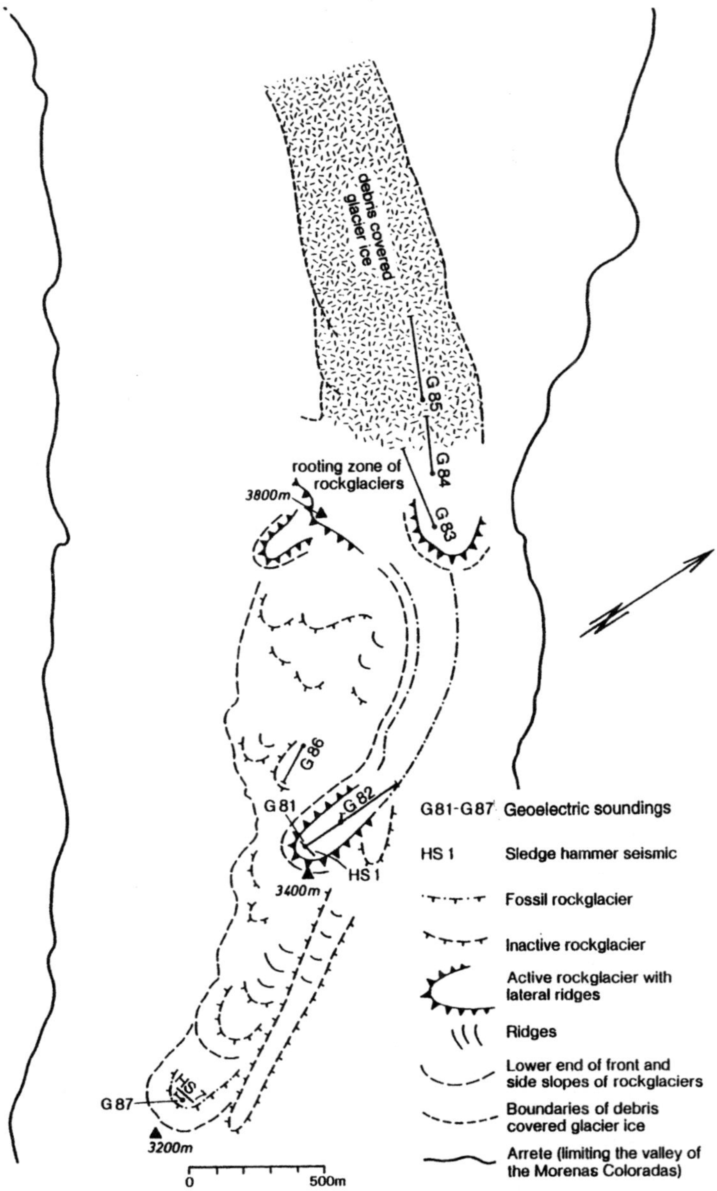

Fig. 8.14. Sketch map of the Morenas Coloradas (Andes W Mendoza, Argentina)

Unfortunately, this hypothesis seems to come up again from time to time (Humlum 1988b; Johnson and Lacasse 1988; Clark et al. 1994a). It is, as Haeberli (1989a) summarizes in a very clear response to Johnson and Lacasse, more a long-standing belief, than "based on any firm evidence from appro-

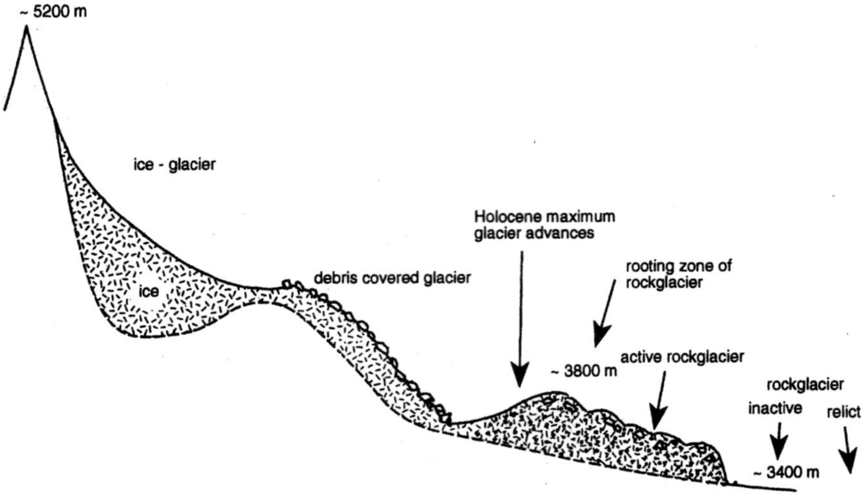

Fig. 8.15. Diagram of the areal connection of debris-covered glaciers and rockglaciers in the Morenas Coloradas (Argentina)

Fig. 8.16. Debris-covered glacier and thermokarst in the Morenas Coloradas (Andes W of Mendoza, Argentina)

priate field measurements or reliable model calculations". Nevertheless, the glacier–rockglacier relationship has to be discussed in some detail.

8.2.2.1 Debris-Covered Glaciers and Thermokarst

The debris-covered glaciers of high semiarid tropical to subtropical mountains are described in many papers. Their existence is always assumed to be due to the high radiation in these latitudes and altitudes. This assumption is correct if the glaciers are stationary or stagnant. Most of the observations available were made between 1930 and 1970 during a period which was unfavorable for glacier advances. Many examples exist from the 1970s and early 1980s which demonstrate this effect: as soon as the glacier advances again due to a positive mass balance, its morainic surface cover disappears in 1–3 years. This was the case on the Horcones glacier (Aconcagua, between Mendoza and Santiago). In January/February 1985 it was still a debris-covered glacier. In February 1986 it was again (as on photographs taken during the 1920/1930 advances) a clear-ice glacier broken into seracs, and advancing in a surge. The same could be observed in the Alps under quite different climatic conditions. The snout of the Steingletscher (Sustenpass, northern Swiss Alps) was completely covered by debris. When the readvance started, the more active left side was devoid of debris in two years. Another two years later, the whole snout was cleared of debris, and the steep profile at the glacier front, which is typical for an advancing glacier, was as perfectly developed as on photographs taken during the advance of 1920/1930.

Therefore, the debris cover on a glacier, can be an indicator of

- A stationary situation,
- A single event (e.g., a bergsturz),
- A high talus production in its environment,
- A climate with high radiation.

The debris cover on a glacier does not display the regular surface relief which is indicative of most rockglaciers. This is especially the case when the so-called thermokarst develops on debris-covered glaciers. It is characterized by melting depressions, by meltwater ponds, or by exposures of the clear glacier ice in places where the debris has been transported by sliding downslope into a depression. Johnson (1992b) describes the downmelting of the glacier ice of the last readvance (probably from the last century) of the Donjek glacier. A typical set of microlandforms and a typical landscape develop, in German called a "*Niedertaulandschaft*", in which stagnant but massive glacier ice is found under a thin veneer of debris/till. The downwasting is normally not so much influenced by conductive heat flow as by meltwater heat transport and by the exposure of the ice due to:

- Sliding of the debris veneer
- Mudflows or till flows originating on steeper slopes
- Collapsing of ice caverns
- Meltwater erosion at the side of shifting channels

Even under 1.5 to 2.5 m of debris the downwasting is about 100 mm a^{-1}. small lakes develop with layered sediments, which are not a response to annual floods and so cannot be used as varves.

All these are normal phenomena in a decaying glacier. It is the situation of a glacier with an extremely negative mass balance. It is unclear how a rockglacier may develop out of this chaotic environment.

The physiognomic differences between a debris-covered glacier and a rockglacier should be obvious to a good observer. This can be demonstrated from the topographic map (1 : 25 000) of the Chomolongma - Mt. Everest (Schneider 1955). The cartographer probably knew nothing about rockglaciers, but he did see and mapped (!) the difference between a debris-covered glacier and a rockglacier. On the map, it is easy to see which of his features are rockglaciers and which are debris-covered glaciers.

A stationary debris-covered glacier is a different system to a rockglacier, despite the plea of Whalley and Azizi (1994) to keep at least the theoretic possibility of a glacier ice core alive. The two systems display different rheologies and belong to quite different realms: the debris-covered glacier is still part of the glacial, the rockglacier part of the periglacial one.

8.2.2.2 Transition from True Glaciers to Rockglaciers?

This view has been discussed in quite a number of papers; it is repeated again and again as an apparently cautious approach to a complex problem, although it is unnecessary (Kozhenikow 1984). In general, there are two possibilities:

1. True debris-covered glaciers are explained as rockglaciers.
2. True debris rockglaciers are explained as debris-covered glaciers.

The first case can be demonstrated, for instance, at glaciers from the tropical Andes (Lliboutry 1965). The author describes true debris-covered glaciers as rockglaciers. As stated before, the debris cover is only of temporary occurrence on a glacier.

The second case is best demonstrated by the so-called ice-cored rock glacier (Potter 1972). The main arguments against this conceptual model have already been discussed for the type locality "Galena Creek" by Barsch (1987b). They can be summarized as follows.

The rockglacier Galena Creek (Fig. 8.17) is a multiunit form. Only the upper two units (II and III) seem to be active today, the lowermost unit is inactive. Sledge-hammer seismic investigation confirms that P-wave velocities (V_2) vary between 2440 and 3350 m s^{-1} below the active layer (bouldery

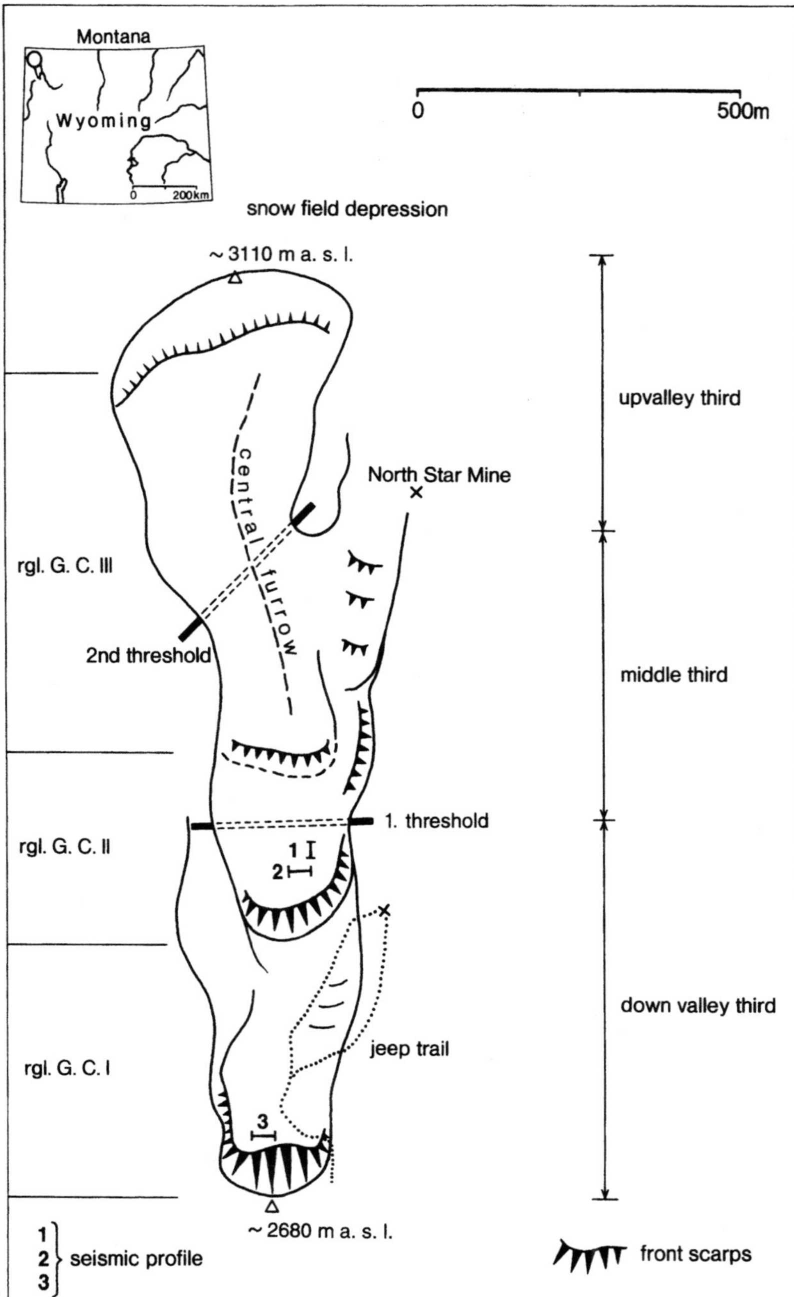

Fig. 8.17. Sketch map of Galena Creek rockglacier (Absaroka Mountains, Wyoming)

mantle), which indicates a very variable frozen material (Sect. 6.3.2). The ice, mentioned by Potter, below the bouldery mantle is "only" the ice-rich (supersaturated) permafrost near the permafrost table on which meltwater flows. All data on the Galena Creek rockglacier, such as water discharge, low sediment concentration in runoff, surface relief, movement velocities, debris mantle thickness etc. are typical for rockglaciers. None of the available data allows the conclusion that this phenomenon is a debris-covered glacier. Because there is only a very limited depression between the rockglacier and the talus below the cirque wall, it is even not very probable that the feature is a debris rockglacier at all. Therefore, Galena Creek rockglacier has to be accepted as a (normal) multiunit rockglacier, which is probably more a talus than a debris rockglacier.

As a consequence, the model of the so-called ice-cored rock glacier has to be abolished. It does not exist at the type locality; up until now, no information has been produced to show, that it exists elsewhere. Whalley and Azizi (1994) try to question this view. They think that the "glacial hypothesis" should be kept in mind as a theoretical possibility. Their argument that the movement of the Galena Creek rockglacier can be sufficiently explained by a "thin" ice layer is not helpful, because – as they themselves state – our rheological information is insufficient. In addition, nothing supports the assumption that the required "thin" layer of ice has to be glacier ice! The Galena Creek rockglacier is situated in a pure Holocene periglacial environment and even the cold periods of the Little Ice Age or others before could not provide a 1.6-km-long glacier at this location.

Quite recently, the glacial hypothesis has been taken up again (Humlum 1988a,b, p. 170 ff). The arguments are neither new, nor are they more convincing today. Because all debris rockglaciers and (with some exceptions) all talus rockglaciers at Disco (Greenland) start at altitudes above the so-called lower envelope curve for the Little Ice Age, Humlum concludes (p.170): "This coincidence suggests a genetic affinity between snow accumulations (glaciers and snow banks) and the majority of rock glaciers within the present study area"; and he repeats: "The rock glaciers classified as glacier-derived should be perceived as true glacial features". Furthermore, he states that the talus rockglaciers "should probably be perceived as somewhat extraordinary glacial features" (p.170). Only the talus rockglaciers which start below the lower envelope of the Little Ice Age equilibrium line are proposed as "true periglacial features" (p.170). This takes up again the old debris-covered glacier hypothesis even for talus rockglaciers. These statements, as well as those from Whalley (1983) and Martin and Whalley (1987a), are not on the present level of discussion (cf. Chap. 6; Wahrhaftig and Cox 1959; Barsch 1969a,b, 1983b, 1992; Haeberli 1985a, 1989a,b; Barsch and King 1989; Vonder Mühll 1993). They neglect all geophysical information collected hitherto. The only information on the interior structures of these rockglaciers is that ice which "appeared" to be glacier ice (p.167) has been found in nine cases. This assumption is based on the ice-crystal size of 1 to 3 cm. No other information on the size of the section, on the ice chemistry, or on the method of crystal-size analysis is given. Because of this lack of information on the material properties and the fact that the thermal conditions which govern the existence of mountain permafrost are not discussed, the proposed hypothesis is not convincing.

A sad highlight in this respect is the paper by Clark et al. (1994a), which has already been strongly questioned by Jakob (1994) who demonstrates that the authors ignore much of the published literature and that they do not discuss the full problem but usually quote only papers which support their view. They do not seem to have been aware that they apparently extrapolate very limited observations from the rooting zone of a rockglacier onto the whole feature without good reasons. Thus, their reconstruction of the ELA has to be disregarded, too. Unfortunately, their response to Jakob (1994) is insufficient and unconvincing.

An interesting example of the wrong interpretation of the observation of ice in the upper part of a rockglacier has been published by Evin (1993). In the valleys of Montgioie and Schiantala (NW Italy) active rockglaciers exist above 2500/2600 m. In the Schiantala valley, ice and thermokarst features have been visible for a long time. Thus, the original topographic map of the valley (published in 1930) presented here a glacier, the Ghiacce di Schiantala. In the neighboring valley of Montgioie neither a glacier nor the rockglacier were mapped. Evin demonstrates clearly that apart from the existing ice between and below the bouldery mantle no glacier existed in the Schiantala valley. The ice–apparently an indicator of the colder conditions of the Little Ice Age – was incorporated into the already existing rockglacier; it cannot be used as a relict of a glacier which did not exist here.

In addition, Evin (1991b) reports a detailed study of a complex of debris-covered glaciers and debris rockglaciers. She develops an interesting model (1991) that clearly demonstrates that debris rockglaciers and debris-covered glaciers are areal neighbors, but are genetically differentiated.

Summarizing, every scientific concept or model carries the risk of being wrong. It has to be tested again and again. However, the tests have to be up to the standards accepted by the scientific community. Unfortunately, the proposed models that explain debris rockglaciers partly or entirely as debris-covered glaciers are not up to this standard because they present neither borehole records, geophysical data, nor any other sound support (e.g., crystallographic analyis of the ice, etc.) for their assumptions (hypothesis). It is clear that debris-covered glaciers exist, but their areal association with debris rockglaciers does not establish an intimate relationship, a simple transition fron one to the other, or a so-called continuum. As stated more than 30 years ago by Clyde Wahrhaftig and Alan Cox (1959): "It is difficult to escape the conclusion that the motion of rockglaciers is the motion of frozen rubble".

8.2.2.3 The Moraine Hypothesis

The Normal Ice-Cored Moraine

It is an often described and a well-known fact that glacial ice is incorporated in end and side moraines. In high cirques solid ice bodies can frequently be observed in moraines reached during the 1850 readvance. The ice has probably been incorporated into the side moraine by shearing processes. Therefore these features, as well as those described by Souchez (1971) from Ellesmere Island or by Johnson (1971), can be named ice-cored moraines. Used in this descriptive sense, an ice-cored moraine is nothing more than an end or lateral moraine containing some glacial ice, an "ice core", or bodies of glacier ice. This fact has been often used to explain an assumed transition from true glaciers to rockglaciers. As already mentioned, this assumed glacier ice core has not yet been found by geophysical means in the rooting zones of debris rockglaciers. In addition the creep of ice-cored moraines (in this descriptive sense) has not been reported so far. The incorporated ice is apparently not thick enough to create the necessary shear stresses.

The Östrem Ice-Cored Moraine Concept

On the other hand, massive ground-ice bodies exist in the ice-cored moraines of the Östrem type. It is unfortunate to use the same term for two different features, therefore, additional information is necessary. The ice-cored moraines (Östrem type) are thought to be end moraine complexes consisting of several ridges which contain ice derived from snowbank accumulations (Östrem 1964a; Östrem and Arnold 1970). In these complexes, the innermost ridge is the highest and oldest according to radiocarbon dates of the ice (Östrem 1965). In Scandinavia, Östrem collected an enormous and valuable set of data on these "ice-cored moraines". However, his conceptual model did not include three very important keywords: permafrost, rockglacier, and push moraine (cf. Sect. 8.2.4).

Haeberli (1979) put forward the idea that some of the ice-cored moraines of the Östrem type are push moraines in mountain permafrost environments, like those from polar permafrost areas (Kaelin 1971; King and Hell 1993; Tatenhove and Dikau 1990). If the stress field generated by the glacier is strong enough and if there are frozen sediments (thick enough) at the glacier's snout, these frozen sediments can be pushed up into one or more frozen moraine ridges as push moraines. If these ridges are high enough and if their ice content is sufficient (ice-supersaturation), the basal layers are plastically deformed and permafrost creep starts. Although, this is possible on flat surfaces, creep will be facilitated on a slope. Thus, extremely ice-rich ridges can be subject to permafrost creep. If they are only high enough to develop the necessary basal shear stress, their basal parts will start to creep out (Fig. 8.18). This has been demonstrated in Ellesmere Island where new ridges are forming due to this creep in front of the push moraine of the Carl Troll glacier. This happens despite the fact that the glacier is no longer in direct contact with its push

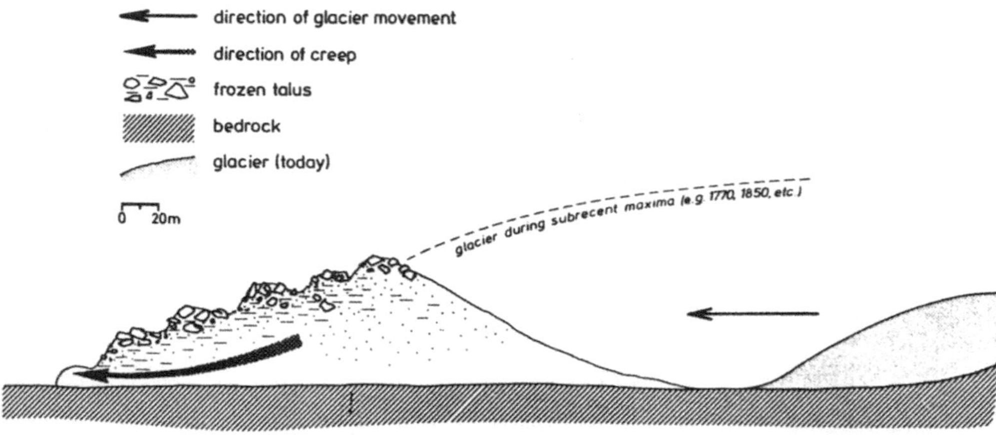

Fig. 8.18. The creep of "ice-cored moraines" (Ostrem type). (Barsch 1977a)

moraine today (Barsch 1981a). This push moraine is built up by frozen, ice-rich fine-layered marine sediments. If frozen moraine ridges like these start to creep on their own for a long time (caused by ice-supersaturation and basal shear stress), they are gradually transformed into debris rockglaciers. Thus, ice-cored moraines (Östrem type) are true rockglaciers (cf. Chap. 8.24; Barsch 1971b).

Special Moraine

Gerhold (1963, 1964, 1965,1967a,b, 1969, 1970, 1971) discussed rockglaciers as special moraines in quite a number of publications. One of his examples is the Glockturm rockglacier (Ötztal, Tirol, Austria). It is a huge rockglacier (1 km long, 500 m wide) with a 180-m-high front slope developed in a narrow valley. At its right-hand side, a number of parallel and longitudinal ridges are developed. Gerhold (1970) believes that these ridges are lateral moraines that were formed during Holocene glacier advances. They incorporated again and again new material into the rockglacier which – according to Gerhold – moves independently by the deformation of its ice content.

Looking at the aerial photographs (Gerhold 1970, Fig. 1, p.109) it becomes obvious that this explanation raises a number of questions. The main problem is caused by the fact that large talus slopes fed talus into the rockglacier at the upper end. Also, the glacier above the rockglacier was restricted to the cirque during the Holocene. Thus, it could only transport material to the end of the cirque where it was deposited on the rooting zone of the rockglacier. Taking this into account, the Glockturm rockglacier is a transition between talus and glacial debris-fed rockglaciers. Because it seems that the talus input is more important, it must be regarded as a talus rockglacier.

Fig. 8.19. Rockglacier Suvretta da S. Murezzan (NE slope of Piz Albana, Upper Engadin, Swiss Alps). The rockglacier is about 1 km long and ends at ca. 2300 m. The lateral ridges in the upper part are extremely well developed. Exposure of the lower part is to the north . Note the recent debris flows on the right lateral slope. (Photograph: W. Haeberli)

Lateral Ridges on Rockglaciers as Moraines?

In the upper part of many active rockglaciers, prominent lateral ridges are developed (Fig. 8.19). Often, they can be traced down to the toe. To the casual observer, they seem to be morainic ridges and, thus, an indication for the existence of a former glacier. It has never been suggested that these lateral ridges are an expression of the cohesive flow (as in debris flows or in a sturzstrom) and that they may have been deposited by the advancing rockglacier. According to the geophysical information in Chapter 6, the ice content is smaller at the sides of an active rockglacier than in the inner part of the core. Therefore, on relict rockglaciers, these lateral outer ridges are more prominent than the inner ones. In addition, they display lower velocities in every cross-sectional profile (cf. Sect. 7.1.6). In the case of the advancing active rockglacier, the ice-rich middle part creeps out and the lateral parts are deposited. Thus, these ridges are connected with the flow of the rockglacier; they are normally not indicators for glacial events in the upper part of the rockglacier.

The Tungsten rockglacier can be cited as an example (Jackson and McDonald 1980). The authors show that the lateral ridges are practically not moving ($V<1$ m over 17 a). Thus, they believe these ridges to be lateral moraines of a former glacier.

8.2.3 The Periglacial (Blockstream) Hypothesis

As already mentioned, the discussion about the inner structure of rockglaciers favours nonglacial (periglacial) model for their genesis. Also in the periglacial realm, landforms (such as blockstreams) exist that might appear (on a first view) as rockglacier-like but which are really quite different from rockglaciers. As a rule, they are smaller or slower by an order of magnitude.

Blockstreams (*Blockströme*) have already been discussed as a terminological problem (Chap. 2). It is today accepted that they were formed during the last glacial period (Würm/Weichsel/Wisconsin) in mountains such as the Smoky Mountains (Appalachians) or the Black Forest (Germany). They are stream-like features consisting of boulders: their thickness does not exceed a few meters. They are (normally) not higher than the adjacent slopes. Today, the assumption is favoured, that they are relicts of bouldery solifluction sheets from which fine-grained material has been washed out (Rother 1965).

Caine and Jennings (1968) use the term blockstream in this sense. Blockstreams in the Toolong Range (New South Wales, Australia) as well as those from Tasmania (Caine 1968a, b) are too thin to cause a shear stress which is sufficient to create a rockglacier-like movement (Caine and Jennings 1968, p.100). These features are an order of magnitude smaller than rockglaciers, and, therefore, should not be considered as those, even if small front slopes and other indicators of an active movement are developed in some cases (Caine 1968a; Braun 1969). Blockstreams belong to another periglacial transport system.

8.3 True Rockglaciers under Wrong Labels

If so many landforms can not be accepted as "true" rockglaciers, the inverse may be equally true: rockglaciers have been identified as other forms. This has already been partly discussed for the "ice-cored-moraines" of Östrem, but this is also true for the so called "protalus ramparts", the "*gehängeschutt-terrassen*" (talus terraces, Barsch 1969c) in the Jura mountains (Switzerland) or some arcuate blocky ridges on talus slopes. Problems related with these wrong interpretations or misconcepts will be discussed in the following sections.

8.3.1 The Östrem Ice-Cored Moraine Concept

This problem has already been discussed in Section 8.2.2 and so needs to be treated only briefly here.

Gunnar Östrem (1959, 1963, 1964a,b, 1965; cf. also Östrem and Arnold 1970) has collected an enormous amount of information on a special type of ice-cored moraine (cf. Sect. 8.2.2). These features are found on level ground in front of small glaciers in Scandinavia. They consist of more or less parallel ridges (cf. Fig. 8.17) with the following characteristics:

1. The outermost ridges are the lowest.
2. The outermost ridges are the youngest.
3. The ice in the ridges is derived from snowbanks and not from glaciers.

According to Östrem, these ridges are formed by the glacier, which pushed debris onto the snowbanks in front. In order to form the outermost ridges, the glacier had to traverse the inner (already existing) ones. Because this is difficult to understand, Barsch (1971b; cf. also Östrem 1971) explained these features as mountain permafrost bodies that are thick (and ice-rich) enough to overcome the critical shear stress at the base. Thus, they start to creep (cf. Fig. 8.18). The ice-cored moraines of the Östrem type are, therefore, nothing other than small debris rockglaciers. This is also true for the examples presented by Östrem and Arnold (1970) from Canada (cf. Fig. 5.14).

8.3.2 The Protalus Rampart Concept

The protalus rampart hypothesis has been discussed during recent years, especially in Britain. According to Ballantyne (1987), a protalus rampart or nivation ridge (German: *Schneehaldenmoräne*) is a ramp or a ridge of coarse debris in front of the foot of a talus slope (cf. Figs. 8.20, 8.21). Since the work of Daly (1912), its development has been explained by the accumulation of talus (produced in the rock walls above) which slides, rolls, or bounces over the snow patch at the foot of the talus; thus it forms the ridge termed "protalus rampart", which is clearly visible as soon as the snowfield has melted (Krebs 1925; Behre 1933; Richmond 1962, 1964; Washburn 1979; Veyret 1981; S.E. White 1981). This hypothesis has some shortcomings:

1. Under present conditions, with a rising snowline, perennial or long-lasting snow patches are more and more restricted. That is, the active formation of

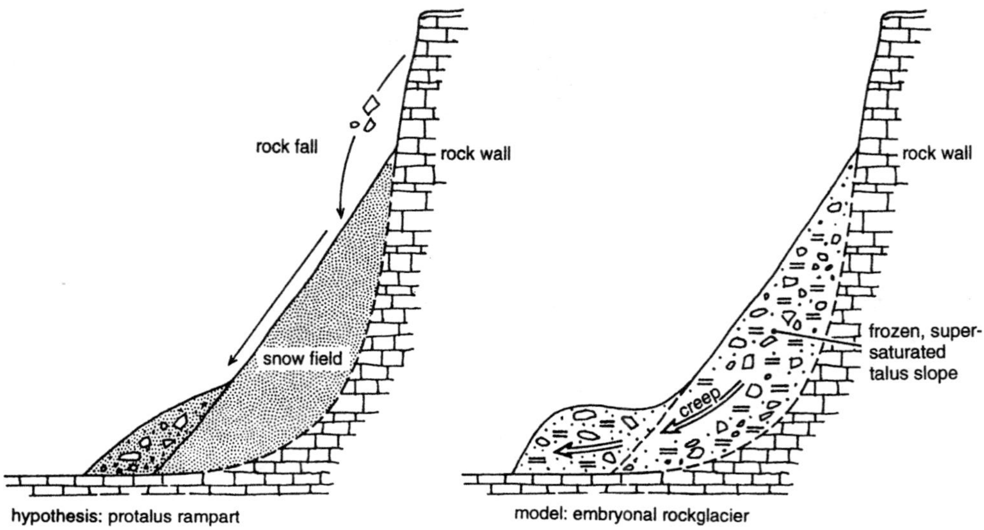

Fig. 8.20. The development of protalus ramparts (Schneehaldenmoranen). *Left* The often accepted hypothesis: Protalus ramparts are though to be formed by talus accumulation at the foot of snowfields by rockfall fragments sliding over the snow surface. *Right* Ridges and terrace-like forms in front of talus slopes are shown as the creep of supersaturated frozen talus which is unstable as soon as certain thresholds (ice volume, shear stress) are overcome

Fig. 8.21. Embryonal rockglacier ("Protalus rampart") near the rockglacier Murtel I (Graubunden, Swiss Alps). The rockglacier already displays the transition to a small lobate talus rockglacier

protalus ramparts (according to this hypothesis) is seldom observed as a recent process (Johnson 1983; Ballantyne 1987).

2. Recent observations demonstrate that clasts and even smaller fragments always lie on the snowfield and not at the foot of it, because the surface of the snow patch is normally too soft. Bouncing over a snow-or firnfield seems only possible if it is ice. In this case we are speaking of an ice patch or ice field, which is normally perennial and thus an indicator of permafrost around and beneath it.

Regarding the first problem, Ballantyne (1987) describes two active protalus ramparts from northern Norway (Lyngen Mountains). The two features studied are situated at altitudes of 760 m (exposure: northwest) and 640 m (facing north). The calculated MAAT is −1.8 °C at 700 m. This indicates the possible existence of discontinuous mountain permafrost, which is assumed above the altitude of MAAT −1.5 °C by King (1984). Based on sedimentary studies four facies are distinguished that Ballantyne (1987 p.167) summarizes as follows:

1. A gentle proximal slope underlain by an openwork mixture of rock-fall boulders and smaller avalanche-transported clasts.
2. A crest over which the frequency of boulders decreases outwards such that the distal part is dominated by avalanche deposits.
3. A steep (up to 43°) unstable and approximately rectilinear distal slope underlain by a diamicton consisting of avalanche deposits form which a larger proportion of clasts has been lost downslope.
4. A basal talus comprising fall-sorted clasts that accumulated mainly by rolling down the distal slope.

The author reports that in only 2 out of 12 rock-fall event clasts reached the rampart. Snow and slush avalanches play an important role in the mass-transport over the firn, where creep is undetectable.

Considering the landform, these modes of transport of talus seem to be of minor importance. The main problem lies in the explanation of the steep, unstable distal slope, which is steeper than the angle of repose. It can only be explained by permafrost creep, that is by the creep of the basal parts of the frozen talus slope which is made unstable by its ice content. According to King (1984) and to the above-cited MAAT calculated by Ballantyne, mountain permafrost is possible at these locations. Therefore, the explanation of these features as small creep phenomena, as small rockglaciers, is not unrealistic.

In addition, looking closely at protalus ramparts in general, the following points support the mountain permafrost hypothesis:

1. The blocky detritus is found on top: on the outer front slope fine-grained material is exposed.
2. In the surroundings of active rockglaciers the active protalus ramparts, as well as the adjacent toe of the slope, are frozen.

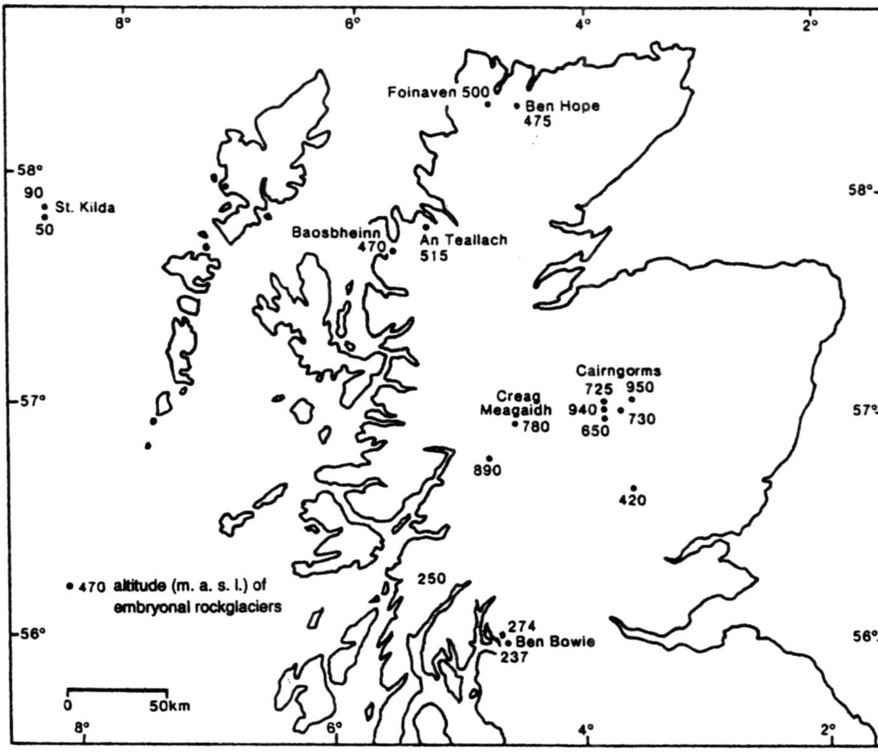

Fig. 8.22. Distribution of Younger Dryas (Loch Lomond Stadial) protalus ramparts in Scotland according to Ballantyne and Kirkbride (1986). The position of the protalus ramparts indicates places with at least discontinuous mountain permafrost during the Younger Dryas in Scotland. The data do not allow reconstruction of the snowline during this time. It is also not possible to reconstruct the lower limit of discontinuous mountain permafrost, but it is possible to demonstrate that the permafrost belt comprised nearly 1000 m during the Younger Dryas. This indicates a very continental climate, in contrast to the present situation

3. In some cases, the depression between the arcuate ridge and the talus slope is quite large. To fill it, a snow patch would have to be of considerable size, probably, almost a glacier. In that case, the deposit should be called a moraine.
4. Frequently, at the foot of adjacent slopes, talus rockglaciers occur. They differ "only" by the fact that they present some more arcuate ridges.

All these points fit well with the definition and description of small but active talus rockglaciers. Even if S.E. White (1981) believes that protalus ramparts are not related to permafrost, there are today many field observations against his hypothesis (Haeberli 1975, 1985a; van Tatenhove and Dikau 1990).

Therefore, "protalus ramparts" can be explained as the visible expression of the creep of mountain permafrost (cf. Fig. 8.20).

Having accepted this hypothesis, a number of difficult points are easier to explain:

- Avalanche activity assists in, but is not critical to, the formation of the landform.
- The explanation is purely process-oriented, and the complicated transport over the snowfield is no longer necessary.
- The difficult differentiation between "protalus ramparts" and talus rockglaciers (the "talus-foot rockglacier" of the authors) is not necessary. This was recognized by Ballantyne and Kirkbride (1986) themselves when they wrote that the dividing line between the two features "may be somewhat arbitrary".

Thus, in most cases, "protalus ramparts" are nothing but embryonal talus rockglaciers (active, inactive, or fossil ones). They can be explained as the creep of mountain permafrost (cf. Fig. 8.20). The base of the frozen talus slope becomes unstable as a consequence of further loading. Creep and deformation follow and cause the development of the first ridge in front of the foot of the talus slope. Thus, the depression between ridge and slope, if it was not occupied once by a glacier, can be explained more easily. In a recent paper Ballantyne and Benn (1994) partly accept this view.

Regarding this discussion, the position of the protalus ramparts allows no information to be obtained on the former snowline, only on the former existence of discontinuous mountain permafrost during the formation of this feature.

Examples.

Relict protalus ramparts are more common than hitherto thought. As already mentioned, many important studies have been done in Britain (Barsch 1993b). The paper by Ballantyne and Kirkbride (1986) displays a map with the altitude of protalus ramparts in upland Britain. The new and much easier interpretation of the protalus ramparts allows a new interpretation of the map itself. According to the authors, these "protalus ramparts" are late Weichsel (Younger Dryas, Loch Lomond stadial). Following the theory that these "protalus ramparts" are the visible expressions of mountain permafrost, the map (Fig. 8.22) shows places with at least discontinuous mountain permafrost during the Loch Lomond stadial (Younger Dryas, 11 000–10 000 ^{14}C years B.P.). This is of paleoclimatic significance, because the position of hte MAAT of about −2 °C isotherm is displayed by the distribution of the small relict talus rockglaciers. The interpretation is discussed in more detail in Section 10.5 (cf. also Barsch 1993b).

Some years ago, Blagbrough and Breed (1967) published a paper on Navajo Mountain (3136 m) in southern Utah. Here a number of features have been described at an altitude of 2600–2700 m in southeastern exposure as protalus ramparts (cf. Fig. 8.10). They too demonstrate the existence of mountain permafrost during Wisconsin time at this altitude.

9 The Age of Rockglaciers

The paucity of good data makes it difficult to provide more than minimum relative ages of active rockglaciers. There is a need for better information, though some initial conclusions are already possible.

Rockglaciers are meso-forms (Barsch 1983a; Thorn and Loewenherz 1987). Based on their size, they belong to the smaller features in this size class (Barsch and Stäblein 1978) which covers areas of 0.01 to 1 km^2. Accepting the fact that a relationship exists between the size and age of landforms, meso-forms are assumed to need a period of several thousand to some 10 000 years in order to develop. For smaller mesoforms such as rockglaciers, an age of several thousand up to 10 000 years seems appropriate. Thus, theoretical considerations lead to ages of some thousand (up to 10 000) years for active rockglaciers. This hypothesis has consequences for the ages of inactive and relict rockglaciers and for their paleoclimatic and paleoecological interpretation.

9.1 The Age of Active Rockglaciers

Direct dating of active rockglaciers is not easy. Organic material has not yet been found in places which are indicative for rockglacier ages. Therefore, relative dating techniques as lichenometry, weathering rinds, and weathering pits on the boulders are used (Birkeland 1973).

Lichenometry has been developed by Beschel (1950, 1957, 1961), and is summarized by Heuberger (1971). It is not without problems and has to face serious criticism (Jochimsen 1966; King and Lehmann 1973; Haeberli et al. 1979). Its value is sometimes overestimated, and lichenometric dates should not be used without cross reference. Dyke (1990 p.30) estimated the oldest rockglaciers to be about 4500 a from lichen (Rhizocarpon) studies. In addition, he found formation times of 3500, 3000, 2200, 1400, 700, 250, 200 and less than 50 B.P. Keeping in mind the general weakness of lichenometry, it seems to be a good assumption that rockglaciers in the Frances Lake Area (Yukon) have formed throughout the Holocene.

Absolute dating seems to be impossible with lichenometry alone. On rockglaciers, this technique has been used, for instance, by Birkeland (1973),

Carrara and Andrews (1973), Heuberger (1974), Miller (1973), Nagl (1971), Haeberli et al. (1979) or Vere and Matthews (1985).

Their results can be summarized as follows. Very young ages assumed for rockglaciers are often based on lichenometric studies. This is normally based on the growth of *Rhizocarpon geographicum* L. It is frequently ignored that with about 1500 species in the Rhizocarpon family the single species is not easily determined, that the taxonomy is difficult (Birkeland 1973), and that the geoecologic situation (including the growth of individual lichens) may be quite different from place to place. Therefore, King and Lehmann (1973) discussed the shortcomings of this method and concluded that this method is only applicable for deposits of the last 300 years. Thus, all lichenometric dates older than 300 years are only acceptable if there is other evidence supporting the given age. Therefore, older ages given by Parson (1987) for active rockglaciers of the Blanca Massif (Audubon, ca. 1800–1000 B.P.) can only be accepted as being tentative. Accepting the difficulties in dating rockglacier deposits, Birkeland (1973), Zielinski (1989), and Calkin et al. (1987) assign mid-Holocene ages to their rockglaciers using, for example, lichenometry, soil development, boulder surfaces, etc.

In addition, present-day movement rates are extrapolated or relationships with Holocene lateral or end moraines are used to determine the ages of active, inactive, or relict rockglaciers. The results from these techniques are sometimes difficult to understand if they suggest rockglacier ages of only a few hundred years. Thus, Boesch (1951) believes that the active rockglaciers in the Swiss Alps have been formed since the beginning of the alpine "Little Ice Age", i.e., since 1500 A.D. Osborn (1975) calculates the maximum age of active rockglaciers in the Banff National Park (Alberta) as being 375 years based on the present-day movement rates. The extrapolation of the present-day movement rates seems inappropriate if we recall the fact (Chap. 6) that the movement of an active rockglacier can change very rapidly from more than 160 cm a^{-1} (around 1918) to zero (1970).

Age assumptions are sometimes also based on questionable relationships. The mid-eighteenth century lateral moraine merges with the uppermost part of the Bukkehol rockglacier (Jotunheimen, Norway). Vere and Matthews (1985) accept this date as a starting point for the rockglacier development. They propose an age of 230 years for this rockglacier. Today, it is known that glaciers in the Alps advanced eight to ten times during the Holocene down to localities of the eighteenth/nineteenth century end moraines (Patzelt 1972, 1977b). The young age given by Vere and Matthews only means that the glacier added material to this moraine and probably to the rockglacier for the last time during the eighteenth or nineteenth century. Therefore, Bukkehol rockglacier is probably much older than 230 years. This can be demonstrated by a simple calculation. Accepting only a mean thickness of 25 m, the volume of debris in the rockglacier (allowing 50% interstitial ice) has to be assumed to be 162 500 m^3 (100 m long, 130 m wide). Allowing the glacier to be stationary for

200 years at this place, it had to transport around 800 m^3 debris per year just to the rooting zone of the rockglacier. This seems impossible!

This calculation can be used generally. The amount of debris incorporated in active rockglaciers (cf. Sect. 10.21) is considerable. It allows the conclusion that relative young ages calculated from velocity measurements of only a few years can be wrong by an order of magnitude. This can be demonstrated by the fact that intensities of frost weathering rates and rock wall retreats based on the present-day debris volume in rockglaciers and on very young ages of these features give unbelievable high results. If these calculated intensities of frost weathering are correct, then it is difficult to accept that high mountains still exist. Therefore, the assumption that huge, active rockglaciers are several thousand years old seems to be appropriate (Evin and Assier 1983a). Thus, on Disko Island (W Greenland), Humlum (1982) believes that the development of the rockglaciers took the whole Holocene period. That these rockglaciers are all inactive (dynamic inactive?) can be doubted. At least they do not appear to be fossil, as Humlum proposes as an alternative to inactive. On the other hand, the hypothesis that the rockglaciers in the Transantarctic Mountains are more than 100 000 years old (Mjagkov 1980) is probably exaggerated.

The White River Ash on the Grizzly Creek rockglacier, S Elias Mountains, Yukon suggests a minimum age of the rockglacier of 1200 to 1500 a (Evin 1992b). Haeberli et al. (1979) estimate an age of 3500 to 10000 years for the Gruben rockglacier (Wallis, Swiss Alps). By this, they oppose the very young age given by Whalley (1974a). This view is supported by new evidence. The pilot analyses of the core from the rockglacier Murtèl I (Haeberli 1989b) opened up new sources of information (Fig. 9.1). At the drilling site the ice in the rockglacier is several thousand years old; at the shear zone at 28 to 32 m depth even early Holocene ice has been assumed. This demonstrates in an exciting way that active rockglaciers formed throughout the Holocene and that individual features are, at least, several thousand years old. From this evidence an instructive model has been proposed (Haeberli et al., in Haeberli 1990). It demonstrates the probable development of the rockglacier during the Holocene. The structured permafrost below the shear zone displays, therefore, decreasing ages from 10 000 to 1000 B.P. from the rooting zone to the toe. The rockglacier itself is oldest (ca. 10 000 years) at the toe. In the borehole, ice is between 5000 (top) and 10 000 (base) years old.

9.2 The Age of Climatic Inactive Rockglaciers

There are two types of inactivity (cf. Chap. 1): the dynamic and the climatic inactive rockglacier. Only the latter will be discussed here. The first has to be accepted as having the same age as active ones; its inactivity is due to topography, to extending flow, or to input changes.

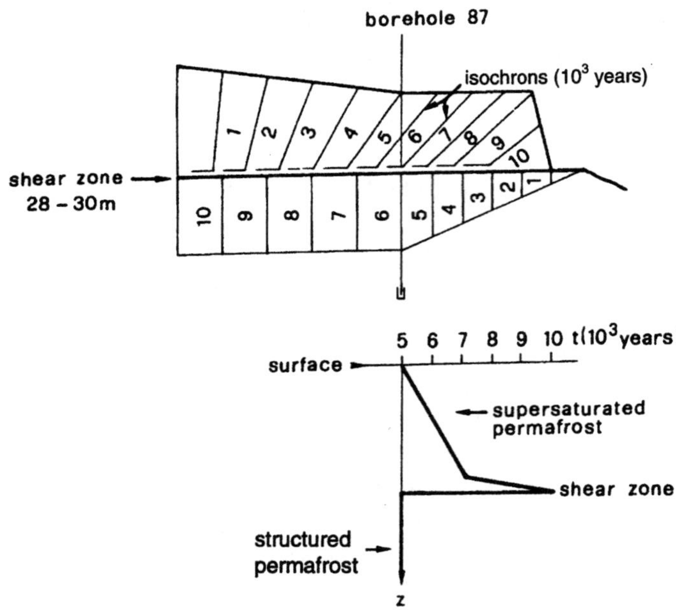

Fig. 9.1. Model of the age of different parts of a rockglacier (case study Murtèl I). (Haeberli et al., in Haeberli 1989b)

Climatic inactive rockglaciers have often been assigned to special periods of the Holocene (Krasnolobodcev 1971; Ivanovskij 1981). Following the same line, Gorbunov (1983) proposes that inactive and active rockglaciers represent two different generations of late-Holocene features. Whereas the inactive ones formed 2500 to 3000 years ago (during a time when the firn line was not more than 100 m lower than today), the active ones started 1500 years ago. This is in accordance with the hypothesis of Gorbunov (1983) that rockglaciers formed during times between Holocene glacier advances. Following the results of the discussion in the foregoing Section 9.1, this view cannot be generally accepted.

In this respect, interesting information is given by P.G. White (1979). He points out that in the San Juan mountains active and inactive rockglaciers end at 3697 and 3516 m a.s.l., respectively. The difference is only 181 m. According to his calculation, the difference in MAAT is less than 1.1 °C (assumed lapse rate: 0.59 °C/900 m). During the Holocene, the elevation of the alpine and nival belts has varied by no more than 200m, i.e., the MAAT differs between warmer and colder periods by ca. 1–1.5°C (Patzelt 1972, 1977b). This is also demonstrated by the altitudinal belt in which inactive rockglaciers occur in the Alps (Fig. 5.2). It seems, therefore, appropriate to accept that the climatic inactive rockglaciers are of the same age as active rockglaciers. Both the active and inactive rockglaciers, which are usually found more or less in the same altitudinal belt, are of Holocene age (cf. Fig. 5.2); on average the climatic inactive forms occupy the lower part of this belt.

Because today's climatic inactive rockglaciers are situated at the lower realm of the present cryogenic belt they have been the first to be affected by the twentieth-century warming. Their inactivity probably developed only during recent decades, i.e., since 1950/1970. Therefore, if the Holocene environment in the upper belts of high mountain systems is discussed, both, the climatic inactive and the active forms should be discussed together. The present-day climatic inactive rockglaciers mark the lower part of the Holocene cryogenic belt where mountain permafrost is not in equilibrium with climate during the warmer periods of the Holocene.

9.3 The Age of Relict (Fossil) Rockglaciers

In contrast to the present-day active rockglaciers, the relict (fossil) rockglaciers today contain no interstitial ice. Their ice-content has been completely melted (cf. Chap. 8); however, they also have to be considered as meso-forms. Thus, their development requires – depending on the intensity of talus production – a period of at least 1000 to some thousand years of activity. Following theoretical considerations, they must have formed during different periods of the Pleistocene. A formation during colder periods of the Holocene, which is often proposed, seems impossible in relation to the Holocene climatic history. Therefore, relict rockglaciers are, if dated accurately, important for chronologies of the younger Quaternary and for climate reconstructions (cf. Sect. 10.5).

In North America, many relict rockglaciers have been accepted as being formed during the earlier periods of the Neoglacial (5000–3000 years B.P.). Using sedimentation in the Temple Lake Valley (Wind River Range, Wyoming), Zielinski (1989) was able to demonstrate that relict rockglaciers, which are in direct contact with the lake shore do not have a Neoglacial, but a Wisconsin age. The study demonstrates the great difficulties that are involved with relative dating techniques on rockglaciers. It seems today that relict (fossil) rockglaciers are much older than active/inactive ones, because they are normally found well below the present cryogenic belt. They were formed during different periods of the last glaciation, and not during the Holocene. Barsch and Updike (1971a,b), therefore, propose a Wisconsinan age for the relict rockglaciers of Kendrik Peak (Arizona) because these features were formed just below the ELA of the last glaciation.

In the Alps, many relict rockglaciers have been found at or below present-day timberline. They must have been formed during late glacial times. Thus, Kerschner (1976, 1978) describes relict rockglaciers from the late Würm (Daun: 13 000–12 000 ^{14}C-years B.P. and Egesen: 11 000–10 000 ^{14}C-years B.P.). Maisch (1981) can demonstrate, for central Graubünden (Swiss Alps), that relict rockglaciers are all within or cross late glacial (Egesen) moraines (ca. 11 000–10 000 ^{14}C-years). They are probably all of the same age; they turned climatically inactive at the Pleistocene/Holocene boundary and had melted completely by 9 000-radiocarbon years B.P. This view is generally accepted in the Alps today (de Jong and Kwadijk 1988; van Tatenhove and Dikau 1990).

Relict (fossil) rockglaciers are, therefore, to be accepted as being older than 11 500 years (i.e., older than the Holocene; cf. Fig. 5.6). They are of Pleistocene age. Dating is therefore related to stratigraphic relationships with Pleistocene or late Pleistocene morainic deposits which are often not simple in alpine environments.

10 Rockglaciers and the High Mountain Environment

The earlier chapters of this book have summarized available information on rockglaciers in an analytical manner; in contrast, this chapter is intended to draw these parts together in a synthesis. The different aspects of rockglaciers will be discussed as an important subsystem (and integral parts) of high mountain environments.

High mountain or high alpine environments are situated above timberline in mountains with a forest belt where they comprise the alpine, the subnival, and the nival belts (Fig. 10.1). In a number of papers, Troll (1941, 1954, 1972, 1973) discussed these belts as *Hochgebirgsstufe* or *Hochgebirgslandschaft* and as being diagnostic of a high mountain system (Barsch and Caine 1984). Only in polar and arid zones, where forests are missing, is this model not applicable.

In midlatitude high mountain systems, the belts above the timberline and below the snowline belong to the alpine periglacial realm. In very continental mountains (like the Rocky Mountains of Alberta) mountain permafrost and active rockglaciers may be found below timberline (Luckman and Crocket 1978; Harris and Brown 1982), and it is a point of discussion whether these areas, within the forest belt, should be included in the periglacial realm (Karte 1978) or not (Barsch 1993a).

In mountains of the arid zones, the determination of cryogenic belts is not simple because arid and cryogenic processes can create similar landscapes. In this case, additional indicators (e.g., active rockglaciers as indicative of discontinuous mountain permafrost) are needed to delimit the present-day periglacial belt. In polar areas, a high mountain belt can only be defined by the relief, which should be characterized by horns, rock walls ($>60°$), and steep mountain slopes ($35–<60°$), because at least discontinuous permafrost is possible anywhere in this zone.

High mountain environments have a cooler and wetter climate than the surrounding lowlands and have often been defined as analogs of polar climates even in tropical mountain systems (von Köppen 1931). However, as Troll (1943; Troll and Paffen 1964) points out, this is incorrect because the climate of high mountain belts represents a lower temperature and usually a higher precipitation version of the climate of the surrounding lowlands with respect to season, day length, precipitation regime, etc.

Despite an increasing human impact on the high alpine environments, our knowledge of these geosytems is still incomplete. Therefore, rockglacier research can be important because it may cast light on the geoecologic state of

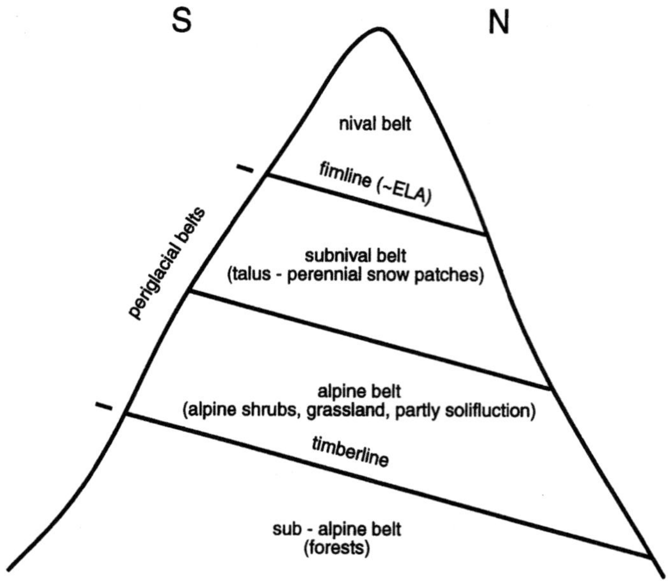

Fig. 10.1. Conceptual diagram of high mountain belts in midlatitude mountains. The MAAT –2 °C isotherm cuts often into the alpine belt and in some mountains even into the subalpine forest belt, e.g. in the continental Rocky Mountains of Alberta (Canada). The lower limit of the cryogenic belt is occasionally associated with this isotherm. The subnival belt corresponds in the zonal concept more or less with the Frostschutt-Zone of Büdel.

high alpine (subnival) environments. It seems appropriate, therefore, to close this volume with a demonstration of what rockglaciers indicate for the geomorphic and geoecologic assessment in high mountain systems.

10.1 Active Rockglaciers and Mountain Permafrost

In contrast to polar regions, permafrost research in high mountains has only a short history (Barsch 1969b, 1977c; Furrer and Fitze 1970; Ives and Fahey 1971; Brown and Pewe 1973). Thus Troll (1944), who compiled much of the information on periglacial processes and forms available at the time, did not believe alpine permafrost to be common in high mountain environments. This illustrates the difficulty of identifying mountain permafrost, a difficulty addressed by the recognition that active rockglaciers are a visible expression of mountain permafrost. They indicate its presence now or in the past. They depend on an energy budget which allows the survival of (discontinuous) permafrost. Although active rockglaciers have long been recognized as bodies of permafrost or underground ice (Wahrhaftig and Cox 1959; Barsch 1969a, b), it is surprising that the first systematic study to define mountain permafrost

Table 10.1. Mountain permafrost in the Grialetch area (Graubünden). (Haeberli 1975)

Belt (m a.s.l.)	Area of permafrost (km^2)	Percentage of the total permafrost area (%)
3000–3200	–	–
2800–3000	0.2	4.7
2600–2800	2.2	59.6
2400–2600	1.2	32.5
2200–2400	0.0	3.2
Σ	3.7	100.0

The amount of water in mountain permafrost bodies of this area is estimated to be 70–80$\times 10^6$ m^3, i.e., about 40% of the water stored in glaciers.

patches in their environments was not published until 1975 (Haeberli 1975; Table 10.1). Following this example, a number of studies have been carried out since (e.g., Figs. 10.2, 10.3).

Using active rockglaciers as permafrost indicators, Haeberli (1975) defined the distribution of mountain permafrost in the Grialetsch area, a continental part of the central Alps. Above 2200/2300 m, mountain permafrost comprises about 25% of the study area (42 km^2). At 2750 m elevation, ca. 50% of the surface is underlain by mountain permafrost. In addition, Haeberli demostrates that topographic situations as the foot of steep slopes, flat surfaces on crests or summits, and slopes with low radiation receipts because of shadowing or northern exposure are favorable for mountain permafrost in alpine relief.

The permafrost map of the Äußeres Hochebenkar (Tirol, Austria) of Haeberli and Patzelt (1982) also demonstrates that mountain permafrost exists outside the active rockglaciers (Fig. 10.3). In the case of the rockglacier Äußeres Hochebenkar, the mountain permafrost is continuous from the rockglacier toe up to the talus slopes and around the small glacierets. It is interesting to note that, above the rockglacier Inneres Hochebenkar, areas with no permafrost are found. Thus, a permafrost connection between the rockglacier and the glacier (Hochebenferner) does not exist. In general, the depressions are permafrost free, whereas the steeper slopes and the areas at the foot of steep slopes are permanently frozen.

Buchenauer (1990, pp. 246, 247) reports for his study area in Austria that discontinuous permafrost comprises about 17 km^2 (Table 10.2). The present periglacial belt (2300–2850 m) covers an area of about 65 km^2. Thus, mountain permafrost influences about 26% of this belt, whereas active and inactive rockglaciers cover only 2.5 km^2, i.e., only 15% of the assumed area of discontinuous permafrost. Thus, intact rockglaciers are real and visible indicators of more widespread permafrost in their neighborhood (cf. Table 10.3).

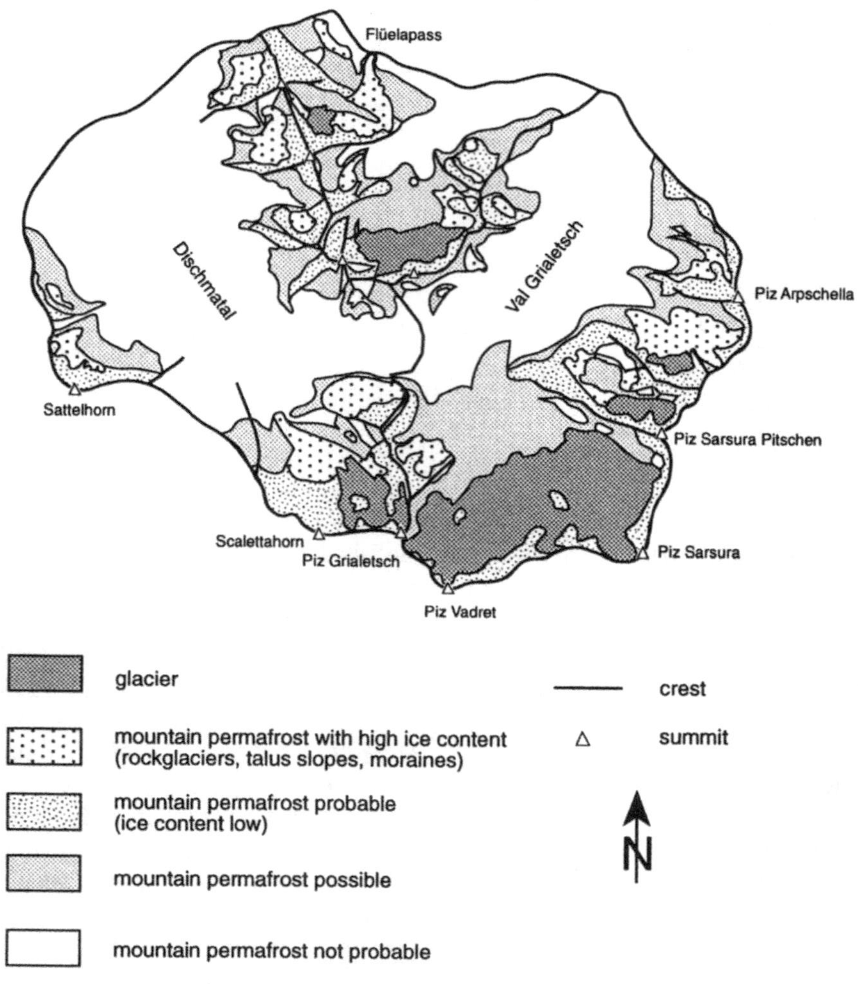

Fig. 10.2. Mountain permafrost distribution in the Grialetsch area (Flüelapass, Graubünden). This is the first map to show alpine permafrost in relation to rockglacier distribution. (Haeberli 1975)

For other parts of the world, Pewe (1983) summarizes mountain permafrost distributions in the contiguous United States; Gorbunov (1988) gives a similar summary for the lands of the former USSR, and Haeberli et al. (1993) provide a recent global perspective on mountain permafrost and global change.

The information collected by Pewe is essentially based on the occurrence of rockglaciers (Fig. 10.4). On a world scale, Fig. 10.5 represents a map with locations of mountain permafrost. The map is based on Gorbunov (1978) and on the simplified version published by Haeberli et al. (1993). The map differentiates between maritime and continental mountain permafrost which is a

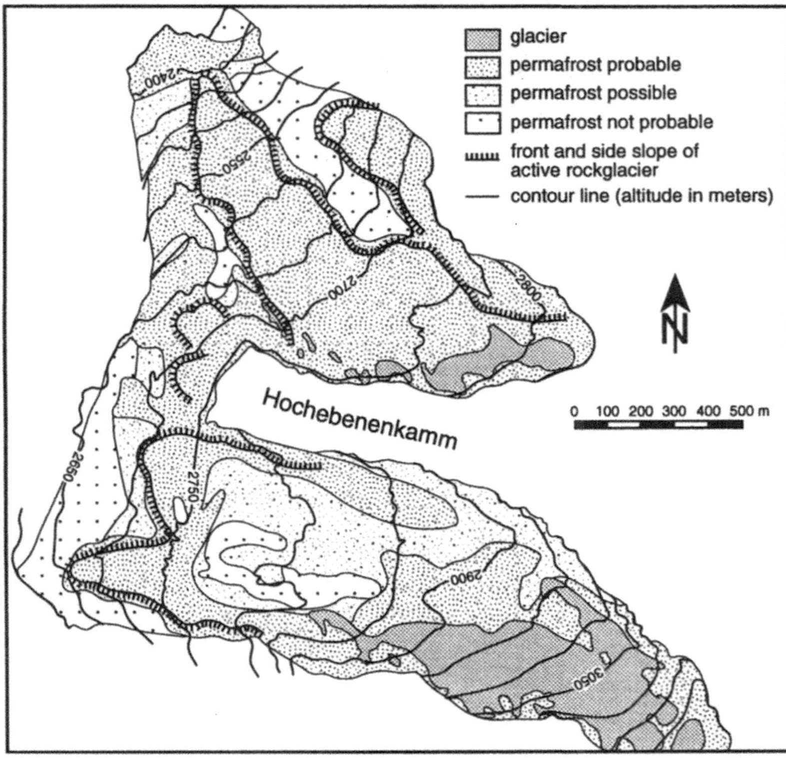

Fig. 10.3. Map of mountain permafrost distribution in the Hochebenkar (Tirol, Austria). (Haeberli and Patzelt 1982)

useful distinction. In maritime climates, high precipitation allows glaciers to exist at relatively warm (low) altitudes, ie., MAAT at the snowline (ELA) is high. Thus, the lower limit of mountain permafrost lies at or even above ELA (i.e., the mountain permafrost belt is very restricted and rockglaciers are rare). In continental climates, on the other hand, ELA is higher because of reduced precipitation. Therefore, the lower limit of mountain permafrost lies well below ELA and rockglaciers are more common. This is especially true for the vast mountain permafrost areas in Asia (Table 10.4). The general relationship between maritime and continental permafrost is discussed in Section 10.6 and summarized in Fig. 10.11. Tropical mountains (e.g., part of the Andes in South America) belong to the maritime mountain permafrost region. They are more strongly influenced by daily than by seasonal temperature variations (Troll 1943), and rockglaciers seem to be rare.

It has often been suggested that the movements of active rockglaciers should be taken into account when defining the distribution of mountain permafrost, since rockglaciers are able to creep into altitudes which are unfavorable for permafrost. Therefore, Humlum (1982) has differentiated auto-

Table 10.2. Area of discontinuous permafrost in the western Schober Mountains outside active rockglaciers

Valley	Area with discontinous permafrost (km^2)
Debanttal	4.30
Neualpl/Schleinitz S	0.05
Catchment Daberbach	1.00
Leibnitztal	2.10
Staniska	2.20
Falwindes	0.30
Lesachtal	4.40
Tschadinalm	0.50
Moostal	1.70
	Total: 16.55

The total area of the periglacial belt (between ELA and lower limit of solifluction) is estimated to be 65 km^2. The belt of discontinuous mountain permafrost comprises about 40 km^2 with ca. 16.5 km^2 of permafrost area (ca. 40%). The lower limit of discontinuous permafrost is assumed at
2400–2450 m in N-exposure,
2650–2700 m in S-exposure and
2500–2550 m in W- or E-exposure.
The lower limits of active rockglaciers is presented in Table 10.3.

chthonous from allochthonous permafrost. The following calculation may demonstrate that this is unnecessary. In the Alps, firnline has risen by ca. 200 m between 1850 and 1990 and MAAT are higher by about 1.5 °C. If we assume a slope of 20° and a rockglacier velocity of 0.3 m a^{-1}, then a rockglacier would need about 2000 years to move this distance. As a consequence, it may be concluded that rockglaciers will be more affected by climatic change over time than by their own downvalley movement into warmer altitudes. The differentiation between autochthonous and allochthonous mountain or rockglacier permafrost is not necessary in this respect.

It seems clear that the exact location of the lower limit of discontinuous mountain permafrost can be evaluated only in a regional sense. On the other hand, active rockglaciers are an easily detected indicator of mountain permafrost in a high mountain range (Barsch 1978). They are also most significant in defining the periglacial belt in arid mountains, where a timberline does not exist. Thus, the regional lower limit of active rockglaciers can be taken (*cum grano salis*) as the approximate lower limit of mountain permafrost although it may be wise to suggest, as did Haeberli (1975), that the true lower limit may be 50 m below this altitude.

Table 10.3. Lower limits of active rockglaciers in the western Schober Mountains in relation to present ELA. (Buchenauer 1990)

Exposure	Ø ELA[a] (1965–1985) (m)	Lower limit of rockglacier (m)	Delta (m)
N	2753	2245	303
NW	2796	2492	304
W	2834	2585	249
SW	2920	2733	187
S	2980[b]	2730	250
SE	2910	2750	160
E	2826[c]	2575[c]	245
NE	2760[c]	2578[c]	182
	Ø 2247	Ø 2512	Ø 235 (NW-N-NE: 263)

[a]Average of all single values in this exposure.
[b]Estimates.
[c]Values including the eastern Schober Mountains.

As Vonder Mühll and Haeberli (1990) mention, permafrost temperature below the ZAA (zero annual amplitude) at the Murtèl I site is about –3 °C (cf. Table 6.9). This is much too low when compared to air temperatures measured at nearby meteorological stations. This discrepancy could be explained by lower temperatures in the permafrost reflecting the cooler climate of the last century, but this was not confirmed by the temperature measurements in the upper 50 m of this deposit (cf. Sect. 6.3.6). Together with the fact that the rockglacier Murtèl is moving onto nonfrozen rock, this demonstrates that

Table 10.4. Estimated areas of mountain permafrost [in km^2] outside the polar zone. (Haeberli et al. 1993)

Mountains in:	Mountain permafrost (km^2)
China, Nepal, India, Pakistan, Afghanistan	2 000 000
former USSR	1 000 000
Mongolia	1 000 000
North America	400 000
South America	270 000
Scandinavia	80 000
Iceland	50 000
Alps	50 000
Caucasus	20 000
New Zealand Alps	10 000
Total	4 880 000

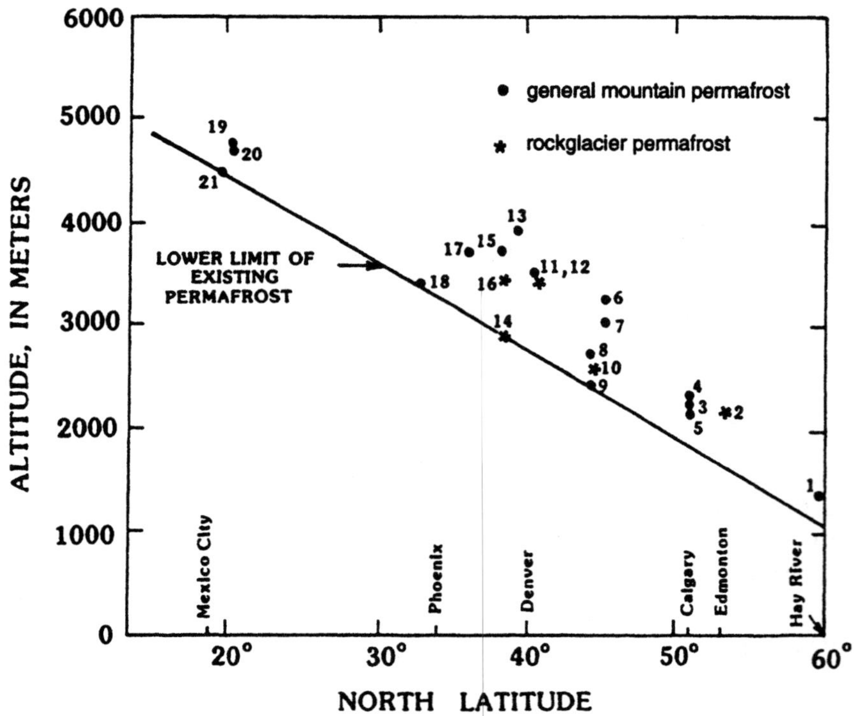

Fig. 10.4. Diagram of the lower limit of mountain permafrost in the contiguous US. (Pewe 1983)

rockglaciers indicate favorable sites for mountain permafrost. This effect is probably magnified by Balch ventilation (Thompson 1962) by which summer warming of the rockglacier surface is prevented by the trapped cold air between the boulders. Furthermore, air cooled at night will tend to replace any warmer daytime air which has perhaps been collected in the upper voids between the boulders. This creates a marked cooling effect that forms part of a positive feedback loop to reduce the melting of rockglacier permafrost.

Differences on a rockglacier surface regarding melting seem to exist. Thus, Humlum (1988) points out that in winter the furrows are normally filled with snow whereas the intervening ridges are windswept and snowfree. These latter areas are, therefore, important cooling places for the rockglacier permafrost, which causes the entire surface to be cooler than the surrounding snow-covered environment. The low temperatures in the furrows are protected by insulation well into the summer. The active layer at these places is often much thinner than that on the ridges (cf. Sect. 6.3).

In summary, active (and inactive) rockglaciers are indicators of mountain permafrost in their neighborhood. They have the advantage of being recognizable landforms that are easily detected on air photographs, and so can be used as a first approximation in defining the distribution of permafrost in all

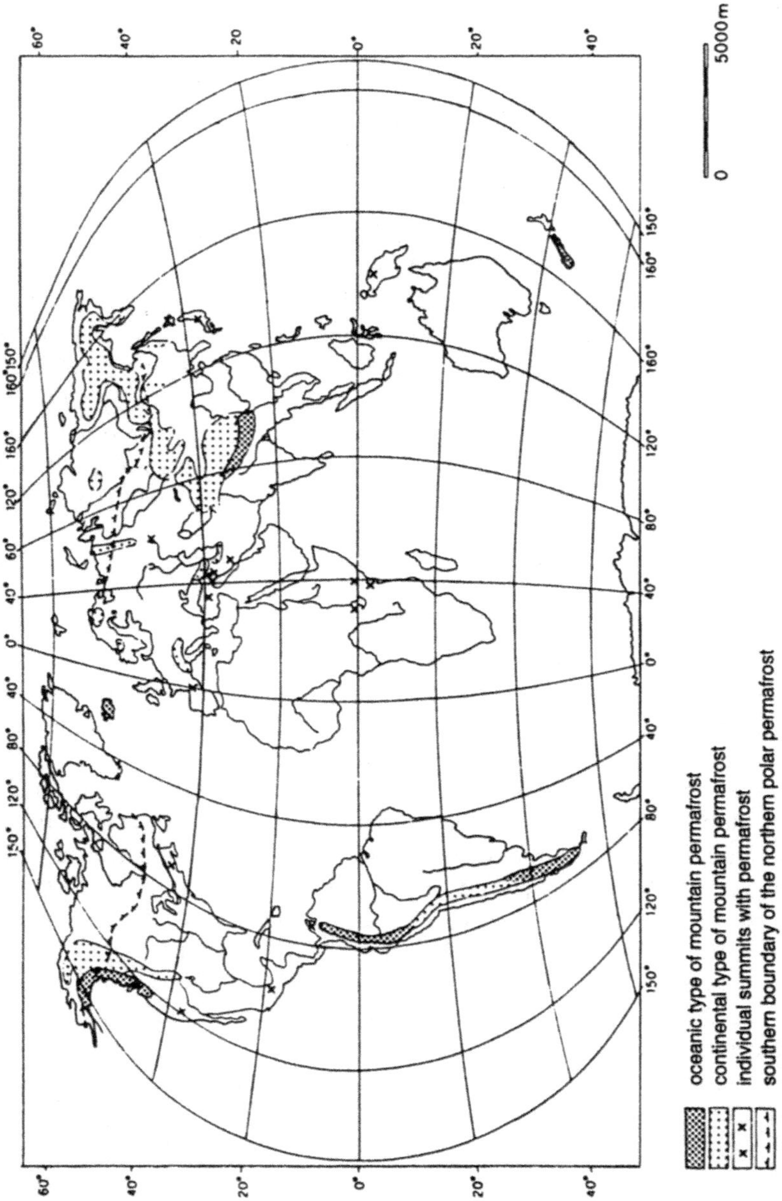

Fig. 10.5. World map of mountain permafrost (Haeberli et al. 1993). This map is a simplified version of the map published by Gorbunov (1978)

mountain systems. Other measurements, e.g., BTS, geophysical soundings etc., and detailed studies form a second step in a geoecologicl assessment.

10.2 Rockglaciers in the Coarse Debris Cycle

The coarse debris system (Caine 1974) forms a very important part of the alpine sediment cascade and active rockglaciers are a prominent element of this system. According to S.E. White (1981), they are noncatastrophic alpine mass-movement forms. Even if the classification of White is not acceptable in all parts, active rockglaciers belong to the slow alpine mass movements that move by creeping of mountain permafrost. On the one side, they are separated from solifluction, where resistance is reduced by the water content in the upper parts of the soil mantle under periglacial, cryogenic conditions. On the other side, they are differentiated from the "*talzuschub*", the large-scale creep of mountain slopes in big valleys under special geologic and geomorphic conditions.

10.2.1 Rockglaciers and Talus Production

Active talus rockglaciers are directly connected with the talus slope and the rock wall above them. Together, they usually form a closed system in the coarse debris cascade (Caine 1974). The amount of talus produced in the rock wall above the rockglacier during a certain period is (in most cases) still found in the talus slope and in the talus rockglacier. The volume of both features (talus rockglacier and talus slope) is, therefore, indicative of talus production from the rock wall during the period of formation, i.e., since the end of the Pleistocene for most active and inactive talus rockglaciers. Thus, the amount of talus in the talus rockglacier and in the talus slope corresponds to the production of talus from the rock walls above the rockglacier during the Holocene, i.e., over the past 10 000 to 11 500 years (maximum assumption).

Wahrhaftig and Cox (1959, 430ff) were the first to use the volumes of active rockglaciers to determine the amount of postglacial denudation in their source area. They calculated a mean denudation rate of 9 to 10 m (30 ft) for the postglacial period. This value (10 m/10 000 a) has to be related to the real surface, i.e., to the real area of the rock walls. This real area of a rock wall is larger by a factor of $1/\cos \hat{A}$ than its horizontal projection on a map. Therefore, the average retreat of the rock walls in the Alaska Range was somewhat less than 1 mm a^{-1} during the Holocene. Wahrhaftig and Cox believe that rockglaciers have not been active throughout the Holocene, and assume higher talus production rates for the times of greatest rockglacier activity.

Higher values have been calculated (Table 10.5) for two well known rockglaciers from the Swiss Alps whose thickness has been determined by seismic refraction. For the calculation of the talus volume, two assumptions were originally used: a 50 and a 40% debris volume in the total rockglacier. Given our present knowledge of the internal composition of rockglaciers, it seems more appropriate to use a value of 30 vol%, which has been added in Table 10.5.

In addition the following assumptions have been used. The source area is taken as the real surface of the rockshed above the talus slope. The volume estimates are conservative; thus, especially in the example of the rockglacier Murtèl I, the estimation of the volume of the talus slope is probably too low. The time during which the talus was produced has been assumed to be either 10 000 or 6 000 years. Thus, four values can be calculated for each source area. The maximum, the minimum, and the average are presented in Table 10.5, which shows an average rock wall retreat above both rockglaciers of between 1.5 and 3.4 mm a^{-1}.

Based on these estimates (Fig. 10.6), an extrapolation to the talus production rate for the entire Swiss Alps suggests a mean rock wall-retreat rate above active rockglaciers to be around 2.5 mm a^{-1} (Barsch 1977d; Tables 10.6, 10.7). All in all, this equals an absolute rate of about 25 m for the last 10 000 years. Even if we accept that the amount of talus in active rockglaciers is generally only 30% (cf. the drilling by Haeberli 1989b; Sect. 6.3.1), rock weathering would still amount to 0.8–1.5 mm a^{-1}. These are relatively high values. They demonstrate the intensity of rock weathering in the subnival belt of the central and southern Alps (Fig. 10.6).

An extrapolation to all high mountain systems of the world is not justified because the environmental differences may be too high as the following estimations may show. In the Zailijskiy Alatau of middle Asia, the Prijamoy shcheli rockglacier is fed from ca. 7 km^2 of rocky slopes (Gorbunov 1983). During times of glacier retreat the source area increases up to 11 or 12 km^2. The amount of talus incorporated in this rockglacier is estimated to be 30 or 50×10^6 m^3. Unfortunately, we are not informed about the slope and the real surface area of the rock walls. For a conservative estimation the given surface is accepted as real and so talus production is very low; 0.4 to 0.7 mm a^{-1} (for 7 km^2 of rock walls) and 0.3 to 0.4 mm a^{-1} (for 11.5 km^2). This seems low for high alpine environments, but it is not as low as rock wall retreat in the Wernecke and Ogilvie Mountains (central Yukon) where Gray (1970) estimated a retreat of 0.1 to 1 m in 10 000 to 14 000 years. It may be that the rocks (quarzites and diorites) are very resistant; on the other hand, this is also a cryogenic environment which allows the existence of active talus rockglaciers, i.e., it is an area with a talus production high enough to allow the existence of rockglaciers.

Considering relict rockglaciers, the same calculations are possible for historical times, and are partly easier because the rockglacier volume corresponds directly to the talus volume. Different values for the rock break-up in relation to present-day data may indicate different climatic conditions. Thus, Buchenauer (1990) estimates the rock wall retreat rate in the western Schobern

Table 10.5. Holocene talus production in the source areas of the debris rockglacier Macun I (Lower Engadin, Graubünden) and the talus rockglacier Murtél I (Upper Engadin, Graubünden). (Barsch 1977d)

Rockglacier	Altitude of toe (m a.s.l.)	Surface of rockglacier ($10^3 m^2$)	Volume ($10^3 m^3$)	Volume talus/debris		Source area ($10^3 m^2$)	Denudation				
				V50 ($10^3 m^3$)	V33 ($10^3 m^3$)		V50 (m)	V33 (m)	max. (mm a^{-1})	min. (mm a^{-1})	average (mm a^{-1})
Macun I	2600	157	6750	3400	2300	ca. 125	27	18	4.5	1.8	3.2
Murtél I	2630	42	2500[a]	1300	830	ca. 100	25	8	4.2	0.8	2.5

Both rockglaciers are on northern exposures.
The volume of talus/debris is assumed to be between 50 vol% (V50) and 33 vol% (V33).
Source area is the estimated true surface.
Denudation is given in two figures as total amount (m) and as denudation per year (mm a^{-1}). For the calculation of the latter, a time of 10 000 and 6000 years has been assumed. Only the maximum, the minimum, and the average are presented.
[a]This figure includes the talus in the talus slope which is estimated to be 0.9×10^6 m^3 and the material beneath the high-viscosity layer at 28–32 m depth (cf. Sect. 6.2.2.5).

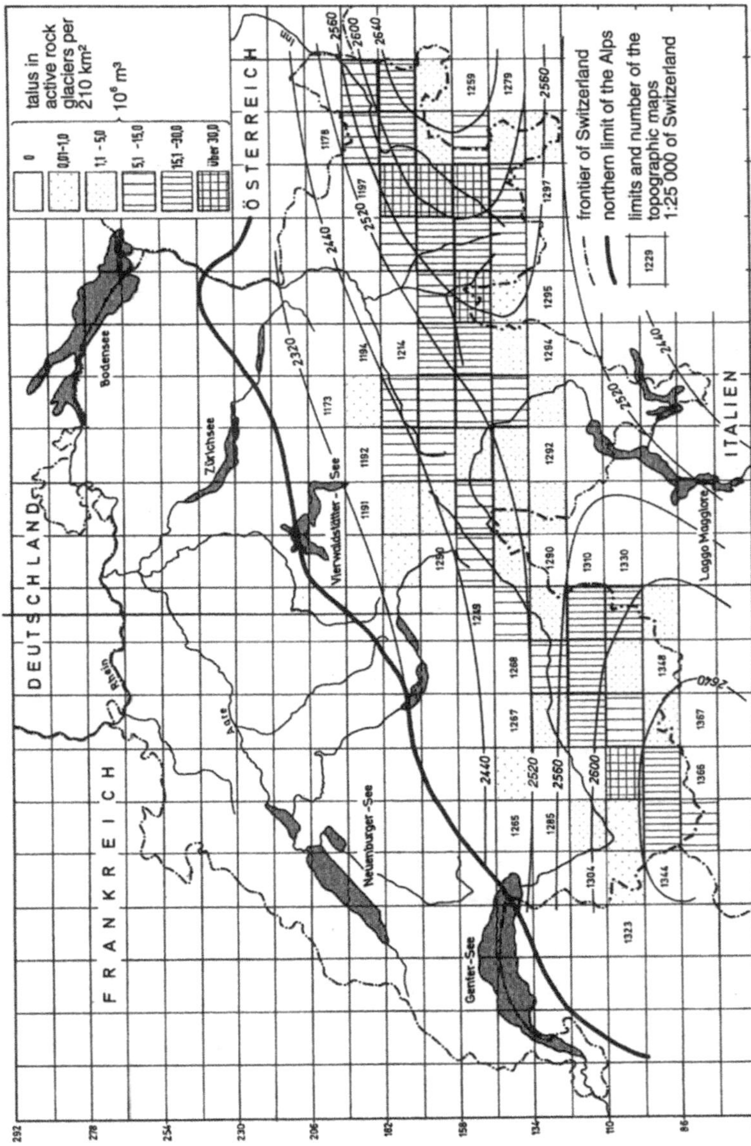

Fig. 10.6. Talus stored in active rockglaciers in the Swiss Alps. (Barsch 1977a)

Table 10.6. Talus production in the source areas of active rockglaciers in the Swiss Alps.(Barsch 1977d)

Rockglacier type	Number [n]	Size [$10^6 m^2$]	Source area[a] [$10^6 m^2$]		Volume[b] [$10^6 m^3$]		Volume of talus[b,c] [$10^6 m^3$]				Denudation (m)			Denudation[d] (mm a^{-1})		
			Model 1	Model 2	Model 3	Model 4	Model 3		Model 4		Max.	Min.	ϕ	Max.	Min.	ϕ
							Vol 50	Vol 40	Vol 50	Vol 40						
1	602	16.1	16.1	48.3	868.1	678.9	433.1	347.3	339.4	271.5	27.0	5.6	10.8	4.5	0.6	2.5
2	392	13.7	13.7	54.8	761.2	595.8	380.6	304.5	297.9	238.3	27.8	4.4	8.9	4.6	0.4	2.5
1+2	994	29.8	29.8	102.7	1629.3	1274.7	814.7	651.8	637.3	509.8	27.3	5.0	9.9	4.6	0.5	2.5

Rockglacier types: 1, talus rockglacier; 2, debris rockglacier.

[a]Model 1: rockglacier surface = source area

Model 2: talus rockglacier: rockglacier surface = 0.33 source area
debris rockglacier: rockglacier surface = 0.25 source area

[b]Model 3: thickness of rockglacier: 30 m (surface < 10 000 m^2)
50 m (surface 10 000–100 000 m^2)
80 m (surface > 100 000 m^2)

Model 4: thickness of rockglacier: 20 m (surface < 10 000 m^2)
40 m (surface 10 000–100 000 m^2)
60 m (surface > 100 000 m^2)

[c]Talus/debris is calculated for 50% ice (Vol 50) and for 60% ice (Vol 40).

[d]Weathering time is assumed to be their 10 000 or 6000 years.

Table 10.7. Holocene talus production above active rockglaciers in different exposures in the Swiss Alps. (Barsch 1977d)

Rockglacier exposure	Number [n]	Size [$10^6 m^2$]	Source area[a] [$10^6 m^2$]		Volume[b] [$10^6 m^3$]		Volume of talus[b,c] [$10^6 m^3$]				Denudation (m)			Denudation[d] (mm a^{-1})		
			Model 1	Model 2	Model 3	Model 4	Model 3		Model 4		Max.	Min.	ϕ	Max.	Min.	ϕ
							Vol 50	Vol 40	Vol 50	Vol 40						
NW–NE	668	21.4	21.4	73.5	1192.0	930.7	596.0	476.8	465.4	372.3	27.9	5.1	10.1	4.6	0.5	2.6
E	85	2.074	2.074	7.461	114.5	90.0	57.3	45.8	45.0	36.0	26.4	4.8	9.6	4.4	0.5	2.5
S	33	0.886	0.886	2.944	47.3	37.0	23.6	19.0	18.5	14.8	26.6	5.0	9.9	4.4	0.5	2.5
W	96	2.664	2.664	9.496	134.7	106.8	67.3	53.9	53.4	42.8	25.3	4.5	8.9	4.2	0.5	2.4

[a]Model 1: rockglacier surface = source area
Model 2: talus rockglacier: rockglacier surface = 0.33 source area
debris rockglacier: rockglacier surface = 0.25 source area
[b]Model 3: thickness of rockglacier: 30 m (surface < 10 000 m^2)
50 m (surface 10 000–100 000 m^2)
80 m (surface > 100 000 m^2)
Model 4: thickness of rockglacier: 20 m (surface < 10 000 m^2)
40 m (surface 10 000–100 000 m^2)
80 m (surface > 100 000 m^2)
[c]Talus/debris is calculated for 50% ice (Vol 50) and for 60% ice (Vol 40).
[d]Weathering time is assumed to be their 10 000 or 6000 years.

Mountains (Austria) to be 5–6 mm a^{-1} above the Oldest Dryas rockglaciers, and 5 mm a^{-1} or more above Younger Dryas rockglaciers. Such a high rate seems only possible under special conditions, in this case a very cold, dry, continental environment. Whether or not large-scale instabilities in the rock walls, following initial deglaciation, contributed to the high rates of talus production cannot be determined at the moment.

Even if active talus rockglaciers indicate rock walls with high weathering rates (at least during part of the Holocene), it is astonishing that so few studies on rock wall retreat have been done in rockglacier environments. Rockglaciers are a trap (or a storage) in the coarse debris cycle and allow a good estimate of the amount of talus production above them. More studies on this aspect would give a worldwide differentiation on rock wall retreat under different climatic, lithologic, and environmental conditions. In any case, rockglaciers, should be used in local and regional studies to evaluate the rockfall risk in high mountain belts.

Under the present climate, rockglaciers in the Alps develop over several thousand years (cf. Chap. 9). On the other hand, many relict rockglaciers have been formed during Egesen time (i.e., during Younger Dryas in the Alps), a period of cold, dry, continental conditions (cf. Sect. 10.3.2; Kerschner 1976). This period lasted probably only 1000 years and must have involved much higher rates of talus production than at present for talus rockglaciers to have formed in such a short interval. Studies on these relationships can lead to a qualitatively better understanding of rock breakup under cold climates and to quite new and more sophisticated information on the paleoenvironment during the Younger Dryas in different mountain systems around the world.

10.2.2 Rockglacier Size and Source Area

Active rockglaciers incorporate a relatively constant proportion (30 to 50%) of rock material (talus/debris). Therefore, the size of a rockglacier should be related to the size of its source area and the weathering rate, i.e., the intensity of talus production From the data in Table 10.5, it may be concluded that the source area of the rockglaciers Macun I and Murtèl I are 1.2 to 2.5 times greater than the rockglacier surface. Measurements in the Alps allow the conclusion that on average the ratio of the size of debris rockglacier to source area is about 1 to 4, and for talus rockglaciers, it is 1:3 (Barsch 1977a, d).

This relationship can be used in defining the paleoenvironment of relict rockglaciers, based on estimates of rockglacier characteristics. This can be demonstrated by computing the necessary size of a source area, if the size of an active rockglacier (width, thickness), its velocity and the incorporated volume of talus are known. According to Table 10.8, the height of the source wall of an active rockglacier is estimated to be between 75 and 450 m. This range demonstrates the possible variability of rockglacier environments. It seems quite

Table 10.8. Possible size and height of source areas of active rockglaciers for different assumptions

Width (m)	Thickness (m)	Velocity ($m\ a^{-1}$)	Volume of talus (%)	Input of talus ($m^3\ a^{-1}$)	Weathering rate ($mm\ a^{-1}$)	Real size of source area (m^3)	Real length of source area (m)	Height of source area (m)
50	15	0.1	20	15	1	15 000	100	ca. 75
		0.3	30	ca. 65	1	ca. 65 000	430	ca. 320
150	20	0.1	20	60	1	60 000	600	ca. 450
		0.3	30	270	1	270 000	135	ca. 100

The real size of the source area is calculated as follows

$$= \frac{\text{input talus}}{\text{weathering rate}}$$

with input of talus = (width × thickness) × veolcity × volume of talus.
The source area is assumed to be three times wider than the rockglacier.
The slope of the source wall is assumed to 50°.

important to use this combined indicator (weathering intensity, flow velocity, size of source area) for an evaluation of the environment of relict, inactive or active rockglaciers.

10.2.3 Rockglaciers as a Debris Transport System

Active rockglaciers are part of the alpine mass movement system. They transport unconsolidated material furnished by glaciers or rockfall processes across distances of 1–2 km. A conceptual model of the system is presented in Fig. 10.7.

The talus or the debris incorporated in active rockglaciers is slowly transported downslope or downvalley. In many cases, it will probably not reach a river in the near future and the debris is often too coarse to be easily transported by headwater streams (Wahrhaftig and Cox 1959). Occasionally, active rockglaciers move directly into the sea (Punt Hennequin, Admirality Bay, King George Islands, Antarctica; cf. Barsch et al. 1985) but this is unusual. Even though a climatic change to renewed glaciation would be required for the transport of rockglacier material from the mountains into their foreland, the present transport capacity of the system is of geomorphic interest. In part, this involves natural-hazard evaluation, because some active rockglaciers move over cliffs where rockfall presents a hazard to lower areas. It also involves the assesment of present geomorphic dynamics in high mountain systems.

In geomorphology, it is only possible to compare different processes if the amounts of material transported and the vertical and horizontal distances of transportation are used. Following Jäckli (1957; cf. the discussion in Rapp 1960a; Barsch 1969c, 1977d, 1993a; Caine 1976b), the mass transport is expressed as $m \cdot t \cdot a^{-1}$ (meters$\times$tonnes$\times$years^{-1}) with a differentiation between the vertical and the horizontal components (VMT, HMT). The first can be expressed in watts (VMT: $1\ m \cdot t \cdot a^{-1} = 311 \times 10^{-6}$ W).

A calculation for the Swiss Alps suggests the following (Barsch 1977d): horizontal mass transport (HMT) by active rockglacier,

$$400 \text{ to } 650 \times 10^6\ m \cdot t \cdot a^{-1};$$

vertical mass transport (VMT) by active rockglaciers,

$$120 \text{ to } 190 \times 10^6\ m \cdot t \cdot a^{-1}\ (= 40 \text{ to } 60 \times 10^3\ W)\ .$$

The ratio of these values to the unit of area (km^2) gives the specific transport rate (Barsch 1981a). In this case, the nonuniform distribution of rockglaciers in the Alps should be kept in mind (cf. Fig. 10.6). Rockglaciers are found in the higher parts of about half of the Swiss Alps (ca. 12000 km^2). If, therefore, the specific mass transport ($m \cdot t \cdot a^{-1} \cdot km^{-2}$; cf. Barsch 1981a) is calculated (for these 12000 km^2), the values are relatively small because only part

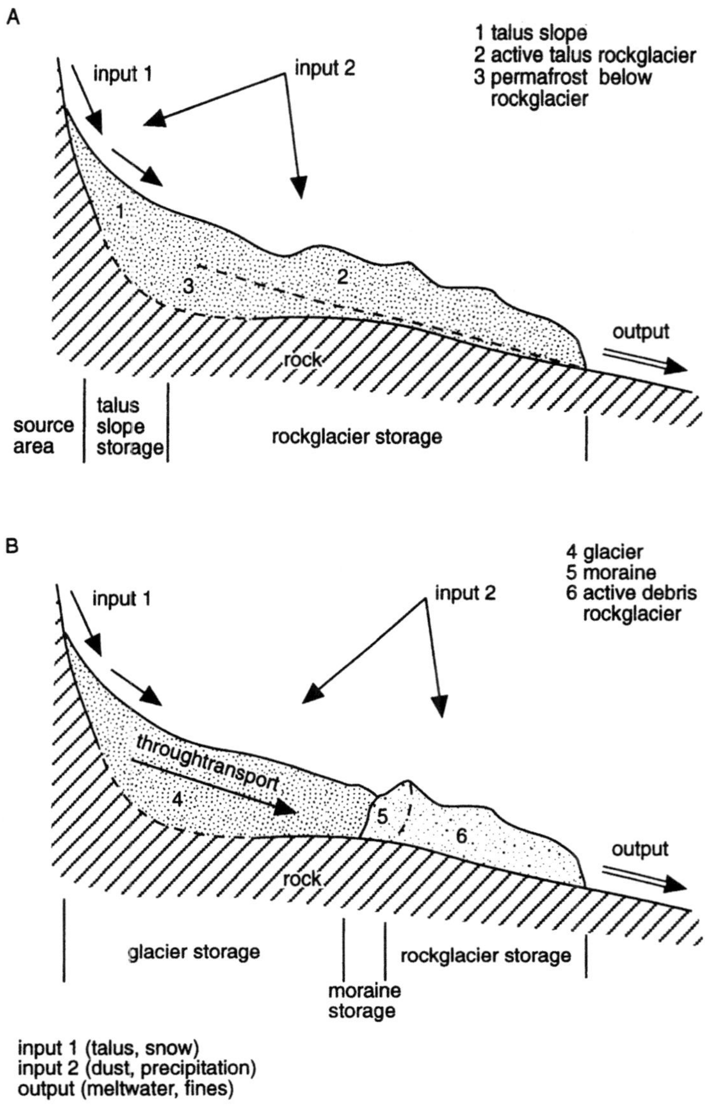

Fig. 10.7. Conceptual model of the rockglacier talus transport system (Modified from Vitek and Giardino 1987)

of the surface is influenced by rockglacier activity. Nevertheless, the values are as follows.

specific HMT, 34 to 54 $\times$ 10^3 $\text{m·t·a}^{-1}\text{·km}^{-2}$;

specific VMT, 10 to 16 $\times$ 10^3 $\text{m·t·a}^{-1}\text{·km}^{-2}$.

In a comparable area, like the Alpine basin of the Rhine South of the Bodensee (Lake Constance), Jäckli (1957) estimated the total transport by mass movements. From Jäckli's, data the specific mass transport by mass movements is calculated as follows:

$$\text{specific HMT}, 247 \times 10^3 \ \text{m}\cdot\text{t}\cdot\text{a}^{-1}\cdot\text{km}^{-2} \ ;$$

$$\text{specific VMT}, 78 \times 10^3 \ \text{m}\cdot\text{t}\cdot\text{a}^{-1}\cdot\text{km}^{-2} \ .$$

If Jäckli's data are typical of the more continental parts of the Swiss Alps, the specific mass transport by active rockglaciers amounts to ca. 15–20% of all mass movements in the Swiss Alps (Barsch 1977d). At first glance, these values seem high, but they have the same magnitude in other regions too. It may, therefore, be assumed that active rockglacier transport comprises about 20% of all mass movements in continental high mountain systems.

Buchenauer (1990) relates the mass transported by active rockglaciers exactly to the periglacial belt (between mean equilibrium line altitude, MELA = 2850 m a.s.l., and lower limit of solifluction, 2300 m a.s.l.), which comprises about 65 km^2 in his study area (eastern Tirolean Alps). He calculates debris transport by active rockglaciers to be 23–30 $\times 10^6$ m^3. Assuming an average rockglacier velocity of 20 cm a^{-1}, an average talus or debris density of 1.8 t m^{-3} and a slope of 15°, the specific mass transport can then be estimated to be:

$$\text{HMT}, 140 \ 10^3 \ \text{m}\cdot\text{t}\cdot\text{a}^{-1} \ \text{km}^{-2} \ ;$$

$$\text{VMT}, 40 \ 10^3 \ \text{m}\cdot\text{t}\cdot\text{a}^{-1} \ \text{km}^{-2} \ ;$$

This corresponds roughly with the regional values from Jäckli, which are about 1.8 to 2 times greater.

The importance of debris removal from the talus by active rockglaciers can be considered from another point of view. As talus rockglaciers transport material from the talus slopes they reduce the height of the talus allowing a larger area of rock wall to be exposed to weathering and a greater rate of talus production. Thus, active rockglaciers play a further important part in the talus transport system of high mountains, especially in continental and arid environments.

10.3 Rockglaciers and Climate

Rockglacier permafrost is dependent on a low energy input. This is normally indicated by low MAAT (<–2 °C). In addition, the distribution of active rockglaciers is also influenced by precipitation. For the Alps, Haeberli (1983a) proposes an annual precipitation of 2500 mm as an upper limit for the existence of rockglaciers. Thus, active rockglaciers are dependent on climate in more than one way.

10.3.1 Rockglaciers and Present Climate

Rockglaciers react to climatic changes either directly by thermal control on their rheology or indirectly through changes in talus or debris input. Direct thermal controls remain poorly understood, because the system response function is nonlinear (cf. Sect. 6.3.6, Fig. 6.18a): the internal energy fluxes are affected by conduction and advection (e.g., meltwater flows) and influenced by phase changes. As creep velocity is partly an indicator for the relationship between climate and rockglacier permafrost, continued high precision monitoring of rockglacier movement in Switzerland (Haeberli and Schmid 1988) will certainly present additional information in future.

Buchenauer (1990) reports that the inactive rockglaciers in the Schober group (eastern Tirol/Kärnten, Austria) have a lower limit that is 90–100 m below that of the active ones (cf. Table 5.5). This seems to be a value which is also true of the Swiss Alps (cf. Fig. 5.2). Because inactive rockglaciers are supposed to have been active during periods of the Holocene with lower temperatures such as the period 1804–1859, which caused the glacier advances of 1820 and 1856; cf. Bircher 1982 or Renner 1982; Table 10.9, Fig. 7.39), the difference in lower limits indicates the region of Holocene fluctuations of mountain permafrost belts. Assuming a gradient of 0.65 °C/100 m, this belt indicates an increase in MAAT from the mid-eighteenth century to today (1985) of about 0.65 °C. This value is reasonable if a time lag of about 30 years [during which the increased temperature (MAAT) has not yet produced a response in active rockglaciers] is accepted.

Table 10.9. Cold and warm summer periods in different millenia after 8000 B.P. The data are based on Renner (1892) cf. also Bircher (1892). Given are the number of years in continuous periods of more than 50 years with higher and lower tree-ring densities. Extreme values are marked. The data are related to summer temperatures and not to precipitation.

Millenium (ka B.P.)	Cold summer period	Warm summer periods
8–7	315	225
7–6	170	–
6–5	280	145
5–4	165	105
4–3	255	430
3–2	205	130
2–1	155	75
1–0	124	60
Σ	1669	1170

10.3.2 Relict Rockglaciers and Paleoclimate Reconstruction

The strong relationship between the climatic environment and active rockglaciers has been demonstrated. Active rockglaciers exist only in areas with alpine permafrost where MAAT is –2 °C or less. A continental climate and precipitation conditions, that are not favorable for glacier development (high radiation and/or limited precipitation) are further requirements for active rockglaciers. Thus, active rockglaciers are indicative of these environmental conditions at present, and relict rockglaciers are evidence of similar conditions at the time of their activity in their present location.

Early attempts to interpret relict rockglaciers paleoclimatically were made 20 years ago when Barsch and Updike (1971a, b) did studies at Kendrick Peak (northern Arizona). They point out that the climate during Wisconsinan times must have been colder, but not necessarily more humid than today. Later, Kerschner (1978, 1983a–c, 1985) presented the following paleoclimatic interpretations for the Tyrolean Alps:

1. Some of the relict rockglaciers have only been active during Egesen (Younger Dryas).
2. They became relict shortly after 10000 radiocarbon years B.P.
3. They are 500–600 m below present-day active rockglaciers; i.e., the contemporaneous lower limit of discontinuous mountain permafrost was 500–600 m below its present position, as well.
4. The climatic snowline (ELA) of Egesen Stade Glaciers was only about 180–220 m lower during the Younger Dryas than the mid-nineteenth century snowline (ELA around 1850), i.e., about 300 to 400 m compared with the present-day ELA (1980).
5. The timberline was about 300 to 400 m below the present one.

In addition, Kerschner assumes:

1. The depression of the lower limit of active rockglaciers during the Egesen stadial is associated with a lower MAAT. The present lower limit of active rockglaciers is at MAAT –2 °C.
2. The depression of the ELA is due to lower mean summer temperatures (T_S: June–September). The influence of precipitation is difficult to define. Because precipitation at the ELA is practically unknown for larger areas, precipitation at 2000 m a.s.l. is used as a substitute. Kerschner (1985) proposes the following equation:

$$P_{2000} = \frac{75^{(0.8\,T_S)} + 400}{1 + 0.015^{(0.8\,T_S)}}\,[\mathrm{mm}] \qquad (10.1)$$

$$\text{with: } 400 < P_{2000} < 5000\,[\mathrm{mm}]$$
$$\text{and: } T_S > 0^\circ\mathrm{c}$$

If no data on T_S are available, Kerschner (1985) uses the relationship proposed by Haeberli (1983a; cf. Fig. 10.11) for MAAT at ELA (T_S) and P_{2000}:

$$P_{2000} = \frac{16150^{(0.7\ T_E)} + 400}{1 + 3.23^{(0.7\ T_E)}} \text{[mm]} \qquad (10.2)$$

3. The lower timberline reflects the level of the 10 °C July isotherm, and, thus, of T_S.

For the climate of the Younger Dryas in the continental Tyrolean Alps, he concludes:

1. Mean summer temperatures were only 1–2 °C lower than today (= average 1931–1960).
2. Mean annual air temperatures (MAAT) have been 3–4 °C lower than today.
3. Hence, mean winter temperatures were roughly 5–6 °C lower than at present.
4. Precipitation (P_{2000}) was about 30% lower than today.

As a consequence, the climate of the Younger Dryas was much more continental, with more marked seasonal contrasts in temperature. In addition, the large differences between the lower limit of discontinuous mountain permafrost and the ELA can only be explained by a greatly reduced precipitation. Thus, the climate had to be much drier than it is today.

These results have since been confirmed by several others (Patzelt 1983; Buchenauer 1990). In particular Buchenauer presented sophisticated information on different late-glacial periods (Table 10.10). During the Older Dryas (Gaimberger–Alm Stade = younger Daun), the difference between the ELA and the lower limit of active rockglaciers was 432 m. About the same value was found for the Lienzer–Hütte Stade (Egesen = Younger Dryas). In relation to the present-day difference of only 235 m, this indicates a dry and continental climate. Summarizing, the distribution of relict rockglaciers in relation to ELA allows an evaluation of the late-glacial climate (Table 10.10). Younger and Older Dryas were both about 2 to 4 °C colder than the cold periods of the Holocene, with annual precipitation of only 50 to 70% of the present. The periglacial belt was much wider than at present and comprised roughly the area between 2000 and 2550 m a.s.l. during Younger Dryas, at present (1969/1985) 2612 to 2847 m a.s.l. It should be kept in mind that this reconstruction of the climate of Older and Younger Dryas says nothing about the interval between them. According to Maisch and Haeberli (1982), a marked climatic jump exists between the two periods; it is assumed that at the end of Daun (Older Dryas) that the climate was probably more humid which resulted in increased glacial activity.

Haeberli (1983a) expanded these possibilities into a discussion of the climate during the last glacial maximum (Würm) around the Alps, when the permafrost limit north of the Alps was about 3000 m lower than at present and ELA was at ca. 1000–1500 m (depression ca.1500 to 2000 m). Thus, in the

Table 10.10. Paleoclimatic reconstruction based on rockglaciers research for eastern Tirol (Austria) (After Buchenauer 1990)

	d/ELA (m)	d/loPF (m)	d/MAAT (°C)	MAAT/ELA (°C)	P_{2000} (mm)	R (%)
Present		0	0	–4.2	1070	100
1850[a]	0	< 150–180	1.0–1.2			
Lower Preboreal	Ca.20	250–380 (?)	2.3–2.5 (?)	–5.7/–5.9	620–650 (?)	60
	75	> 350–400	> 2.3–2.6	–5.3/–5.7	< 660–730	< 65
Younger Dryas[b]	120	> 450	> 3	–5.7	< 660	< 60
	200	550–600	3.6–3.9	–5.8/–6.1	600–650	< 60
Alleröd/ Bölling	< 110	?	?	?	?	?
Older Dryas	250	650–700	4.2–4.6	–0.6/–6.3	580–620	55
	280[c]					70
	420	650–700 (?)	4.2–4.6 (?)	–5.0/–5.4	750–800	< 70
	570[d]	800–?	5.2–?	–5.1/?	< 7000–780	?
	650[e]	?	?	?	?	

Present MELA (1969–1985) is situated at 2847 m a.s.l.
MELA around 1850 is calculated to 2746 m a.s.l.
Mean lower limit of active rockglaciers (1985) lies at 2612 m a.s.l.
The lower limit of discontinous mountain permafrost is assumed to be around 100 m below the lower limit of active rockglaciers.
The lower limit of present solifluction lies at ca. 2300 m a.s.l.
d/ELA, depression of the equilibirum line on glaciers against 1850;
d/loPF, depression of the lower limit of permafrost based on rockglaciers studies against the present situation (1985);
d/MAAT, depression of the mean annual air temperature;
MAAT/ELA, mean annual air temperature at ELA;
P_{2000}, precipitation at 2000 m a.s.l.;
R, relation of former to present preciption.
[a]1850 and similar Holocene climatic depressions during glacier advances.
[b]Egesen readvances.
[d]Daun maximum readvance.
[d]Gschnitz readvnace.
[e]Steinach readvance.

northern Alps, MAAT depression was about 15 °C, and mean annual temperature at ELA (T_E) is estimated to –10 to –15 °C. Precipitation was very low, probably only 30% or less than that at present.

Ballantyne and Kirkbride (1986; cf. Sect. 8.2.4; Fig 8.22) published an interesting map showing the altitudes of late-glacial (Younger Dryas, Loch Lomond Stadial) protalus ramparts in upland Britain. From this they reconstruct the altitude of the late-glacial snowline. However, as discussed in Section 8.2.4, the late-glacial protalus ramparts may be an indicator of discontinuous mountain permafrost. MAAT at these sites was –2 °C or lower. Given protalus ramparts in Baosbheinn and a temperature gradient of 6 °C km^{-1}, MAAT in the Cairngorms would have been around –5 °C at 950 m a.s.l., which suggests a situation close to continuous mountain permafrost in the Cairngorms during late-glacial times.

At this time the climate in Scotland must have been very continental. This is demonstrated by the existence of protalus ramparts formed during the Loch Lomond Stadial on St. Kilda at only 50–90 m a.s.l. The island has today an extremely maritime climate with a MAAT of around 8.5 °C (summer ca. 12.5 °C, winter ca. 5 °C, according to the nearby station at Scalpay). If the age of the protalus rampart is correct, the temperature difference in MAAT between the late-glacial and present-day climates must have been around 10 °C. This is a very great temperature depression for late-glacial times and it suggests a climate with very low winter temperatures (around –6.5 °C) and summer temperatures of ca. 2.5 °C.

The fossil rockglacier in the central Appenines (Maiella Mtns., Italy) at 43 °N is probably a Wurmian feature (Dramis and Kotarba 1992). If this is correct, it marks mountain permafrost near Rome for this time at ca. 2300 m a.s.l. and a last-glacial MAAT of –2 °C, too.

Lewis reports relict protalus ramparts from the Drakensberg (South Africa) from which he reconstructs the last-glacial snowline (ELA) at 2000 m a.s.l. If these features are really protalus ramparts, then they do not indicate snowline but discontinuous mountain permafrost. At the moment, a final decision is not possible because these features are questioned by Grab (pers. comm.); otherwise it would have been fascinating to discuss the consequences of the existence of discontinuous mountain permafrost during the last glaciation in South Africa at 2000 m a.s.l.

Heine (1994) evaluates rockglaciers in the Cordillera Neovolcanica in Mexico (Fig. 10.8). Talus and debris rockglaciers occur at different altitudes. The youngest ones are supposed to be inactive and are related to the Little Ice Age (350 to 100 a B.P). Older ones are from the period M IV (2000–3400 B.P.), the period M III (8500–10 000 B.P.) and M II (before 11 500 B.P.). There are three surprising facts:

1. The inactive rockglaciers dated as Little Ice Age were probably not formed during such a short period of only 200 to 300 years. It is more likely that this was only the last time of activity. Abnormally intense rock weathering would be required to produce the large amount of talus and debris necessary to produce them in only a few centuries.
2. There are no active rockglaciers reported, despite the fact that a discontinuous mountain permafrost belt exists at roughly between 4600 and 5000 m. Heine (1994) explains this by very low rates of talus production at present. It is difficult to accept this explanation. Perhaps it would be better

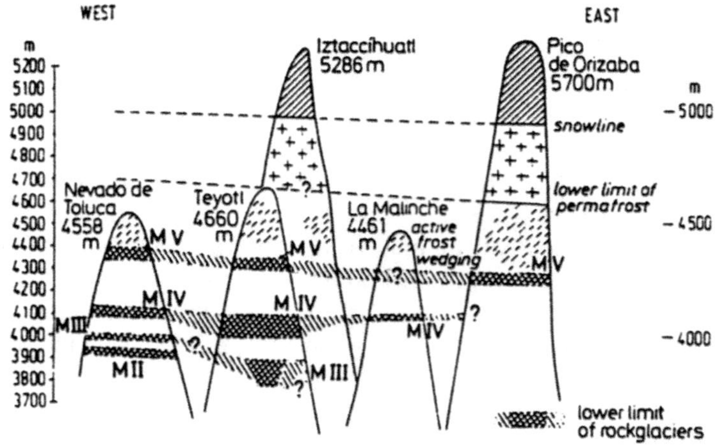

Fig. 10.8. Lower limit of rockglaciers during different periods in the Cordillera Neovolcanica (Mexico) *MII* Late glacial; *MIII* 10 000–9000 B.P.; *MIV* 3500–2000 B.P.; Little Ice Age. (Heine 1994)

to check if topographic controls prevent the formation of rockglaciers today.

3. The fossil rockglaciers are reported to be of Holocene age. This is difficult to accept since, in most other mountain systems on Earth, only one generation of rockglaciers (active and inactive) has been formed during the Holocene. There are two solutions to this problem:
   - The Holocene climatic history of Mexico is quite different to anywhere else; or
   - The fossil rockglaciers of the period M IV to M III are older than reported, i.e., of Pleistocene age.

Taking the Iztaccihuatl as an example, if the reconstruction by Heine (1994) is correct, then the MAAT for different altitudes and periods with different lower limits of discontinuous mountain permafrost can be calculated (Table 10.11). As a consequence, it can be said that the MAAT at the present lower limit of discontinuous permafrost (4600 m) was during M III (10 000 – 8500 B.P) nearly 7 °C lower than today. Such a temperature depression is found nowhere else in the world during the early Holocene.

10.3.3 Reactivation of Inactive or Relict Rockglaciers

Increased melting of permafrost leads to reduced movements of active rockglaciers and finally to their inactivity, as observed during periods of climatic warming like that since 1926. Similarly, climatic change after late-glacial times

Table 10.11. Mean annual air temperatures (MAAT) at different altitudes of Iztaccihuatl (Sierra Neovulcanico, Mexico) calculated according to information by Heine (1994). Temperatures are computed for the lower limit of mountain permafrost at different times

	MAAT (°C)				
	T(m a.s.l.)	At present	I.I.A.	MIV	M III
Permafrost of present	4600	−2.0	−2.8	−5.0	−6.8
Permafrost Little Ice Age (L.I.A.)	4300	+0.2	−2.0	−3.2	−5.0
Permafrost during stage MIV	4100	+1.4	−0.8	−2.0	−3.8
Permafrost during stage M III	3800	+3.2	+1.0	+0.2	−2.0

MAAT at the lower limit of permafrost is assumed to be −2 °C, lapse rate to be 0.6 °C/100 m.
MIV is 2000–3400 B.P., M III 8500–10 000 B.P.

(Daun, Egesen) led to many rockglaciers of late Pleistocene age becoming relict.

The possibility of an inactive, or even a fossil rockglacier, being reactivated during times of cooling with increasing permafrost has often been discussed on field trips. Answers remain speculative: an inactive rockglacier still contains a frozen core and so, if climate permits, meltwater from different sources may be added to it as ice. Thus, reactivation of an inactive rockglacier following climatic cooling should be possible. On the other hand, a relict rockglacier that has lost its entire ice content and often its finer-grained materials, seems unlikely to be reactivated, though the possibility cannot be definitely ruled out.

10.3.4 Rockglaciers and Climatic Change

In recent years, the discussion of global climate change has influenced all of the environmental sciences, as well as politics. The present warming period between 1850 and 1950 and especially after 1980 is documented in many mid-latitude mountains, although its cause, whether an anthropogenically increased greenhouse effect or a more intense heat transfer from lower to higher latitudes, remains open to debate. In any case, it is important to observe sensitive geosystems closely.

Up until now in high mountains, glaciers have been widely used as a general climatic indicator, even if their value in this regard is sometimes questionable. In Chapter 7 it was demonstrated that rockglaciers (depending on altitude and exposure) display velocity changes which are probably connected with climate and which lead to a climatic inactive status. In a more general view, Haeberli (1992b) discusses the effects of climate warming on mountain permafrost. Both discontinuous mountain permafrost in general and rockglacier permafrost exist at temperatures just below zero (–1 to –3 °C). A slight increase in MAAT, initiating a greater heat transfer into the soil would start a tendency to melt the ice in these permafrost bodies. Thus, a number of problems regarding stability (slides, bergsturzes, debris flows, etc.) would probably be triggered.

An increase in surface temperature of mountain permafrost bodies will be first found at the lower limit of active rockglaciers, where the unfrozen rockglacier mantle becomes thicker (more than ca. 10 m) and finally forms an obstacle to further creep (cf. Sect. 8.1.2; Barsch 1973). Thus, the rockglacier Val Sassa (cf. Table 7.5, 7.7), which is situated at only 2100 m a.s.l., became (climatic) inactive between ca. 1950 and 1970. Others, such as Murtèl I (cf. Table 7.10) have displayed a marked decrease in velocity since 1955, which can be explained as a consequence of warming. However, there remain a number of unresolved questions:

1. The heat flow into the ground and the subsequent melting of permafrost can be impeded by a thicker snow cover (Haeberli 1985a; cf. Sect. 6.2.2.5).
2. The reaction time of active rockglaciers is neither well known nor well understood, but appears to be relatively long (i.e., decades; cf. Sect. 7.1.5).
3. The reaction of rockglacier permafrost to increased temperature is not yet known. It could include any of the following:

   - A general increase in the thickness of the active layer (cf. Sect. 6.3),
   - An increase in the thickness of the unfrozen layer at the toe in summer to 10 m and more (cf. Sect. 8.1.2),
   - Transfer of heat deeper into rockglaciers by circulating meltwater.
   - Different creep rates (strain–temperature relation).

Information on these points is necessary for the development of physically based models of the future development of active rockglaciers under a warmer atmosphere. Even if much information is still missing, it seems obvious that active rockglaciers form a geosystem with a very high indicator value that could be used in environmental monitoring. The borehole measurements on the rockglacier Murtèl I (Haeberli 1989b) and the rockglacier observation program is Switzerland (Haeberli and Schmid 1988; Haeberli 1992b) present very valuable examples. They should be extended to a larger international project with examples in other mountain ranges under different climatic conditions, e.g., by the IPA working group on mountain permafrost.

It seems that a general warming will cause more active rockglaciers to become inactive. Thus, the belt of inactive rockglaciers, which comprises today in the Swiss Alps about 100 m, gives an indication of the Holocene fluctuations of the lower limit of discontinuous mountain permafrost. At present, it is not possible to define the highest Holocene limit of former inactive rockglaciers. This has to be a task of the future; to determine the possible former positions of the upper limit of former inactive rockglaciers.

10.4 Rockglaciers in the Alpine Hydrological Cycle

The alpine or mountain hydrological cycle is greatly influenced by a variety of storage features: winter snowcover, perennial or nearly perennial snow patches, glaciers and permafrost, in addition to the more usual storage as groundwater, soil moisture surface lakes, and reservoirs. The first set of these is dependent on climate, but not all in the same way. The nival or glacial regime of alpine rivers, in which the more obvious influences of snow and glacier ice melt are dealt with, are well known. The influence of alpine permafrost on the hydrological system has not received so much attention, apart from the work of Corte (1976b, 1978; cf. also Schrott 1993, 1995). Three areas of concern may be suggested:

1. How much water is presently stored in alpine permafrost or in active and inactive rockglaciers? (*Storage*)
2. What is the present outflow from these sources; alpine permafrost, active and inactive storage? (*Output*)
3. What changes occur at present, regarding the input/output ratio in alpine permafrost or active/inactive rockglaciers regarding short- and long-term fluctuations? (*Budget*)

A diagram developed for the semiarid Andes near Mendoza may illustrate the problem (Fig. 10.9).

10.4.1 Rockglaciers as a Water Store

Active and inactive rockglaciers are bodies of alpine (mountain) permafrost. Often more than 60% of the volume of active rockglaciers is ice. Thus, they are important water storage features. Very few studies on this topic have been performed. Even today, only first-order approximations are possible (Haeberli 1985a). Corte (1976b, 1978) tackles this problem hypothetically in relation to the water supply for the Rio and the city Mendoza. Unfortunately, not many data are available.

In the Swiss Alps, the total amount of water stored as ice in rockglacier permafrost is estimated to be 0.8 km^3 (Barsch 1977c; Table 10.12). Though this represents only 1.5% of the water stored in Swiss glaciers, the large amount of permafrost water can be of great importance in the water balance of high mountains. Perhaps the storage itself is not so important, but the changes due to the variability of the climate can be sensitive to the mountains and their surrounding lowlands. During warmer summers, more water will be released due to increased permafrost melting. During colder years, more water will especially be stored in rockglaciers. This has been observed at the rockglacier Murtèl I (Engadin, Swiss Alps), where the voids between the boulders in cold and rainy summers are filled to a higher level with ice, than during warm ones

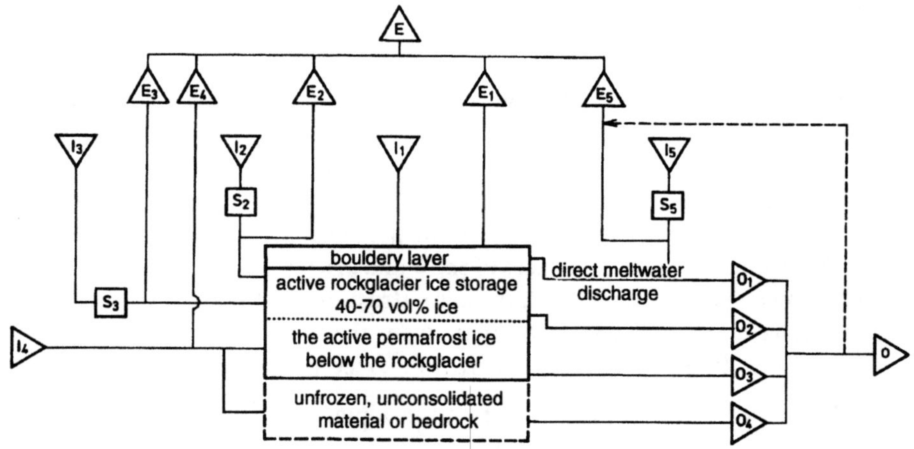

I1 direct precipitation on the rockglacier surface
I2 direct input from side slopes (as water flow or as avalanches)
I3 direct input from the head slopes (as water flow or as avalanches)
I4 meltwater input from glaciers, glacierets or snowfields upvalley from the rockglacier
I5 possible inputs below the rockglacier and above the gauging station
E_1- E_5 evaporation output
S_1- S_5 local storage
O1 output and throughput from the active layer
O2 output and throughput from the rockglacier permafrost
O3 output and throughput from the permafrost beneath the rockglacier
O4 output and throughput from the underlying, non-frozen layer

Fig. 10.9. Conceptual model of rockglacier hydrology

Table 10.12. Water storage in rockglacier permafrost in the Swiss Alps. (Barsch 1977d, 1977c)

Active rock-glaciers (n)	Volume of active rockglaciers (km^3)	Water[a] in active rock-glaciers (km^3)	Discontinuous[b] mountain permafrost (km^2)	Volume[c] of discontinuous mountain permafrost (km^3)	Water[d] in discontinuous mountain permafrost (km^3)
994	1.0–1.6	0.6–1.0	965	20	6

[a]The amount of water is assumed to be 60% of the total rockglacier volume.
[b]The belt between 2600 and 300 m a.s.l. comprises in the Swiss Alps about 3221 km^2. The area with discontinuous mountain permafrost is estimated to be about 30% of this belt.
[c]The average thickness of mountain permafrost is assumed to 20m.
[d]The water volume in this mountain permafrost is estimated to 30%.

Table 10.13. Water storage in mountain permafrost of the western Schober Mountains (eastern Tirol, Austria) in comparison to the water stored in present-day glaciers. (Buchenauer 1990)

Discontinous mountain permafrost	Total area (km^2)	Permafrost area (%)	Ø Thickness (m)	Ø Ice (%)	Ice volume ($10^6 m^3$)	Water ($10^6 m^3$) volume
Patchy	23	20–30	20[a]	30[a]	28–42	100–115
Areal	16	50	30[a]	30[a]	72	
Active						
Rockglaciers	1.3		35–45[c]	50[a]	23–30	21–27
	(2.2)[b]				(38–50)	(34–45)
Inactive						
Rockglaciers	1.2		30–35[c]	50[a]	18–21	16–19
	(1.5)[b]				(22–26)	(20–23)
Glaciers	2		5/10/15[d]		14	13
	(5.3)				(43)	(39)

[a]Estimates following Barsch (1977a–d).
[b]Rockglaciers of the eastern Schober Mountains from Lieb (1987).
[c]Estimates following Barsch (1977b, 1977 d).
[d]Estimates following Müller et al. (1976): for the glaciers Schober-, Horn- and W-Gradenkees and the glacier W Glödis und NE Roter Knopf the average thickness was assumed to be 10 m; for the Gößnitzkees, 15 m.
The values in parentheses refer to the eastern and western Schober Mountains combined.

(Barsch 1973). This is a negative feedback, which is quite different from the pattern found in lowland rivers where an increased input and a reduced evapotranspiration lead to higher runoff volumes. In areas of mountain permafrost, the same conditions lead to an increase in the amount of water storage, which may then be released later.

For the Schober Mountains (Austria), Buchenauer (1990; cf. Table 10.13) estimates, according to the method proposed by Barsch (1977a–d), the water volume in active rockglaciers to be $34–45 \times 10^6$ m^3. This corresponds to the water volume stored in the glaciers of this mountain group. If – according to Buchenauer – the water stored in mountain permafrost ($100–115 \times 10^6$ m^3 for the western Schober group only) and in inactive rockglaciers ($20–23 \times 10^6$ m^3) is added, the total water storage is considerable ($154–183 \times 10^6$ m^3).

In semiarid and continental dry mountains, this storage should be even more important. In these areas, the water volume stored in rockglacier permafrost is usually greater than that stored in glaciers. The balance of this storage, i.e., the release of permafrost water during warm dry periods and the increased storage during more humid, cooler decades can be of great geoecologic and economic importance (Schrott 1993, 1995).

10.4.2 Discharge from Rockglacier Permafrost

Water discharge from springs at the foot of active rockglaciers is generally estimated to be 5–250 l s^{-1} (Table 10.14); a relationship with the rockglacier surface is normally not discussed. The water itself can be derived from very different sources (Fig. 10.9). It can be the outflow generated by:

- Direct input of precipitation as rain,
- Seasonal storage of the previous winter's snow cover,
- Summer melting at the permafrost table of ice that has been stored longer,
- Through-flow over the permafrost table from higher sources,
- Meltwater circulation through or beneath the rockglacier permafrost.

Even though only a few studies have been done on rockglacier hydrology, some points are worth discussion:

1. Long-term discharge from active rockglaciers is always clear and does not contain the high suspended-sediment concentrations of glacial streams. Its temperature and chemical composition (electrical conductivity and pH) are very stable without daily fluctuations. Normal discharge is on the order of 8–151 s^{-1} or often less, in individual sources.

2. Mineralization of rockglacier water. The water of active rockglaciers has a higher ion content than glacier water (Harris et al. 1994). This has to be expected because it is permafrost water which generally has been in longer contact with minerals than water from glacier ice (Souchez and Lorrain 1991). The data from Harris et al. (1994) show that the pH values vary by less than 0.4 units. Temperature is normally below 2 °C.

The mineralization of rockglacier water is an indicator of the local situation (regarding rock types, climate, etc). Thus, solute content in the Andes of Mendoza is around 204 to 209 mg l^{-1} (Corte 1978) or 168– 171 mg/l in glacier meltwater. In the French Alps, the major ion concentrations in rockglacier water amount to only 77–92 mg l^{-1} (Evin 1993).

In the drainage from the Slim valley rockglacier, conductivity (as a measure of dissolved ions) amounts to 900 μS cm^{-1} during snowmelt (Harris et al.

Table 10.14. Discharge from rockglacier permafrost

Discharge (l/s)	Rockglacier	Area	Reference
5–20	Tungsten	N.W.T., Canada	Jackson and MacDonald (1980)
2–50	Gruben	Wallis, Swiss Alps	Haeberli (1985a)
90–270	Hilda	Rocky Mts., Alberta	Gardner and Bajewski (1987)
50–200	Slimvalley	St. Elias Mts., Yukon	Harris et al. (1994)

The Hilda rockglacier seems to be inactive (Luckman and Crocket 1978). According to Harris et al. (1994), it probably discharges not only the rockglacier permafrost but an additional groundwater spring.

1994). This is probably due to atmospheric deposition on the rockglacier in winter. During summer, conductivity decrease to 700–800 μS cm^{-1}.

3. Meltwater peak flow. First peak flow occurs at the beginning of snowmelt in early summer (Harris et al. 1994). It is of the same magnitude as storm responses discussed below. This observation is confirmed by Johnson (1978), Jackson and McDonald (1980), Evin and Assier (1983d), Haeberli and Patzelt (1983), Haeberli (1985a).

4. Storm discharge. Despite the "filtering effect" of a rockglacier (see above), it seems that a direct connection between summer storm precipitation and discharge exists. Both Johnson (1981) and Haeberli (1985a) report the direct influence of summer storms on rockglacier discharge. Most of the additional input is immediately released. Figure 10.10 shows that, during storm events, peak flow is more than four times higher than normal (ca. 3.5 compared to 0.8 l s^{-1}).

In contrast, Harris et al. (1994) suggest that the discharge responds only 10–14 days after storm events. If the precipitation occurs with air temperatures above 15 °C, runoff from the rockglacier is increased by additional melting due to the increased energy input. The direct storm response and the increased discharge after warm precipitation is easily understood. The first is the runoff over the permafrost table where the normal discharge often heard on rockglaciers has frequently been reported. Increased melting during and after warm rain is to be expected at the permafrost table; this presents the same effect as rain-on-snow events, it is the relationship between the heat of water to the latent heat of fusion. On the other hand, the long reaction time reported by Harris et al. (1994) after normal storms is astonishing. It can only be explained by an increased storage either due to freezing (e.g., below the boulders icicles are observed during or after early summer rains) or to infiltration. In both cases, the temperature and intensity of the precipitation have to fulfill the necessary boundary conditions. Regarding the time lag reported by Harris et al. (1994), flow velocity has to be about 10^{-4}m s^{-1}. This is only acceptable for intra-or subrockglacier permafrost flow (see below).

5. Summer melting has only been qualitatively observed. Tritium analysis of water samples from rockglacier springs suggests a mixed origin of the water, which also allows the conclusion that the water has been stored for some time in the rockglacier (cf. Haeberli 1985a, after data from Barsch). The same has been observed by Moore and Friedman (1991) in Colorado.

6. Water circulation on the permafrost table. The observed water flow during normal and storm runoff over the permafrost table (below the active layer) always seems to be fast. It is apparently not much delayed in relation to normal surface flow. Velocity is around 0.5 to 1.3 m s^{-1} (Johnson 1978), 0.01 to 0.1 m s^{-1} (Evin and Assier 1983d) or 0.06 m s^{-1} (Tenthorey 1992).

7. Water circulation through and below rockglacier permafrost has been tentatively assumed by Haeberli (1985a). This is based on the experience that permafrost contains (depending on temperature) up to 10 vol% of water (liquid state). There have also been experiences which demonstrate water flow

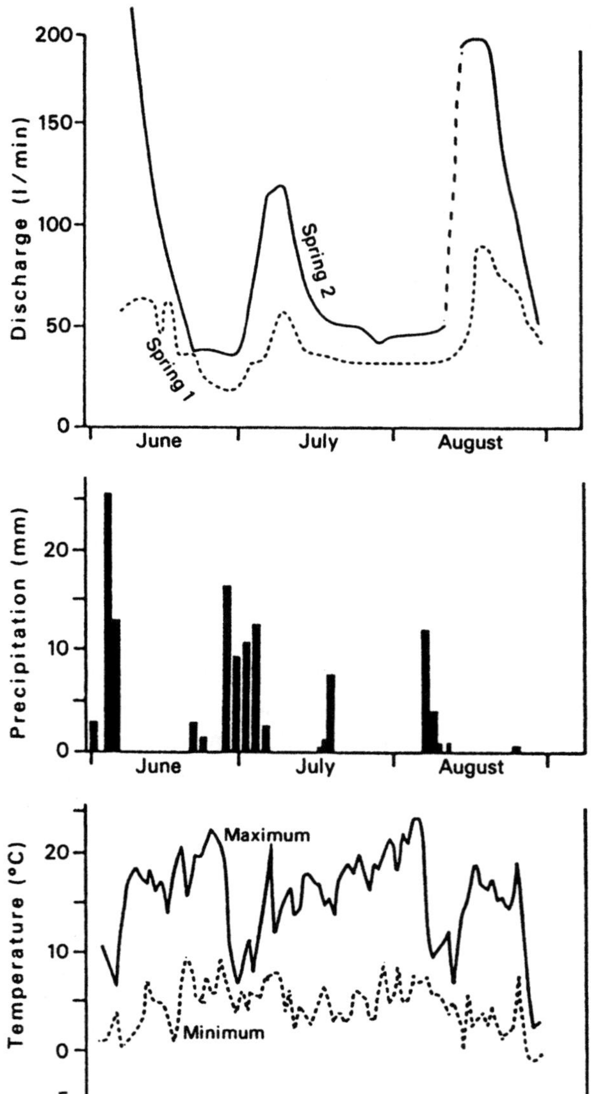

Fig. 10.10. Discharge from the rockglacier Slim river (St. Elias Mountains, Yukon) from (Harris et al. 1994). The discharge is measured at two springs. For comparison, precipitation and temperature are given

beneath or through active rockglaciers. For example, the rockglacier Macun I can be cited. The glacier above the rockglacier has been melting since ca. 1856. It was estimated that about 30 000 $m^3\ a^{-1}$ had to pass through or under this rockglacier.

Evin and Assier (1983d) report tracer experiments at the rockglacier Marinet (French Alps). The glacial meltwater disappeared into the rockglacier

3 to 20 h before reappearing at the rockglacier front. This corresponds to a flow velocity of 8 to 80×10^{-3} m s^{-1}, i.e., an order of magnitude lower than the water flow over the permafrost table.

The borehole through the rockglacier Murtèl I (Vonder Mühll 1993) demonstrated the existence of intra-rockglacier water flow at greater depth. Regarding distances and times, the travel velocites seem to be larger than the 10^{-3} to 10^{-4} m s^{-1} which Haeberli (1985) estimated for the Gruben rockglacier.

The velocity of flow below a rockglacier has been determined from tracer experiments. Fluorescene was introduced at the rooting zone of the rockglacier Becs de Bosson (Wallis) (Tenthorey 1992). The distance of 750 m was travelled in a time of between 1 month to less than 1 year. This corresponds to a velocity of 2.5×10^{-4} to 2.5×10^{-5} m s^{-1}, which are the best available for water flow below rockglacier permafrost and may be taken as characteristic of flows in that environment.

8. Discharge from inactive rockglaciers. Inactive rockglaciers may behave quite differently because they contain degrading permafrost and a thick unfrozen layer, at least over the toe. The study by Gardner and Bajewski (1987) displays results from the inactive Hilda rockglacier in the rocky Mountains of Alberta, which seems to have a higher discharge than active ones. This is unlikely and may be explained by the suggestion of Harris et al. (1994) that the Hilda rockglacier masks a groundwater spring. Therefore, it cannot be used as a good example for inactive rockglaciers.

10.4.3 Fluctuations in Rockglacier Permafrost Storage

The problem itself has to be discussed with climatic change and with the problems of melting of permafrost. Unfortunately, no long-term hydrological data on rockglaciers are available. Thus, an ergodic approach, based on the existence of inactive rockglaciers which have lost about half of the ice (i.e., ca. 30 to 35% of their volume) from their lower parts in the last 50 years, may be necessary. Even this would yield only very rough estimates at present because of the lack of good data on the water balance of different rockglacier systems.

10.5 Rockglaciers as Hazards in Alpine Environments

At a time when resource development in high-altitude alpine environments is very fast and when access to remote mountain parts is increasing from year to year, the problems of hazard management are considerable. Most hazards arise because engineers and planners are not aware of the geoecologic system they are working in. Thus, Giardino and Vick (1987) state, in one of the few papers that discuss the geomorphic hazard of rockglaciers, that the most important problem is to recognize a rockglacier in the field. If this has been done, the potential hazard is manageable. The next decision is related to the question of activity or inactivity. This is easily done by a geomorphologist acquainted with rockglaciers and using geophysical methods (Chap. 6).

On active rockglaciers, fixed installations, e.g., pillars, etc., should be avoided because of the risk of movement or subsidence due to melting. Other problems may arise in front of the rockglacier by overriding, by talus slides at the steep front slope or by boulder fall. Yet more widespread hazards arise where an active rockglacier moves over a cliff or onto a steeper slope. In this case its velocity increases, crevasses open and – if the stress exceeds strength thresholds – large-scale failure may occur, as can be expected at the rockglacier Hochebenkar near Obergurgl (Austria). There, the formation of crevasses due to the increased stress has been observed for several decades (cf. Chap. 6).

On inactive rockglaciers where there is no movement, subsidence may still occur as permafrost continues to degrade, not only by ice melting but also by compaction of the unconsolidated material and perhaps due to the loss of fine-grained sand and silt following internal erosion (piping). At the fronts of inactive rockglaciers, rockfall and slumping may increase because the front slope is steeper than the angle of repose and, therefore, unstable. Further, inactive rockglaciers may be reactivated during a cold phase in the future, and thus start moving again, which may be a hazard to nearby installations.

Active and inactive rockglaciers are sometimes used as sources for construction material. Normally, excavation stops when the frozen material is reached. Examples are quoted by Haeberli (1975) or Vitek and Giardino (1987). Another example can be seen from the road just south of the Flüela Pass (Graubünden, Swiss Alps). Here, hazards are associated with the created instabilities at the excavation sites.

Buchenauer (1990) reports an example of a catastrophic debris flow in the eastern Tirolean Schober Mountains (Austria). During an intense thunderstorm in the summer of 1986, fast runoff started on the permafrost table of an active rockglacier. This water (probably including additional meltwater from rockglacier permafrost) triggered the debris flow in unconsolidated materials below the rockglacier. Following the catastrophic events of the summer and early fall of 1987 in the Swiss Alps, it was found that huge debris flows were released from the lower part of the discontinuous mountain permafrost belt and even from rockglacier side slopes (Haeberli et al. 1990).

10.6 The Environment of Active Rockglaciers

According to the available information, active rockglaciers are limited to areas with mountain permafrost and with a certain amount of relief. Thus, they are phenomena of the subnival belt of Carl Troll, which is situated between the

ELA and the lower limit of perennial or quasi-perennial snow patches and which is part of the periglacial or cryogenic belt. The existence of the periglacial belt (between firnline and timberline in midlatitude mountains) is, according to Troll (1941, 1954, 1972, 1973; Barsch and Caine 1984), indicative for the definition of high mountains. Thus, active rockglaciers form a special subsystem in the hierarchy of high mountain systems.

Haeberli (1983a) discusses these belts in the Swiss Alps in relation to mean annual air temperature (MAAT) and mean annual precipitation, both of which influence the regional distribution of glaciers, mountain permafrost, and rockglaciers and, thus, the extension of the above-mentioned high mountain environment. The conceptual model is displayed in a schematic way in Fig. 10.11. The most important information is given by the equilibrium line altitude (ELA) on glaciers as a function of mean annual air temperature and mean annual precipitation. It separates the nival and the cryogenic belts. According to many data, the lower limit of discontinuous mountain permafrost is defined by a MAAT of −1 to −2 °C. In the diagram this is accepted as the lower limit of the cryogenic belt, which is correct for a thermal discussion, even if the landscape belt (cf. Fig. 10.1) may cut across this isotherm. The cryogenic and the nival belts are divided by the MAAT isotherm of −8 °C: at lower temperatures, continuous mountain permafrost is expected in the cryogenic belt of the Alps, and cold firn in the nival belt. At an annual precipitation of about 2250 mm, the equilibrium line cuts the −1 °C MAAT, i.e., there is no cryogenic belt in areas with higher precipitation. As is the case for many mountain systems, the Alps are characterized by more humid outer ranges and a more continental and drier interior. Thus, an area like Engelberg in the northern Alps has an annual precipitation of nearly 2000 mm, with a MAAT at the equilibrium line of −1 to −2.5 °C at an elevation of about 2600 m a.s.l. The vertical extension of the cryogenic belt is limited in this area and so discontinuous mountain permafrost and active rockglaciers are very rare. Perennial frozen ground is restricted to special situations in the nival belt such as horns, aretes, and rock walls above the glaciers. In contrast to this area, the Mischabel (Wallis) has an annual precipitation of less than 1000 mm, and the MAAT at the ELA (3000 m and more) is between −5 and −7 °C. In this area the potential vertical extension of discontinuous permafrost is around 770 m, which greatly favors rockglacier development. The calculation for the vertical extension of the cryogenic belt (H_c) is as follows:

$$H_c = \frac{(|T_E| - 1) \times 100}{T_G} \text{ [m]} \tag{10.3}$$

where T_E = MAAT at the ELA, T_G = air temperature gradient (assumed to be 0.65 °C/100 m).

The result (Fig. 10.11) demonstrates that rockglacier distribution is dependent on climatic parameters. It also shows that rockglaciers indicate a relatively continental, dry climate. They are indicative of intense rock weathering and of sites with cool microclimates and with low radiation receipts.

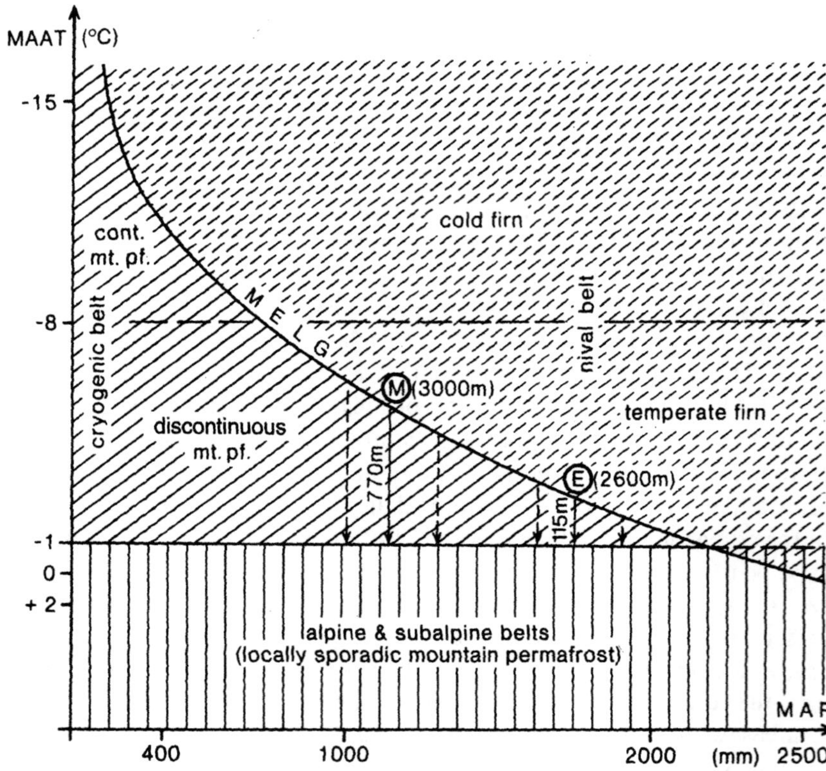

MAAT	Mean Annual Air Temperature	Ⓔ	Engelberg
MELG	Mean Equilibrium Line on Glaciers	Ⓜ	Mischabel
MAP	Mean Annual Precipitation		

Fig. 10.11. The diagram (after Haeberli 1983a) displays the relationship between the nival and the cryogenic belt as a function of temperature *(MAAT)* and precipitation *(MAP)* for the Swiss Alps; cf. text

Thus, patches of discontinuous permafrost around them, more or less intense seasonal and daily rockfall, a lack of surface water in late fall, a special hydrologic budget, and certain hazards with stable mountain permafrost (creep and rockfall at the front slopes) or with decaying mountain permafrost (debris flows and subsidence) may be expected in their environment. All construction problems connected with permafrost may occur in this belt. Thus, active and inactive rockglaciers can be used as a first indicator for the geoecologic potential of a high mountain environment. If development is planned in or near an area with rockglaciers, the recognition of their presence may help to direct further detailed studies to ensure a sustainable development. Knowledge about rockglacier environments is, thus, not "l'art pour l'art" but essential for the sensible use of high continental mountain systems.

11 Summary and Outstanding Problems

Active rockglaciers are the visible expression of the creep of mountain permafrost. Thus, they are truly periglacial features despite suggestions, still repeated, that they belong at least partly in the glacial realm. This view cannot be confirmed and adds only confusion to basic research. It is derived from a failure to differentiate between thermal conditions, material properties and rheology as defining characteristics. Permafrost, including mountain permafrost and its creep, is generally defined by its temperature; below 0 °C for more than 1 year (or two winters and the intervening summer). In contrast, glaciers are primarily defined by their movement and not by material properties (petrology and structure of the ice contained in them). If ice patches do not move they are called glacierets. Both, rockglaciers and glaciers, are, therefore, quite different features. The first are determined by thermal boundary conditions, the second by a surples of snow accumulation versus melting and sublimation of snow at the surface.

Since rockglaciers are landforms created by processes of slow mass movement, they display a great variability of forms, shapes, and sizes and this has often been used to define "genetic types". In this book, only two main classes of rockglaciers are used:

- the talus rockglacier,
- the debris rockglacier.

They are differentiated by the origin of the rock material transported in them: in other respects they are similar; they move by the same process of slow mass wasting at rates of 1 to 200 cm a^{-1}.

Active rockglaciers are a two-layered phenomenon. Below a blocky mantle (normally 2–5 m thick), they contain an ice-supersaturated core (normally 30–50 m thick) with mainly fine-grained material (sand and silt). The ice content is on average about 50–70% of the entire mass, though it may aproach 100% by volume locally. In the classification of Shumskii (1964), this was originally not sedimentary ice (vulgo: glacier ice) but mainly congelation ice, i.e., ice which has been either syngenetically formed below the surface as the rockglacier developed or added to the rockglacier core later by the refreezing of percolating water. The material contrast between the mantle and the core is not reflected in a temperature contrast: the active layer, i.e., the near-surface part which thaws in summer, may be thicker or thinner than the blocky mantle.

Despite a century of scientific research, many problems regarding rockglacier movement remain unresolved and require further studies:

1. There is little known of short-term and seasonal patterns of movement.
2. The same is true for the variations in vertical displacement of rockglaciers since almost all studies have resolved motion to the horizontal, or slope, plane only.
3. Although they are relatively common, long-term studies of horizontal surface displacements in active rockglaciers are usually not well enough documented or of sufficient duration to allow an empirical correlation with climatic changes or other environmental conditions. In some cases, the transition from activity to inactivity has been documented and related to climatic warming.
4. A systematic analysis of velocity variations along the flow lines of rockglaciers and the correspondence of these to changes in thickness are also not possible at the moment. Thus, the reported pulses in rockglacier movement, the observed harmonica-like movement of rockglaciers (e.g., rockglacier Gruben), and the changes in thickness of the rockglaciers Ölgruben (Sect. 7.1.3) or Äußeres Hochebenkar (Sect. 7.1.7) are not understood at all.
5. This leads to the still open questions of internal strain distribution in active rockglaciers, which should be related to surface structures, to thickening or thinning, and to movement in general.

These, and other, questions regarding the movement of active rockglaciers will be important for the development of better physically based, rheologic models of rockglacier creep. Answers to them would also contribute much to the theoretical discussion of rockglacier movement and history.

The lack of good data on the ages of both active and inactive rockglaciers has already been noted. Dating of the ice in different parts and different depths of active rockglaciers would be a critical test of models for rockglacier movement, e.g., in the interpretation of the rockglacier Murtèl I. It would also provide a more detailed model of rockglacier development during the Holocene and of the environment in which that development has occurred.

The hydrological importance of active rockglaciers in high mountains was suggested years ago by Corte but still remains poorly studied. In particular, long-term trends are not known, as is generally true for mountain permafrost despite its importance for many mountain systems. Further, the mass budgets (of both water and debris) of active rockglaciers remain insufficiently studied. This leads to unresolved questions regarding the dynamics of active rockglaciers, e.g., the reaction time to changes in input and climate, i.e., from active to dynamic or climatic inactive rockglaciers. The change from activity to climatic inactivity probably takes several decades, but has never been quantitatively defined for even a single case.

This is unfortunate because it hampers the development of possible uses that rockglaciers present for better indicators of discontinuous alpine permafrost and its future development, for estimation of rock wall retreat rates under different climatic conditions, and for paleoclimatic and geoecologic interpretation of former and present high mountain environments. Furthering the development of these potentials has been one of the main purposes of this book.

Although these problems (and others) remain open and unsolved, active, inactive, and relict rockglaciers indicate the present or former high mountain environment in the belt of their occurence. Properly used, they provide information on the geomorphic situation and the geoecologic potential in their neighborhood. A better knowledge of this geomorphic system will facilitate more detailed studies of these environments and may prevent poor decisions with regard to their future development.

References

Ackert R (1984) Ice-cored lateral moraines in Tarfala valley, Swedish Lapland. Geogr Ann 66A:79–88

Aguado C (1983) Comparación del inventario de glaciares de la cuenca del Río de los Patos con otros inventarios de los Andes Centrales de Argentina conénfasis en glaciares de escombros Acta Reunión Grupo Periglacial Republica Argentina 1:3–11

Ahumada AL (1992) Periglacial climatic conditions and vertical form associations in Quebrada Benjamin Matienzo, Mendoza, Argentina. Permafrost and Periglacial Processes 3:221–224

Alley NF, Young GK (1978) North central British Columbia, environmental significance of geomorphic processes. Ministery of Environment, Province of British Columbia, Victoria, RAB Bull 3:21–53

Anderson DM, Morgenstern NR (1973) Physics, chemistry, and mechanics of frozen ground: a review. In: 2nd Int Conf on Permafrost, North American Contribution, Natl Acad Sci, Washington DC, pp 257–288

Anderson LW, Anderson DS (1981) Weathering rinds on quartz-arenite clasts as a relative age indicator and the glacial chronology of Mount Timpanogos Wasatch Range, Utah. Arct Alp Res 13:25–31

André MF (1989) Vitesses d'accumulation des debris rocheux au pied des parois supraglaciaires du nord-ouest du Spitsberg. Z Geomorphol 32:351–373

André MF (1992) Les glaciers rocheux du Spitsberg central et nord-occidental. Rev Géomorphol Dyn 41:47–63

André M-F (1993) Les versants du Spitsberg. Presse Univ Nancy, 361 pp

André M-F (1994) Rock glaciers in Svalbard. Geogr Ann 76A:235–245

Andrews JT, Williams LD, Carrara PE (1976) Ecological overview II: the geomorphological overview. In: Steinhoff HW, Ives JD (eds) Ecological impact of snowpack augmentation in the San Juan Mountains of Colorado, Final Report. Colorado State Univ, Fort Collins, pp 87–104

Angely AG (1967) Anciens glaciers rocheux dans l'Est des Pyrenées centrales. Rev Geogr Pyrenées et du Sud-Ouest 38:5–28

Angenheister G (1980) Structures and movements of the rock masses under the Eastern Alps. Eclogae Geol Helv 2:399–424

Annan AP, Davis JL (1976) Impulse radar sounding in permafrost. Radio Sci 11:383–394

Annan AP, Davis JL (1978) High frequency electrical methods for the detection of freeze-thaw interfaces. In: 3rd Int Conf on Permafrost, Natl Res Council Can, Ottawa, pp 496–500

Antevs E (1932) Alpine zone of Mt. Washington Range. Auburn, Maine, 118 pp

Anton Burgos FJ (1982) Glacial and nivo-glacial morphology in the Iberian System (Spain): morphogenesis and deposits. Reunión Regional Sudamericana INQUA, Univ Nac Comahue, pp 3–5
Arpat E, Oezguel N (1972) Rock glaciers in Geyik Dagi area, central Taurus. Maden Tetkik ve Arama Enstitüsü, Ankara, 78:28–32
Assier A (1981) Glaciers et glaciers rocheux du Queyras. Paris, 250 pp
Assier A (1992) Evolution des marges proglaciaires en conditions de pergélisol, Vallon de Grizzly Creek (Yukon, Canada). Commission pour l'étude des phénomènes périglaciaires, Réunion annuelle de la Commission à Fontainebleau-Avon, 18 et 19 Janvier 92
Atwood WW, Mather KF (1932) Physiogeography and quaternary geology of the San Juan Mountains, Colorado. US Geol Surv Prof Pap 166:162–163
Azizi F (1989) Primary creep of polycrystalline ice under constant stress. Cold Regions Sci Technol 16:159–165
Bajewski I, Gardner JS (1989) Discharge and sediment-load characteristics of Hilda rock-glacier stream, Canadian Rocky Mountains, Alberta. Phys Geogr 10:295–306
Baker D, Escher-Vetter H, Moser H, Oerter H, Reinwarth O (1982) A glacier discharge model based on results from field studies of energy balance, water storage and flow. In: Hydrological aspects of alpine and high mountain areas. Int Assoc Hydrol Sci Publ 138:103–112
Ballantyne CK (1984) The late Devensian periglaciation of upland Scotland. Quat Sci Rev 3:311–343
Ballantyne CK (1986) Protalus rampart development and the limits of former glaciers in the vicinity of Baosbheinn, Western Ross. Scot J Geol 22:13–25
Ballantyne CK (1987) Some observations on the morphology and sedimentology of two active protalus ramparts, Lyngen, northern Norway. Arct Alp Res 19:167–174
Ballantyne CK, Benn DI (1994) Glaciological constraints on protalus rampart development. Permafrost and Periglacial Processes 5:145–153
Ballantyne CK, Kirkbride MP (1986) The characteristics and significance of some lateglacial protalus ramparts in upland Britain. Earth Surface Processes and Landforms 11:659–671
Ballantyne CK, Kirkbride MP (1987) Rockfall activity in upland Britain during the Loch Lomond Stadial. Geogr J 153:86–92
Barnes D (1965) Geophysical methods for delineating permafrost. In: Proc 1st Int Conf on Permafrost, Natl Acad Sci, Washington DC, pp 349–355
Barrere P (1952) Le relief des massifs granitiques du Néouvielle, de Cauterets et de Panticosa. Rev Géogr Pyrenées et du Sud-Ouest 23:69–98
Barsch D (1969a) Studien und Messungen an Blockgletschern in Macun, Unterengadin. In: Kaiser K (ed) Glazialmorphologie. Z Geomorphol NF (suppl) 8:11–30
Barsch D (1969b) Permafrost in der oberen subnivalen Stufe der Alpen. Geogr Helv 24:10–12
Barsch D (1969c) Studien zur Geomorphogenese des zentralen Berner Juras. Basler Beitr Geogr 9:221
Barsch D (1970) Fossile Blockgletscher als Klimaindikatoren der letzten Kaltzeit. Eiszeitalter Gegenwart 21:194
Barsch D (1971a) Neuere Untersuchungen an Blockgletschern in den Schweizer Alpen. Verh Schweiz Naturforsch Ges 151:122
Barsch D (1971b) Rock glaciers and ice-cored moraines. Geogr Ann 53A:203–206

Barsch D (1973) Refraktionsseismische Bestimmung der Obergrenze des gefrorenen Schuttkörpers in verschiedenen Blockgletschern Graubündens, Schweizer Alpen. Z Gletscherk Glazialgeol 9:143–167

Barsch D (1977a) Nature and importance of mass-wasting by rock glaciers in alpine permafrost environments. Earth Surface Processes 2:231–245

Barsch D (1977b) Ein Permafrostprofil aus Graubünden, Schweizer Alpen. Z Geomorphol NF 21:79–86

Barsch D (1977c) Alpiner Permafrost: ein Beitrag zur Verbreitung, zum Charakter und zur Ökologie am Beispiel der Schweizer Alpen. In: Poser H (ed) Formen, Formengesellschaften und Untergrenzen in den heutigen periglazialen Höhenstufen der Hochgebirge Europas und Afrikas zwischen Arktis und Äquator. Akad Wiss Gött Math Phys Kl Folge 3, 31:118–141

Barsch D (1977d) Eine Abschätzung von Schuttproduktion und Schutttransport im Bereich aktiver Blockgletscher der Schweizer Alpen. In: Wirthmann A (ed) Hangformen und Hangprozesse. Z Geomorphol Suppl 28:148–160

Barsch D (1977e) Das geomorphologische Symposium der Akademie der Wissenschaften in Göttingen vom 19.-23.9.76: Ein Beitrag zu einer ökologischen Geomorphologie der rezenten Periglazialgebiete Europas und Afrikas. Z Geomorphol NF 21:223–227

Barsch D (1978) Active rockglaciers as indicators of discontinuous permafrost: an example from the Swiss Alps. In: 3rd Int Conf on Permafrost, Natl Res Cen, Natl Res Council Can, Ottawa, 349–352

Barsch D (1980) Die Beziehung zwischen der Schneegrenze und der Untergrenze der aktiven Blockgletscher. In: Jentsch C, Liedtke H (eds) Höhengrenzen in Hochgebirgen. Arb Geogr Inst Univ Saarlandes 29:119–133

Barsch D (1981a) Studien zur gegenwärtigen Geomorphodynamik im Bereich der Oobloyah Bay, N-Ellesmere Island, N.W.T., Kanada. Heidelb Geogr Arbe 69:123–162

Barsch D (1981b) Terrassen, Fluß arbeit und das Modell der exzessiven Talbildungszone im Expeditionsgebiet Oobloyah Bay, N-Ellesmere Island, N.W.T., Kanada. Heidelb Geogr Arb 69:269–284

Barsch D (1983a) Blockgletcher-Studien, Zusammenfassung und offene Probleme. In: Poser H, Schunke E (eds) Mesoformen des Reliefs im heutigen Periglazialraum. Abh Akad Wiss Gött Math Phys Kl Folge 3, 35:133–150

Barsch D (1983b) Problems of the delimination of geomorphic belts on the east slopes of the Cordillera Frontal. Acta Criogénica 1 (Mendoza):18–21

Barsch D (1984) Geomorphologische Untersuchungen zum periglazialen Milieu polarer Geosysteme. In: Stäblein G (ed) Grundvorstellungen und Entwicklungen in der Geomorphologie. Z Geomorphol Suppl 50:107–116

Barsch D (1987a) Rock Glaciers: an approach to their systematics. In: Giardino JR, Shroder JF, Vitek JD (eds) Rock glaciers. Allen & Unwin, London pp 41–44

Barsch D (1987b) The problem of the ice-cored rock glacier. In: Giardino JR, Shroder JF, Vitek JD (eds) Rock glaciers. Allen & Unwin, London pp 45–53

Barsch D (1988) Rockglaciers. In: Clark MJ (ed) Advances in periglacial geomorphology. Wiley, Chichester, pp 69–90

Barsch D (1992) Permafrost creep and rockglaciers. Permafrost and Periglacial Processes 3:175–188

Barsch D (1993a) Periglacial Geomorphology in the 21st century. Binghamton Symp. Geomorphology 7:141–163

Barsch D (1993b) Schneehaldenmoränen (Protalus Ramparts). Ein falsches Modell behindert die paläoklimatische Deutung. Würzb Geogr Arb 87:257–267

Barsch D, Caine N (1984) The nature of mountain geomorphology. Mountain Res Dev 4:287–298

Barsch D, Happoldt H (1984) Blockgletscherbildung und holozäne Höhenstufengliederung in den mendozinischen Anden, Argentinien. Zentralbl Geol Paläontol Teil 1:1625–1632

Barsch D, Hell G (1975) Photogrammetrische Bewegungsmessungen am Blockgletscher Murtèl I Oberengadin, Schweizer Alpen. Z Gletscherk Glazialgeol 11:111–142

Barsch D, Jakob M (1993) Active rockglaciers and the lower limit of discontinuous alpine permafrost in the Khumbu Himalaya, Nepal. Proc 6th Int Conf on Permafrost, Beijing, 5–9 July 1993, 1:17–31

Barsch D, King L (1975) An attempt to date fossil rock glaciers in Grison, Swiss Alps. Quaest Geogr 2:5–14

Barsch D, King L (1989) Origin and geoelectrical resistivity of rockglaciers in semi-arid subtropical mountains (Andes of Mendoza, Argentina). Z Geomorphol NF 33:151–163

Barsch D, Mäusbacher R (1986) Beiträge zur Vergletscherungsgeschichte und zur Reliefentwicklung der Südshetland Inseln. Z Geomorphol NF Suppl 61:25–37

Barsch D, Stäblein G (1978) EDV-gerechter Symbolschlüssel für die geomorphologische Detailaufnahme. In: Stäblein G (ed) Geomorphologische Detailaufnahme. Berl Geogr Abh 30:63–78

Barsch D, Treter U (1976) Zur Verbreitung von Periglazialphänomenen in Rondane/Norwegen. Geogr Ann 58A:83–93

Barsch D, Updike RG (1971a) Late pleistocene periglacial geomorphology (rock glaciers and blockfields) at Kendrick Peak, northern Arizona. Ariz Geol Soc Digest 9:225–243

Barsch D, Updike RG (1971b) Periglaziale Formung am Kendrick Peak in Nord-Arizona während der letzten Kalzeit. Geogr Helv 26:99–114

Barsch D, Zick W (1991) Die Bewegungen des Blockgletschers Macun I von 1965–1988 (Unterengadin, Graubünden, Schweiz). Z Geomorphol NF 35:9–14

Barsch D, Fierz H, Haeberli W (1979) Shallow core drilling and bore-hole measurements in permafrost of an active rock glacier near the Grubengletscher, Wallis, Swiss Alps. Arct Alp Res 11:215–228

Barsch D, Blümel WD, Flügel WA, Mäusbacher R, Stäblein G, Zick W (1985) Untersuchungen zum Periglazial auf der König-Georg-Insel, Südshetlandinseln, Antarktika. Ber Polarforsch 24:75

Behre CH (1933) Talus behavior above timber in the Rocky Mountains. J Geol 41:622–625

Belloni S, Pelfini M, Smiraglia C (1988) Morphological features of the active rock glaciers in the Italian Alps and climatic correlations. In: 5th Int Conf Permafrost in Trondheim, Norway, Proc 1:678–682

Benedict JB (1968) Recent glacial history of an alpine area in the Colorado Front Range, U.S.A: II. Dating the glacial deposits. J Glaciol 7:77–87

Benedict JB (1973) The origin of rock glaciers. J Glaciol 12:520–522

Benedict JB, Benedict RJ, Sanville D (1986) Arapaho rock glacier, Front Range, Colorado, U.S.A: a 25-year resurvey. Arct Alp Res 18:349–352

Bennet LP, French HM (1991) Solifluction and the role of permafrost creep, eastern Melville Island, NWT, Canada. Permafrost Periglacial Processes 2:95–102

Besancon J (1978) Remarques sur le périglaciaire du Liban. In: Colloq sur le periglaciaire d'altitude du domaine mediterraneen et abords, Strasbourg, May 12–14, 1977, Assoc Géol Alsace, pp 231–241

Beschel RE (1950) Flechten als Altersmaßstab rezenter Moränen. Z Gletscherkd Glazialgeol 1:152–161, see also Arct Alp Res 5:303–309

Beschel RE (1957) Lichenometrie im Gletschervorfeld. Jahrb Schutze Alpenpflanz Tiere 22:164–185

Beschel RE (1961) Dating rock surfaces by lichen growth and its application to glaciology and physiography (lichenometry). In: Raasch GO (ed) Geology of the Arctic. 1st Int Symp on Arctic geology, Proc 2, Calgary 1960, Univ Toronto, Toronto, pp 1044–1062

Bindschadler R, Harrison WD, Raymond CF, Crosson R (1977) Geometry and dynamics of a surge-type glacier. J Glaciol 18:181–194

Binggeli V (1965) Der Blockstrom im Val Cadlimo. Regio Basiliensis 6:149–154

Bircher W (1982) Zur Gletscher- und Klimageschichte des Saastales. Physische Geogr Zürich 9:233

Birkeland PW (1973) Use of relative age-dating methods in a stratigraphic study of rock glacier deposits, Mt Sopris, Colorado. Arct Alp Res 5:404–416

Birkeland PW (1974) Rock glaciers. Off Belay 17:2–10

Birkeland PW (1982) Subdivision of Holocene glacial deposits, Ben Ohau Range, New Zealand, using relative-dating methods. Geol Soc Am Bull 93:433–449

Birkenmajer K (1982) Talus moraines in South Spitsbergen and comparison with East Greenland. Bull Acad Polonaise Sci Sér Sci Chim Géol Géogr 7:29–38

Birman JH (1964) Glacial geology across the crest of the Sierra Nevada, California. Geol Soc Am Spec Pap 75:80

Birnie R, Thom G (1982) Preliminary observations on two rock glaciers in South Georgia Islands Dependencies. J Glaciol 28:377–386

Blagbrough JW (1967) Cenozoic geology of the Chuska Mountains. In: 18th Field Conference. New Mexico Geol Soc, pp 70–77

Blagbrough JW (1976) Rock glaciers in the Capitan Mountains, south-central New Mexico. Geol Soc Am Abstr Progr 8:570–571

Blagbrough JW (1984) Fossil rock glaciers on Carrizo Mountain, Lincoln County, New Mexico. N M Geol 6:65–68

Blagbrough JW, Breed WJ (1967) Protalus ramparts on Navajo Mountain, southern Utah. Am J Sci 265:759–772

Blagbrough JW, Breed WJ (1969) A periglacial amphitheater on the northeast side of Navajo Mountain, southern Utah. Plateau 42:20–26

Blagbrough JW, Brown HG III (1983) Rock glaciers on the west slope of South Baldy, Magdalena Mountains, Socorro County, New Mexico. N M Geol Soc Guidebook, 34th Field Conf, Socorro Region II:299–303

Blagbrough JW, Farkas SE (1968) Rock glaciers in the San Mateo Mountains, south-central New Mexico. Am J Sci 266:812–823

Blake W (1977) Holocene glacier fluctuations and sea level changes in arctic Canada: a reply. Geogr Ann 39A:257–260

Blankenhorn M (1896) Theorie der Bewegung des Erdbodens. Z Dtsch Geol Ges 48:382–400

Blumstengel WK (1986) Case study of an active rock glacier, Slims River valley, Kluane National Park, Yukon Territory. Ann Meet Can Assoc of Geographers, Programs and Abstract, 57 pp

Blumstengel W, Harris SA (1988) Observations on an active lobate rock glacier, Slims River Valley, St. Elias Range, Canada. In: 5th Int Conf on Permafrost in Trondheim, Norway, Proc 1:689–694

Bodere JC, Moign A, Simon T, Carre F (1982) Cronique arctique: Les glaciers rocheux. Norois 29:258

Boesch H (1951) Beiträge zur Kenntnis der Blockströme. Die Alpen 27:1–5

Bogrodoskiy VW, Trepov GV, Federov BA, Khokhlov GP (1978) Radar probing of ice and the possibility of the sounding of permafrost. In: 2nd Int Conf on Permafrost, USSR Contrib, Natl Acad Sci, Washington DC, pp 488–491

Boulton GS (1987) Progress in glaciology during the last fifty years. J Glaciol Spec Issue:25–32

Boulton GS, Jones AS (1979) Stability of temperate ice caps and ice sheets resting on beds of deformable sediments. J Glaciol 23:464–466

Boulton GS, Dickson JH, Nichols H, Nichols M, Short SK (1976) Late Holocene glacier fluctuations and vegetation changes at Maktak Fiord, Baffin Island. N.W.T., Canada. Arct Alp Res 8:343–356

Braun U (1969) Der Felsenberg im Odenwald. Heidelb Geogr Arb 26:175

Broscoe AJ, Clissold RJ (1967) Rock glaciers, Maligne Range, Jasper National Park, Alberta. Master's Thesis, Univ Alberta, Canada

Brown WH (1925) A probable fossil glacier. J Geol 33:464–466

Brown RJE, Pewe TL (1973) Distribution of permafrost in North America and its relationship to the environment: a review 1963–1973. In: 2nd Int Conf on Permafrost, N Am Contrib, Natl Acad Sci, Washington DC, pp 71–100

Brukhanda VI (1976) Glacial pulsations and their interaction with the rock glaciers of Kavkaz and Pamir-Altai. Res Materials Glaciers, Chronicle, Discussions 27:63–70

Bryant B (1971) Movement measurements on two rock glaciers in the Eastern Elk Mountains, Colorado. US Geol Surv Prof Pap 750-B:108–116

Buchenauer HW (1990) Gletscher- und Blockgletschergeschichte der westlichen Schobergruppe (Osttirol). Marb Geogr Schr 117:376

Buk EM (1983a) Sobre las condiciones hidrológicas del glaciar de escombros El Salto, Potrerillos Mendoza. Inst Argentino de Nivologia y Glaciologia, Mendoza, Informe inédito, 40 pp

Buk EM (1983b) Glaciares de escombros y su significación hidrológica. In: Actas 1 ra Reunión del grupo periglacial Argentino. IANIGLIA-CONICET 83:22–28

Busch G (1982) Zur heutigen Kenntnis der Verbreitung und Entstehung von Blockgletschern in den Alpen. Wiss Arbeit im Rahmen der 1. Staatsprüfung für das Lehramt Sekt II, Bonn, 110 pp

Cailleux A (1947) Charactères distinctifs des coulèes de blocailles liées au gel intense. C R Somm Séances Soc Géol Fr 16:323–325

Cailleux A, Taylor G (1954) Cryopédologie: Etude des sols gelées. Exp Pol Fr 4:219

Caine N (1968a) The blockfields on northeastern Tasmania. Aust Natl Univ, Canberra, Res School of Pacific Studies, Dep Geogr, Publ G6:127

Caine N (1968b) The fabric of periglacial blockfield material on Mt. Barrow, Tasmania. Geogr Ann 50A:193–206

Caine N (1969) A model for alpine talus slope development by slush avalanching. J Geol 77:92–100
Caine N (1971) A conceptual model for alpine slope process study. Arct Alp Res 3:319–329
Caine N (1972) Air photo analysis of blockfield fabrics in Talus Valley, Tasmania. J Sediment Petrol 42:33–48
Caine N (1974) The geomorphic processes of the alpine environment. In: Ives JD, Barry RG (ed) Arctic and alpine environments. Methuen, London, pp 721–748
Caine N (1976a) A model for alpine talus development by slush avalanching. In: King CAM (ed) Periglacial processes. Methuen, London, pp 354–360
Caine N (1976b) A uniform measure of subaerial erosion. Geol Soc Am Bull 87:137–140
Caine N (1978) Climatic geomorphology in mid-latitude mountains. Davies JL, Williams MAJ (eds) Landform evolution in Australasia. Aust Natl Univ Press, Canberra, pp 113–127
Caine N (1986) Sediment movement and storage on alpine slopes in the Colorado Rocky Mountains. In: Abrahams A (ed) Hillslope processes. Allen, Unwin, London, pp 115–137
Caine N, Jennings JN (1968) Some blockstreams of the Toolong Range Kosciusko State Park, New South Wales. R Soc New South Wales, J Proc 101:93–103
Calkin PE, Haworth LA, Ellis JM (1987) Rock glaciers of central Brooks Range, Alaska, U.S.A. In: Giardino JR, Shroder JF, Vitek JD (eds) Rock glaciers. Allen & Unwin, London, pp 65–82
Capello CF (1947) Le "pietraie semoventi" delle Alpi occidentali. Natura 38:17–23
Capello CF (1959) Prime richerche sulle "pietraie semoventi" del settore montuoso del Gran Paradiso. Club Alpino Italiano Rivista mensile 78:294–300, 371–376
Capello CF (1963) Le morfologie crionivali (periglaciali) nelle Alpi Graie meridionali italiani. Publ Ist Geogr Alp 3:126
Capello CF (1969) Le morfologie tardo-glaciali e crionivali della val Maira (Piemonte). Stud Morfol Crionivali Turin 2:130
Capps SR (1910) Rock glaciers in Alaska. J Geol 18:359–375
Capps SR (1919) The Cantishna region, Alaska. US Geol Surv Bull 687:1–116
Carrara PE (1975) Transition from shear moraines to rock glaciers. J Glaciol 12:149
Carrara PE (1979) Rock Glaciers – some comments about age. J Glaciol 22:411
Carrara PE, Andrews JT (1973) Problems and applications of lichenometry to geomorphic studies, San Juan Mountains, Colorado. Arct Alp Res 5:373–384
Carrara PE, Andrews JT (1975) Holocene glacial/periglacial record, Northern San Juan Mountains, Southwestern Colorado. Z Gletscherkd Glazialgeol 11:155–174
Carton A, Dramis F, Smiraglia C (1988) A first approach to the systematic study of the rock glaciers in the Italian Alps. In: 5th Int Conf on permafrost in Trondheim, Norway, Proc 1:712–717
Casati P (1988) I Ghiacciai di pietre (rock glaciers). Boll Annu Com Sci Cent Club Alp Italiano 87:30–33
Catalano LR (1923) Contribucion al conocimiento de los fenomenos geofisicos atmosfericos. Dcion Gral de Minas, Geol Hidrol 24:78 pp
Cazenave-Piarrot F, Tihay JP (1983) Eboulis, formations morainiques et glaciers rocheux dans le Massif de l'Ardidien (Pyrenées Centrales). In: Actes du Colloque tenu le 8 Janvier 1983 à Paris lors de la séance de l'Association de Géographes Francais pp 121–135

Cazenave-Piarrot F, Tihay JP (1985) Glaciers rocheux dans les pyrenées occidentales et centrales. Notes C R groupe travail "régionalisation du Périglaciaire", Comité Natl Fr Géogr 10:31–43
Chaix A (1919) Coulées de blocs (rock-glaciers, rock streams) dans le Parc National Suisse de la Basse-Engadine. C R Séances Soc Phys Hist Nat Genève 36:12–15
Chaix A (1923) Les coulées de blocs du Parc National Suisse d'Engadine (Note préliminaire). Le Globe 62:1–35
Chaix A (1943) Les coulées de blocs du Parc National Suisse: Novelles mésures et comparison avec les "rock streams" de la Sierra Nevada de Californie. Le Globe 82:121–128
Chamberlin TC, Salisbury RD (1906) Geology III. In: Earth history. New York, 474 pp
Chandler JR (1973) The inclination of talus, arctic talus terraces, and other slopes composed of granular materials. J Geol 81:1–14
Chardon M, Riser J (1981) Formes et processus géomorphologiques dans le Haut Atlas marocain. Rev Géogr Alp 69:561–582
Chattopadhya GP (1984) A fossil valley-wall rock glacier in the Cairngorm Mountains. Scott J Geol 20:121–125
Chelius C (1896) Die Bildung der Felsenmeere im Odenwald. Z Dtsch Geol Ges 48:644–651
Cheng G (1983) Vertical and horizontal zonation of high-altitude permafrost. In: Proc 4th Int Conf on Permafrost, National Academic Press, Washington, Univ Alaska, Fairbanks, 136–141
Cheng G, Dramis F (1992) Distribution of Mountain permafrost and climate. Permafrost Periglacial Processes 3:83–91
Chueca J (1989) Caracteristicas de los glaciares rocosos del área meridional del batolito de Panticosa (Pirineo oscense). Geographicalia 26:61–74
Chueca J (1991a) Análisis de la distribución espacial de los glaciares rocosos en el Pirineo Central oscense. Geographicalia 28:85–99
Chueca J (1992) Estimación de paleotemperaturas durante el Pleistoceno final: Pirineo Central espanol. Estud Geogr 53(207): 241–263
Chueca J (1992) A statistical analysis of the spatial distribution of rock glaciers, Spanish Central Pyrenees. Permafrost Periglacial Processes 3:261–265
Church M, Stock RF, Ryder RM (1979) Contemporary sedimentary environments on Baffin Island, N.W.T., Canada: debris slope accumulations. Arct Alp Res 11:371–402
Clapperton CM (1975) Further observations on the stone runs of the Falkland Islands. Biul Peryglacjalny 24:211–217
Clapperton CM, Sugden DE, Birnie RV, Hanson JD, Thom G (1978) Glacier fluctuations in South Georgia and comparison with other island groups in the Scotia Sea. In: Zinderen Bakker EM van (ed) Antarctic glacial history and world palaeoenvironments. Int Counc Sci Unions, Sci Comm Antarctic Res, Proc Symp Rotterdam, pp 94–105
Clark DH, Clark MM, Gillespie AR (1994a) Debris covered glaciers in the Sierra Nevada, California, and their implications for snowline reconstruction. Quat Res 41:139–153
Clark DH, Clark MM, Gillespie AR (1994b) Reply to comment by M Jakob on debris covered glaciers in the Sierra Nevada, California, and their implications for snowline reconstruction. Quat Res 42:359–362

Clark GM, Ciolkosz EJ (1988) Periglacial geomorphology of the Appalachian Highlands and interior Highlands south of the glacial border: a review. Geomorphology 1:191–220

Clark GM, Schmidlin TW (1992) Alpine periglacial landforms of eastern North America: a review. Permafrost Periglacial Processes 3:225–230

Clark R (1972) Periglacial landforms and landscapes in the Falkland Islands. Biul Peryglacjalny 21:33–50

Clark R (1981) Cold climate features of the Cordillera Cantabrica, northern Spain. Biul Peryglacjalny 28:5–13

Colhoun EA (1981) A protalus rampart from the western Mourne Mountains, Northern Ireland. Ir Geogr 14:85–90

Collins DN (1978) Hydrology of an alpine glacier as indicated by the chemical composition of the meltwater. Z Gletscherkd Glazialgeol 13:219–238

Corte AE (1955) Contribución a la morfología periglacial especialmente criopedológica de la Repblica Argentina. Acta Geogr 14:83–102

Corte AE (1976a) Rock glaciers. Biul Peryglacjalny 26:175–197

Corte AE (1976b) The hydrological significance of rock glaciers. J Glaciol 17:157–158

Corte AE (1978) Rock glaciers as permafrost bodies with a debris cover as an active layer: a hydrological approach in the Andes of Mendoza, Argentina. In: 3rd Int Conf on Permafrost, Edmonton, Alberta, Proc, Nath Res Council, Ottawa, Canada, 1:263–269

Corte AE (1980) Glaciers and glaciolithic systems of the central Andes. In: World Glacier Inventory. Proc Worksh, Riederalp Switzerland, Int Assoc Hydrol Sci, IAHS-AISH 126, pp 11–24

Corte AE (1986) Delimitation of geocryogenic (periglacial) regions and associated geomorphic belts at 33°S, Andes of Mendoza, Argentina. Biul Periglacjalny 31:31–34

Corte AE (1987a) Rock glacier taxonomy. In: Giardino JR, Shroder JF, Vitek JD (eds) Rock glaciers. Allen & Unwin, London, pp 27–39

Corte AE (1987b) Central Andes rock glaciers: applied aspects. In: Giardino JR, Shroder JF, Vitek JD (eds) Rock glaciers. Allen & Unwin, London, pp 289–303

Corte AE (1988) Geocryology of the Central Andes and rock glaciers. In: Proc 5th Int Conf on Permafrost 1, Tapir, Trondheim, pp 718–723

Corte AE, Blanca B (1963) Relationship between four ground patterns, structure of the active layer, and type and distribution of ice in permafrost. Biul Peryglacjalny 12:7–90

Corte AE, Espizua LE (1981) Inventario de glaciares de la cuenca del rio Mendoza. Inst Argentino de Nivologia y Glaciologia, Mendoza, 64 pp

Corte AE, Figueroa A (1970) Quarternary and present mass wasting features in Sierra de la Ventana (Argentina). In: Macar P (ed) New contributions to slope evolution. Z Geomorphol Suppl 9:138–145

Corte AE, Trombotto D (1984) Quartz grain surface textures in laboratory experiments and in field conditions of rock glaciers. Microsc Electron Biol Cel 8:71–79

Cotta BV (1843) Karte von Sachsen; geognostische Konstitutionn des Fichtelgebirges. N Jahrb Mineral Geognosie Geol Petrefaktenkd 1843:171–178

Coutard JP, Francou B (1989) Rock temperature measurements in two alpine environments: implications for frost shattering. Arct Alp Res 21:399–416

Coutard JP, Helluin M, Helluin-Quinejure E, Journaux A, Lautridou JP, Ozouf JC, Pellerin J, Galert P, Igout M, van Vliet-Lanoe B, Levant M, Lecoustumer MN (1980) Observations sur quelques formes et processus périglaciaires dans le Massif du Chambeyron (Alpes de Haute-Provence). Rev Géogr Alp 68:349–382

Coutard JP, van Vliet-Lanoe B, Auzet AV (1988) Frost heaving and frost creep on an experimental slope: results for soil structures and sorted stripes. Z Geomorphol NF 71:13–23

Cross CW, Howe E (1905) Geography and general geology of the quadrangle in Silverton folio. US Geol Surv Folio 120:1–25

Cui Z (1982) Periglacial landforms and regional characteristics on Quinghai-Xizang (Tibet) Plateau. Beijing. vol II. Gordon and Breach, New York

Cui Z (1983) An investigation of rock glaciers in the Kunlun Shan, China. In: 4th Int Conf on Permafrost, Fairbanks, Alaska, Proc National Academic Press, Washington, pp 208–211

Cui Z, Zhu CH (1988) Rock glaciers in the source region of Urumqi River, middle Tian Shan, China. In: 5th Int Conf on Permafrost in Trondheim, Norway, Proceedings 1:724–727

Currey RR (1968) Neoglaciation in the mountains of the southwestern United States. PhD Thesis, Dep Geogr, Univ Kansas 171 pp

Curry RR (1968) Quaternary climatic and glacial history of the Sierra Nevada, California. PhD Thesis, Univ of California, Berkeley, 238 pp

Curry RR (1969) Quaternary climatic and glacial history of the Central Sierra Nevada, California. In: Schumm SA, Bradley WC (eds) United States contribution to quaternary research. Geol Soc Am Spec Pap 123:47

Czurda KA (1983) Freezing effects on soils: comprehensive summary of the ISGF 82. Cold Regions Sci Technol 8:93–107

Dahl R (1966) Block fields, weathering pits and tor-like forms in the Narvik Mountains, Nordland, Norway. Geogr Ann 48A:55–85

Daly RA (1912) Geology of the North American Cordillera at the forty-ninth parallel. Can Geol Surv Mem 38:857 pp

Dawson AG (1977) A fossil lobate rock glacier in Jura. Scott J Geol 13:37–42

Dawson AG, Matthews JA, Shakesbury RA (1986) A catastrophic landslide in Verkilsdalen, Rondane National Park, southern Norway. Geogr Ann 68A:77–87

De Martonne E (1920) Le rôle morphologique de la neige en montagne. La Géographie 34:255–267

De Robin GQ (1975) Radio-echo sounding: glaciological interpretations and applications. J Glaciol 15:49–64

Demangeot J (1978) L'étage périglaciaire au Gran Sasso (2912m), Italie Centrale. In: Colloq sur le Périglaciaire d'Altitude du Domaine Méditerranéen et Abords. Assoc Géogr Alsace, Strasbourg 1977, pp 209–220

Denton GH, Stuiver M (1966) Neoglacial chronology and northeastern St. Elias Mountains, Canada. Am J Sci 264:577–599

Dieling K (1973) Studien zur Verteilung der Blockgletscher in den Schweizer Alpen. Staatsexamensarbeit Univ Kiel, 175 pp

Dingwall PR (1973) Rock glaciers in the Canadian Rocky Mountains. In: 9th Congr Int Union Quat Res Abstr, Agricultural Development and Advisory Service (ADAS), London, pp 80–81

Domaradzki J (1951) Blockströme im Kanton Graubünden. Ergeb Wiss Untersuch Schweiz Nationalparks 3:177–235
Dona F (1956) Osservazioni sulla morfologia periglaciale della Valle Aurina. In: Studi sui fenomeni crionivali nelle Alpi Italiane, Fondazione per i Problemi Montani dell'Arco Alpino, Milano 11:105–113
Dowdeswell JA (1982) Relative dating of late Quaternary deposits using cluster and discriminant analysis, Audubon cirque, Mt. Audubon, Colorado Front Range Boreas 11:151–161
Dramis F, Kotarba A (1992) Southern limit of relict rock glaciers, Central Apennines, Italy. Permafrost Periglacial Processes 3:257–260
Dramis F, Kotarba A (1994) Geomorphological evidence of high mountain permafrost in central Apennines. Geogr Fisica Dyn Quat 17:29–36
Dramis F, Govi M, Guglielmin M, Mortara G (1995) Mountain permafrost and slope instability in the Italian Alps: the Val Pola landslide. Permafrost periglacial Processes 6:73–88
Dresch J (1978) Rôle des montagnes dans la morphogénèse actuelle et quarternaire au Moyen-Orient. Colloq sur le périglaciaire d'altitude du domaine Méditerranéen et abords, Strasbourg, Assoc Géogr Alsace, Strasbourg 245–252
Dresch J (1982) A propos des glaciers rocheux en pays méditerranéens arides. In: Hommages à René Raynal Réch Géogr Strasbourg 16–17:107–110
Dyke AS (1990) A lichenometric study of Holocene rock glaciers and neoglacial moraines, Frances Lake Map Area, southeastern Yukon Territory and Northwest Territories. Geol Surv Can 394:1–33
Dzierzek J, Nitychoruk J (1986) Types of fossil rock glaciers in the Polish High Tatra Mountains. Bull Pol Acad Sci Earth Sci 34:409–418
Dzierzek J, Lindner L, Nitychoruk J (1986) Late Quarternary deglaciation of the eastern Polish Tatra Mountains. Bull Pol Acad Sci Earth Sci 34:395–407
Ellis JM (1982) Holocene glaciation of the central Brooks Range, Alaska Diss Univ New York, Buffalo 396 pp (microfilm)
Ellis JM, Calkin PE (1979) Nature and distribution of glaciers, neo-glacial moraines and rock glaciers, east-central Brooks Range, Alaska. Arct Alp Res 11:403–420
Ellis JM, Calkin PE (1981a) Climatic reconstruction and environments of cirque glaciers, central Brooks Range, Alaska. Geol Soc Am Abstr Progr 13:445–446
Ellis JM, Calkin PE (1981b) Insolation control of glacial and periglacial landforms. Geol Soc Am Abstr Progr 13:131
Ellis JM, Calkin PE (1983) Environments and soils of Holocene moraines and rock glaciers, central Brooks Range, Alaska. In: Evenson EB, Schlüchter C, Rabassa J (eds) Tills and related deposits. Balkema, Rotterdam, pp 315–328
Ellis JM, Calkin PE, Bruen MP (1983) Rock glaciers of the east central Brooks Range, Alaska. In: Proc 4th Int Conf on Permafrost, Univ Alaska, Fairbanks, National Academic Press, Washington, p 322
Ellis JM, Parker EC (1979) Nature and distribution of glaciers, neoglacial moraines and rock glaciers, east-central Brooks Range, Alaska. Arct Alp Res 11:403–420
Embleton C, King CAM (1968) Glacial and periglacial geomorphology. Arnold, London, 608 pp
England J (1976) Late Quaternary glaciation of the eastern Queen Elizabeth Islands, N.W.T., Canada: Alternative models. Quat Res 6:185–202

Eskenasy MD (1978) The origin of the King Ravine Rock Glacier in the Presidential Range of the White Mountains of New Hampshire. MS Thesis, Univ Massachusetts, Amherst, 85 pp

Eskenasy MD (1979) The origin of the King Ravine rock glacier, Presidential Range, New Hampshire. Geol Soc Am Abstr Progr 11:11

Espizua L (1983a) Glacier and moraine inventory on the eastern slopes of Cordon del Plata and Cordonn del Portillo, Central Andes, Argentina. In: Evenson EB, Schlüchter CH, Rabassa J (eds) Tills and related deposits. Elsevier, Rotterdam, pp 381–395

Espizua L (1983b) Differencia altitudinal del límite inferior de los glaciares de escombros activos entre laderas norta y sur de los cordones del Plata y Portillo, Provincia de Mendoza. In: Acta Criogénica, 1ra Reunión 79–87

Espizua de Bengochea L (1982) Inventory of glaciers and moraines of the Tunuyan River basin, eastern Cordones del Plata and Portillo, Mendoza, Argentina. In: Rabassa J: Reunión Regional Sudamericana INQUA, Univ Nac Comahue, Neuguen, pp 21–22

Eugster H (1973) Bericht über die Untersuchungen des Blockstroms in der Val Sassa im Schweiz. Nationalpark (GR) von 1917–1971. Ergeb Wiss Untersuch Schweiz Natlparks 11:368–384

Evin M (1982) Mise en évidence de mouvements sur le glacier rocheux du Pic d'Asti (Queyras, Alpes du Sud, France). Rev Géomorphol Dyn 31:127–136

Evin M (1983a) Présence et signification morphoclimatic de sédiments gelés à l'amont des glaciers rocheux. In: Actes du Coll de l'Association des Géographes francais, Paris 8.1.83, Eboulis et environnement géographique passé et actuel, pp 137–142

Evin M (1983b) Structure et mouvement des glaciers rocheux des Alpes du Sud. Univ Grenoble I, Inst Géogr Alp, Thèse de 3e cycle, 343 pp

Evin M (1984a) Glaciers rocheux actuels et limites du pérgelisol alpin: l'example des Alpes du Sud. Communication à la Commission pour l'Etude des Phénomènes Périglaciaires, Fontainebleau-Avon, 14–15 Janv 1984

Evin M (1984b) Characteristiques physico-chimiques des eaux issues des glaciers rocheux des Alpes du Sud (France). Z Gletscherkd Glazialgeol 20:27–40

Evin M (1994c) Sondages électriques sur quelques glaciers rocheux et moraines des Alpes du Sud (France). Rev Géomorphol Dyn 34:127–137

Evin M (1985a) Contribution à l'étude de la macrofissuration dans les parvis situées à l'amont des glaciers rocheux Alpes du Sud (France, Italie). Rev Géomorphol Dyn 34:17–30

Evin M (1985b) Les glaciers rocheux: une mise au point. In: Notes et comptes rendus du Groupe de Travail "Régionalisation du Périglaciaire". Com Natl Fr Géogr 10:19–29

Evin M (1987a) Dynamique, repartition et âge des glaciers rocheux des Alpes du Sud. Thèse de doctorat d'état, Poitiers, 309 pp

Evin M (1987b) Lithology and fracturing control of rock glaciers in southwestern Alps of France and Italy. In: Giardino JR, Shroder JF, Vitek JD (eds) Rock glaciers. Allen & Unwin, London, pp 83–106

Evin M (1988) Propositions de terminologie concernant les glaciers rocheux. Notes et comptes-rendus du groupe de travail "Régionalisation du périglaciaire", Com Natl Fr Géogr, Comm pour l'étude des phénomènes périglaciaires 13:12–20

Evin M (1991a) Glaciers et glaciers rocheux sur le versant nord-est du Viso (Alpes Cottiennes, Italie). Rev Géomorphol Dyn 40:131–147

Evin M (1991b) Les rélations existant entre la moraine de refoulement, le glacier rocheux et le glacier du Petit Age Glaciaire dans le Haut Vallon d'Asti (Queyras, Alpes du Sud, France), In: Sesto Convegno Glaciologico Italiano, Gressoney St Jean/La Trinité, Valle d'Aosta, Italia, 26–27–28 Settembre 1991

Evin M (1992a) Glaciers rocheux et sondages géoélectriques sur le versant nord est du Viso (Alpes Cottiennes, Italie). In: Com Natl Fr Géogr Comm pour l'étude des phénomènes périglaciaires, Réunion annuelle de la Comm à Fontainebleau-Avon, 18–19 Janvier 92

Evin M (1992b) Mésures de résistivité électrique sur les glaciers rocheux de Grizzly Creek, Territoire du Yukon, Canada. Rapp de mission, Services Culturels de l'Ambassade du Canada, Janvier 92

Evin M (1993) Glaciers et glaciers rocheux dans les vallons de Mongioie et de Schiantala (Haute Stura di Demonte, Italie). Z Gletscherkd Glazialgeol 27/28:1–10

Evin M, Assier A (1982) Mise en évidence du mouvement sur le glacier rocheux du Pic d'Asti, (Queyras, Alpes du Sud, France). Rev Géomorphol Dyn 31:127–136

Evin M, Assier A (1983a) Glaciers et glaciers rocheux dans le Haut-Vallon du Loup (Haute-Ubaye, Alpes du Sud, France). Z Gletscherkd Glazialgeol 19:27–41

Evin M, Assier A (1983b) Mésures d'orientation de blocs sur quelques accumulations des Alpes du Sud (Brianconnais, Queyras, Embrunais, Ubaye, Argentera-Mercantour. France et Italie): Etablissement de coefficients permettant l'étude des glaciers rocheux. Z Gletscherkd Glazialgeol 19:107–126

Evin M, Assier A (1983c) Mise en évidence de mouvements sur le glacier rocheux de Sainte-Anne (Queyras, Alpes du Sud, France): le rôle du pergélisol alpin. Rev Géogr Alp 71:165–178

Evin M, Assier A (1983d) Relations hydrologiques entre glacier et glaciers rocheux: l'example du cirque de Marinet (Haute-Ubaye, Alpes du Sud). In: Communication, Section de Glaciologie de la Societé hydrotechnique de France, Grenoble, 5 pp

Evin M, Assier (1984) Sondages électriques sur quelques glaciers rocheux et moraines des Alpes du Sud (France). Geomorphology 3:57–71

Evin M, Assier A, Fabre D (1990) Les glaciers rocheux du Marinet (Haute-Ubaye, France). Rev Géomorphol Dyn 39:139–155

Evin M, Fabre D, Assier A, Guillain C (1993) Les glaciers rocheux du Roure (Haute Ubaye). Soc Hydrotech Fr, Grenoble, 11–12 Mars 1993, (unpubl report) 6 pp

Eyles N (1978) Rock glaciers in Esjufjoll nunatak area, southeast Iceland. Jökull 28:53–56

Fabre D, Evin M (1990) Prospection électrique des milieux à très forte résistivité: le cas du pergélisol alpin. 6th Int Congr Int Assoc Engin Geol, The Netherlands, pp 927–934

Faure-Muret A (1949) Les 'rock-streams' ou 'pseudo-moraines' du Massiv de l'Argentera-Mercantour. C R Bull Soc Géol Fr 7:118–120

Ferrians OJR, Hoson GD (1973) Mapping and predicting permafrost in North America: a review 1963–1973. In: 2nd Int Conf on Permafrost, N Am Contrib, Natl Acad Sci, Washington DC, pp 479–498

Finsterwalder S (1928) Begleitworte zur Karte des Gepatschferners. Z Gletscherkd 16:20–41

Fisch W sen, Fisch W jun, Haeberli W (1978) Electrical DC resistivity soundings with long profiles on rock glaciers and moraines in the Alps of Switzerland. Z Gletscherkd Glazialgeol 13:239–260

Fischer K (1967) Erdströme in den Alpen. Mitt Geogr Ges München 52:231–246

Fish AM (1983) Thermodynamic model of creep at constant stresses and constant strain rates. Cold Regions Res and Engineering Lab, CRREL Rep 83, 33:24

Flakstad N, Sollid JL, Tolgensbakk J (1985) Nordre Andoya, Kvartaergeoologi og geomorfologi 1:50 000. Geogr Inst, Univ Oslo

Flint RF, Denny CS (1958) Quaternary geology of Boulder Mountain, Aquarius Plateau, Utah. US Geol Surv Bull 1061-D:103–164

Foster HL, Holmes GW (1965) A large transitional rock glacier in the Johnson River Area, Alaska Range. US Geol Surv Prof Pap 525-B:B112–B116

Fraedrich R (1979) Spät- und postglaziale Gletscherschwankungen in der Ferwallgruppe (Tirol, Vorarlberg). Düsseld Geogr Schr 12:161

Francou B (1979) Formes d'éboulis dans le Brianconnais. Rev Géogr Alp 65:64–73

Francou B (1982) Chutes de pierres et éboulisation dans les parois de l'étage périglaciaire. Rev Géogr Alp 70:279–300

Francou B (1983a) Régimes thermiques de sol à l'étage périglaciaire et leurs consequences géomorphologiques: Example de la Combe de Laurichard. Alpes du Brianconnais. France. Géogr Phys Quat 37:27–38

Francou B (1983b) Géodynamique des dépots de pied de parois dans l'étage périglaciaire. Rev Géol Dyn Géogr Phys 24:411–424

Francou B (1988) L'Eboulisation en Haute Montagne. Caen 2:696

Francou B, Hetu B (1989) Eboulis et autres formations de pentes hétérométriques Contribution à une terminologie géomorphologique. Régional Périglaciaire 14:11–69

Francou B, Reynaud L (1992) 10 years surficial velocities on a rock glacier (Laurichard, French Alps.) Permafrost Periglacial Processes 3:209–213

French HM (1976) The periglacial environment. Longman, London, 309 pp

French HM, Egginton P (1973) Thermokarst development, Banks Islands, western Canadian Arctic. In: 2nd Int Conf on Permafrost, N Am Contrib, Natl Acad Sci, Washington DC, pp 203–212

Frich P, Brandt E (1985) Holocene talus accumulation rates and their influence on rock glacier growth: a case study from Igpik, Disko, West Greenland. Geogr Tidsskr 85:32–43

Fritz P (1976) Gesteinsbedingte Standorts- und Formendifferenzierung rezenter Periglazialerscheinungen in den Ostalpen. Mitt Österr Geogr Ges 118:237–273

Fritz P (1976) Vergleichende Untersuchung der rezenten periglazialen Region im Kristallin und Kalk der Ostalpen (Schneeberger Alpen, Hochschwab, Dachstein, Hochwesel, Schladminger Tauern, Sadnig-Gruppe). Diss Phil Univ Wien, 406 pp

Fritzsche W, Osterer F (1978) Elektronische Messungen mit Puls-Echo in der Glaziologie. Z Gletscherkd Glazialgeol 13:275–283

Fujii Y, Higuchi K (1978) Distribution of alpine permafrost in the Northern Hemisphere and its relation to air temperature. In: Proc 3rd Conf on Permafrost, Natl Res Counc Can, Ottawa, pp 366–371

Funk M, Hoelzle M (1992) A model of potential direct solar radiation for investigating occurrences of mountain permafrost. Permafrost Periglacial Processes 3:139–142

Furrer G (1954) Solifluktionsformen im schweizerischen Nationalpark. Ergebnisse der wissenschaftlichen Untersuchungen des schweizerischen Nationalparks 4 (Neue Folge), pp 201–275

Furrer G, Fitze P (1970) Beitrag zum Permafrostproblem in den Alpen. Vierteljahrssch Naturforsch Ges Zür 115:353–368

Gagne RM, Hunter JA (1975) Hammer seismic studies of surficial materials, Banks Island, Ellesmere Island, and Boothia Peninsula, N.W.T. Geol Surv Can Rep Activities 75, 1B:13–17

Gamper M (1985) Morphochronologische Untersuchungen an Solifluktionszungen, Moränen und Schwemmkegeln in den Schweizer Alpen. Phys Geogr Zür 17:115

Gardner JS (1972) Recent glacial activity and some associated landforms in the Canadian Rocky Mountains. In: Slaymaker O, McPherson HJ (eds) Mountain geomorphology. Allen & Unwin, Vancouver, pp 52–62

Gardner JS (1978) Wenkchemna Glacier: ablation complex and rock glacier in the Canadian Rocky Mountains. Can J Earth Sci 15:1200–1204

Gardner JS (1980) Frequency, magnitude, and spatial distribution of mountain rockfalls and rockslides in the Highwood Pass area, Alberta, Canada. In: Coates DR, Vitek JD (eds) Thresholds in geomorphology. London, pp 267–295

Gardner JS, Bajewski I (1987) Hilda rock glacier stream discharge and sediment load characteristics, Sunmwapta Pass area, Canadian Rocky Mountains. In: Giardino JR, Shroder JF, Vitek JD (eds) Rock glaciers. Allen & Unwin, London, pp 161–174

Garleff K (1970) Verbreitung und Vergesellschaftung rezenter Periglazialerscheinungen in Skandinavien. Gött Geogr Abh 51:7–66

Garleff K (1977) Höhenstufen der Argentinischen Anden in Cuyo, Patagonien und Feuerland. Gött Geogr Abh 68:150

Garleff K, Stingl H (1984) Höhenstufen und ihre neuzeitlichen Veränderungen in den argentinischen Anden. Zentralbl Geol Paläontol 1:1701–1707

Garleff K, Stingl H (1988) Geomorphologische Untersuchungen in der nivalen und subnivalen Stufe der argentinischen Anden: Ergebnisse und Paläoklimatische Interpretation. Tagungsber Wiss Abh Dtsch Geographentag München 46:419–425

Gerber E (1994) Geomorphologie und Geomorphodynamik der Region Lona Sassenaire (Wallis, Schweizer Alpen) unter besonderer Berücksichtigung von Lockersedimenten mit Permafrost. Dissertation, Univ Freiburg/Schweiz, Math Naturwiss Fakultät, 330 pp

Gerber E, Scheidegger AE (1966) Bewegungen in Schuttmantelhängen. Geogr Helv 21:20–31

Gerhold N (1963) Zwei berühmte Blockgletscher in der Umgebung des Gepatschhauses. Nachrichtenbl DAV (Deutscher Alpenverein) Sekt Frankfurt 32:3–4

Gerhold N (1964) Die Blockgletscher – eine besondere Moränenform? PhD Dissertation, Geogr Inst Univ Innsbruck, 200 pp

Gerhold N (1965) Die Blockgletscher-eine besondere Moränenform? Jahresber Bischöfl Gymansiums Paulinum Schwaz 32:3–24

Gerhold N (1967a) Die Kare des nordöstlichen Kaunergrates. Jahresber Bundesgymanasiums Bundesrealgymnasiums Innsbruck, pp 1–4

Gerhold N (1967b) Zur Glazialgeologie der westlichen Ötztaler Alpen unter besonderer Berücksichtigung des Blockgletscherproblems. Veröff Museum Ferdinandeum Innsbruck 47:5–52

Gerhold N (1969) Zur Glazialgeologie der westlichen Ötztaler Alpen: Unter besonderer Berücksichtigung des Blockgletscherproblems. Veröff Museum Ferdinandeum Innsbruck 49:46–78

Gerhold N (1970) Blockgletscher im Ötztal. Tiroler Heimatbl 45:107–114

Gerhold N (1971) Glaziale und periglaziale Ablagerungen in den westlichen Ötztaler Alpen. Tiroler Heimatbl 46:31–35

Giardino JR (1979) Rock glacier mechanics and chronologies, Mount Mestas, Colorado. PhD Diss, Univ Nebraska, Lincoln, 244 pp

Giardino JR (1980) Rock glacier mechanics. Am Quat Assoc Abstr 83

Giardino JR (1983) Movement of ice cemented rock glaciers by hydrostatic pressure: an example from Mount Mestas, Colorado. Z Geomorphol NF 27:297–310

Giardino JR, Shroder JF (1978) The origin of the term "Rock glacier". J Glaciol 20:446

Giardino JR, Vick SG (1987) Geologic engineering aspects of rock glaciers. In: Giardino JR, Shroder JF, Vitek JD (eds) Rock glaciers. Allen & Unwin, London, pp 265–287

Giardino JR, Vitek JD (1985) A statistical study of the fabric of a rock glacier. Arct Alp Res 17:165–177

Giardino JR, Vitek JD (1988a) Interpreting the internal fabric of a rock glacier. Geogr Ann 70A:15–25

Giardino JR, Vitek JD (1988b) Rock glacier rheology: a preliminary assessment. In: 5th Int Conf on Permafrost in Trondheim, Norway, Proc 1:744–748

Giardino JR, Vitek JD (1988c) The significance of rock glaciers in the glacial – periglacial landscape continuum. J Quart Sci 3:97–103

Giardino JR, Shroder JF, Lawson MP (1978) Movement mechanisms for ice-cemented rock glaciers. Am Quat Assoc Abstr 202

Giardino JR, Lawson MP, Shroder JF (1979) Fabric analysis of rock glaciers. Assoc Am Geogr, AAG Progr Abstr 79:191

Giardino JR, Shroder JF, Lawson MP (1984) Tree-ring analysis of movement of a rock-glacier complex on Mount Mestas, Colorado, U.S.A. Arct Alp Res 16:299–309

Giardino JR, Shroder JF, Vitek JD (eds) (1987) Rock glaciers. Allen & Unwin, London, 355 pp

Giardino JR, Vitek JD, DéMorett JL, (1992) A model of water movement in rock glaciers and associated water characteristics. In: Dixon J, Abrahams A (eds) Periglacial geomorphology. Wiley, Chichester, pp 159–184

Giardino JR, Vitek JD, Marston RA (1984) Surface fabric characteristics of an ice-cemented rock glacier. Geol Soc Am Abstr Progr 16:518

Giardino JR, Vitek JD, Hoskins ER (1985) Rheology of rock glaciers: A preliminary assessment. Geol Soc Am Abstr Progr 17:591

Girsperger S (1973) Bericht über die Klinometermessungen im Val Sassa. Unpubl Rep to the Schweizerische Nationalparkkommission, 7 pp

Glazovskiy AF (1978) Rock glaciers (main problems). In: Gorbunov AP (ed) Cryogenic occurrences of high mountains. Sci Acad USSR, Siberial Chapter, Novosibirsk, pp 85–92

Glazovskij AF (1978a) Kamennye gletcery (sostojanie problemy). Kryogennye jarlenija vysokogomj, Novosibirsk, pp 59–72

Glazovskij AF (1978b) Kamennye gletcery bassejna r bol'soj almatinki. Kryogennye jarlenija vysokogomj, Novosibirsk, pp 85–92

Glen JW (1952) Experiments on the deformation of ice. J Glaciol 2:111–114

Gold LW, Lachenbruch AH (1973) Thermal conditions in permafrost: A review of North American literature. In: 2nd Int Conf on Permafrost, N Am Contrib, Natl Acad Sci, Washington DC, pp 3–25

Goolsby JE (1972) East rock glacier of Lone Mountain, Madison County, Montana. Geol Soc Am Abstr Progr 4:377–378
Gorbunov AP (1978) Permafrost investigations in high-mountain regions. Arct Alp Res 10:283–294
Gorbunov AP (1983) Rock glaciers of the mountains of middle Asia. In: Proc 4th Int Conf on Permafrost, Fairbanks, Alaska. National Academic Press, Washington, pp 359–362
Gorbunov AP (1988) The alpine permafrost zone of the USSR. In: Proc 5th Int Conf on Permafrost 1. Tapir, Trondheim, pp 154–158
Gorbunov AP, Titov SN (1986) Kamenne gletcery. Priroda 5:73–77
Gorbunov AP, Titkov SN (1992) Dynamics of rock glaciers of the Northern Tien Shan and the Djungar Ala Tau, Kazakhstan. Permafrost Periglacial Processes 3:29–39
Gordon JE, Birnie RV, Timmis RJ (1978) A major rockfall and debris slide on the Lyell Glacier, South Georgia. Arct Alp Res 10:49–60
Goudong C (1983) Vertical and horizontal zonation of high altitude permafrost. In: Proc 4th Int Conf on Permafrost, Fairbanks, Alaska. National Academic Press, Washington, pp 136–141
Goughnour RR, Andersland OB (1968) Mechanical properties of a sand–ice system. J Soil Mechanics Foundation Division (Proc Am Soc Civil Engin) 94:923–950
Graf WL (1983) Rock glaciers: a case for a change in nomenclature. Geogr Ann 65A:401–410
Gray JT (1970) Mass wasting studies in the Ogilvie and Wernecke Mountains, Central Yukon territory. Geol Surv Can Pap 70-1(A):192–195
Gray JT (1972) Processes and rates of development of talus slopes and protalus rock glaciers in the Ogilvie and Wernecke Mountains, Yukon Territory. PhD Thesis, Mc Gill Univ, Montreal, 285 pp
Greenstein LA (1983) An investigation of mid-latitude alpine permafrost on Niwot Ridge, Colorado, Rocky Mountains, U.S.A., In: Proc 4th Int Conf on Permafrost, Fairbanks, Alaska. National Academic Press, Washington, pp 380–383
Griffey NJ, Whalley WB (1979) A rock glacier and moraine-ridge complex, Lyngen peninsula, north Norway. Nor Geogr Tidsskr 33:117–124
Griffiths TM (1958) A theory of rock stream morphology. Ann Assoc Am Geogr 48:265–266
Grötzbach E (1965a) Beobachtungen an Blockströmen im afghanischen Hindukusch und in den Ostalpen. Mitt Geogr Ges München 50:175–201
Grötzbach E (1965b) Die Blockgletscher – eine besondere Moränenform? Bischöfl Gymnasium Schwaz 32:3–24
Grötzbach E, Rathjens C (1969) Die heutige und die jungpleistozäne Vergletscherung des afghanischen Hindukusch. In: Kaiser K (ed) Glazialmorphologie. Z Geomorphol Suppl 8:58–75
Grogger PK (1979) The rock glaciers of the high Uintas primitive area, Uinta Mountains, Utah. Geol Soc Am Abstr Progr 11:273
Gross G, Kerschner H, Patzelt G (1978) Methodische Untersuchungen über die Schneegrenze in alpinen Gletschergebieten. Z Gletscherkd Glazialgeol 12:223–251
Grout M (1979) Amount and rate of Holocene cirque erosion, northern Sangre de Cristo Mountains, Colorado. Geol Soc Am Abstr Progr 11:273–274
Grout M (1981) Rockwall erosion in the northern Sangre de Cristo Mountains, Colorado. Master's Thesis, Univ Colorado, Boulder, 165 pp

Grout M (1982) Sourcewall relief chiefly influences the current supply of rockfall debris to rock glaciers, northern Sangre de Cristo Mountains, Colorado. Geol Soc Am Abstr Progr 14:313

Grout M (1984) Rates of rock glacier source wall erosion, northern Sangre de Cristo Mountains, south-central Colorado. Am Quat Assoc Progr Abstr 8:54

Grove JM (1988) The Little Ice Age. Methuen, Cambridge, 498 pp

Gruppo Nazionale Geografia Fisica E Geomorfologia CNR (1986) Ricerche geomorfologiche nell'alta Val di Peio (Gruppo del Cevedale). Geogr Fis Din Quat 9:137–191

Gruppo Nazionale Geografia Fisica E Geomorfologia (1987) Nuovi dati per lo studio dei rock-glaciers del gruppo Ortles-Cevedale (Alpi). Riv Geogr Ital 94:425–450

Guglielmin M, Lozej A, Tellini C (1994) Permafrost distribution and rock glaciers in the Livigno area (northern Italy). Permafrost Periglacial Processes 5:25–36

Guiter V (1972) Une forme montagnarde: le rock-glacier. Rev Géogr Alp 60:467–487

Gullentops F, Huysmans L, Crovnenborghs B, Willemsen H (1985) Rock glaciers and nivation (abs) In: INQUA Eurosiberian Subcommission for the Study of the Holocene IGCP Projects 158, Paleohydrology of the temperate zone in the last 15,000 years. Bern, 24.6–1.7.1985, 11 pp

Gutierrez M, Pena JL (1981) Los glaciares rocosos y el modelado acompanante en el área de la Bonaigua (Pirineo de Lérida). Bol Geol Minero 92:11–20

Haeberli W (1973) Die Basis-Temperatur der winterlichen Schneedecke als möglicher Indikator für die Verbreitung von Permafrost in den Alpen. Z Gletscherkd Glazialgeol 9:221–227

Haeberli W (1975) Untersuchungen zur Verbreitung von Permafrost zwischen Flüelapass und Piz Grialetsch (Graubünden). Mitt Versuchsanst Wasserbau Hydrol Glaziol ETH Zür 17:221

Haeberli W (1976) Eistemperaturen in den Alpen. Z Gletscherkd Glazialgeol 11:203–220

Haeberli W (1978) Special aspects of high mountain permafrost methodology and zonation in the Alps. In: 3rd Int Conf on Permafrost, Natl Res Cen, Ottawa, 1:378–384

Haeberli W (1979) Holocene push-moraines in alpine permafrost. Geogr Ann 61A:43–48

Haeberli W (1980) Morphodynamische Aspekte aktueller Gletscherhochwässer in den Schweizer Alpen. Regio Basiliensis 21:58–78

Haeberli W (1981) Ice motion on deformable sediments. J Glaciol 27:365–366

Haeberli W (1982) Klimarekonstruktionen mit Gletscher-Permafrost-Beziehungen. Basler Beitr Physiogeogr 4:9–17

Haeberli W (1983a) Permafrost–glacier relationships in the Swiss Alps – today and in the past. In: Proc 4th Int Conf on Permafrost, Fairbanks, Alaska. National Academic Press, Washington, pp 415–420

Haeberli W (1983b) Frozen ground: rock glacier studies. Ice 72/73:14

Haeberli W (1985a) Creep of mountain permafrost: internal structure and flow of alpine rock glaciers. Mitt Versuchsanst Wasserbau Hydrol Glaziol ETH Zür 77:142

Haeberli W (1985b) Permafrostkriechen in den Alpen. Phys Geogr 16:43–48

Haeberli W (1989a) Glacier ice-cored rock glaciers in the Yukon Territory, Canada. J Glaciol 35:294–295

Haeberli W (1989b) Pilot analysis of permafrost cores from the active rockglacier Murtel I, Piz Corvatsch, Eastern Swiss Alps. Versuchsanst Wasserbau Hydrol Glaziol ETH Zür Arbeitsh 9:38

Haeberli W (1990a) Glacier and permafrost signals of 20th-century warming. Ann Glaciol 14:99–101

Haeberli W (1990b) Scientific, environmental and climatic significance of rock glaciers. Mem Soc Geol Ital 45:823–831

Haeberli W (1991) Zur Glaziologie der letzteiszeitlichen Alpenvergletscherung. In: Frenzel B (ed) Klimageschichtliche Probleme der letzten 130.000 Jahre. Fischer, Stuttgart, pp 409–419

Haeberli W (1992a) Construction, environmental problems and natural hazards in periglacial mountain belts. permafrost Periglacial Processes 3:111–124

Haeberli W (1992b) Possible effects of climatic change on the evolution of alpine permafrost. Catena Suppl 22:23–35

Haeberli W, Epifani F (1986) Mapping the distribution of buried glacier ice – an example from Lago Delle Locce, Monte Rosa, Italian Alps. Ann Glaciol 8:78–81

Haeberli W, Fisch W (1984) Electrical resistivity soundings of glacier beds: a test study on the Grubengletscher, Wallis, Swiss Alps. J Glaciol 30:373–376

Haeberli W, Funk M (1991) Borehole temperatures at the Colle Gnifetti core-drilling site (Monte Rosa, Swiss Alps). J Glaciol 37:37–46

Haeberli W, Lautridou JP (1992) Permafrostkartierung im Gebiet der Hochebenkar-Blockgletscher, Obergurgl, Ötztaler Alpen. Z Gletscherkd Glazialgeol 18:127–150

Haeberli W, Röthlisberger H (1976) Beobachtungen zum Mechanismus und zu den Auswirkungen von Kalbungen am Grubengletscher (Saastal, Schweiz). Z Gletscherkd Glazialgeol 11:221–228

Haeberli W, Schlüchter CH (1987) Geological evidence to constrain modelling of the Late Pleistocene Rhonegletscher (Switzerland). In: The physical basis of ice-sheet modelling. Proc Vancouver Symp, August 1987, IAHS 170:333–345

Haeberli W, Schmid W (1988) Aerophotogrammetrical monitoring of rock glaciers. In: Proc 5th Int Conf on Permafrost in Trondheim, Norway, 1:764–769

Haeberli W, Schweizer J (1988) Rhonegletscher 1850: Eismechanische Überlegungen zu einem historischen Gletscherstand. Versuchsanst Wasserbau Hydrol Glaziol ETH Zür Mitt 94:59–70

Haeberli W, King L, Flotron A (1979) Surface movement and lichen-cover studies at the active rock glacier near the Grubengletscher, Wallis, Swiss Alps. Arct Alp Res 11:421–441

Haeberli W, Wächter H-P, Schmid W, Sidler C (1982) Erste Erfahrungen mit dem US Geological Survey – Monopuls Radioecholot im Firn, Eis und Permafrost der Schweizer Alpen. Versuchsanst Wasserbau Hydrol Glaziol ETH Zür Arbeitsh 6:1–23

Haeberli W, Rellstab W, Harrison WD (1984) Geothermal effects of 18 ka BP ice conditions in the Swiss Plateau. Ann Glaciol 5:56–60

Haeberli W, Huder J, Keusen HR, Pika J, Röthlisberger H (1988) Core drilling through rock glacier permafrost. In: Proc 5th Int Conf Permafrost in Trondheim, Norway, 2:937–942

Haeberli W, Rickernmann D, Zimmermann M (1990a) Investigation of 1987 debris flows in the Swiss Alps: general concept and geophysical soundings. In: Hydrology in mountainous regions II – artificial reservoirs; water and slopes. Proc of two Lausanne Symposia, August 1990, IAHS 194:303–310

Haeberli W, Zimmermann M, Kienholz H (1990c) (eds) Morphogenesis and Morphodynamics in high mountain environments – Fieldtrip D4. 2nd Int Conf Geomorphol Frankfurt 1989. Versuchsanst Wasserbau Hydrol Glaziol ETH Zür Arbeitsh 14:65

Haeberli W, Evin M, Tenthorey G, Kensen HR, Hoelzle M, Keller F, Vonder Mühll D, Wagner S, Pelfini M, Smiraglia C (1992) Field Report. permafrost research sites in the Alps: excursions of the international workshop on permafrost and periglacial environment in mountain areas. Permafrost Periglacial Processes 3:189–202

Hagedoorn JG (1959) The plus-minus method of interpreting seismic refraction sections. Geophys Prospect 8:158–182

Hamilton SJ, Whalley WB (1995) Rock glacier nomenclature: re-assessment. Geomorphology 14:73–80

Happoldt H, Schrott L (1989) Globalstrahlung und Bodentemperaturen in der periglazialen Höhenstufe am Aconcagua, Argentinische Hochanden. In: Hüser K, Kleber A, Stingl H (eds) Beiträge zur Geomorphologie: Quartärmorphologie Relief und Tektonik. Bayreuther Geowiss Arb 14:35–45

Happoldt H, Schrott L (1992) A note on ground thermal regimes and global solar radiation at 4720 m a.s.l., High Andes of Argentina. Permafrost Periglacial Processes 3:241–245

Harris C (1986) Some observations concerning the morphology and sedimentology of a protalus rampart, Okstindan, Norway. Earth Surface Proc Landforms 11:673–676

Harris SA (1979) Ice caves and permafrost zones in southwest Alberta. Erdkunde 33:61–70

Harris SA (1981a) Distribution of active glaciers and rock glaciers compared to permafrost landforms, based on freezing and thawing indices. Can J Earth Sci 18:376–381

Harris SA (1981b) Climatic relationship of permafrost zones in areas of low winter snow cover. Arctic 34:64–70

Hariss SA (1981c) Distribution of zonal permafrost landforms with freezing and thawing indices. Can J Earth Sci 18:376–381

Harris SA (1982a) Cold air drainage west of Fort Nelson, British Columbia. Arctic 35:537–541

Harris SA (1982b) Identification of permafrost zones using selected permafrost landforms. In: Proc 4th Can Permafrost Conf Calgary 1981:49–58

Harris SA (1986) The permafrost environment. Croom Helm, London, 276 pp

Harris SA (1988) The alpine periglacial zone. In: Clark MJ (eds) Advances in periglacial geomorphology Wiley, Chichester, pp 369–413

Harris SA (1994) Climatic zonality of periglacial landforms in mountain areas. Arctic 47:184–192

Harris SA, Brown RJE (1982) Permafrost distribution along the Rocky Mountains in Alberta. In: Proc 4th Can Permafrost Conf, Calgary 1981:59–67

Harris SA, Corte AE (1992) Interactions and relations between mountain permafrost, glaciers, snow and water. Permafrost Periglacial Processes 3:103–110

Harris SA, Wayne K, Blumenstengel D, Cook H, Krouse R, Whitley G (1994) Comparison of the water drainage from an active near-slope rock glacier and a glacier, St. Elias Mountains, Yukon Territory. Erdkunde 48:81–91

Harrison WD (1975) Temperature measurements in a temperate glacier. J Glaciol 14:23–30
Harry DG (1988) Ground ice and permafrost. In: Clark MJ (ed) Advances in periglacial geomorphology. Wiley, Chichester, pp 113–149
Hartman EL, Rottman ML (1986) The vascular flora of five rock glaciers in the San Juan Mountains, Colorado. Phytologia 60:225–235
Hassinger JM (1981) Characteristics, dynamics and paleoclimatic significance of rock glaciers in southern Victoria Land, Antarctica. Master's Thesis, Univ New Hampshire, 134 pp
Hassinger JM, Mayewski PA (1983) Morphology and dynamics of rock glaciers in Southern Victoria Land, Antarctica. Arct Alp Res 15:351–368
Haynes FD (1978) Strength and deformation of frozen silt. In 3rd Int Conf on permafrost, Natl Res Cent, Ottawa, 1:656–661
Heald WF (1946/47) Palisade Glacier Survey, Sierra Nevada. Am Alp J 6:332–339
Heim A (1882) Der Bergsturz von Elm. Z Dsch Geol Ges 34:74–115
Heim A (1898) "Campo/Ravona". Vierteljahresschr Naturforsch Ges Zür
Heim A (1932) Bergsturz und Menschenleben. Fretz, Wasmuth, Zürich, 218 pp
Heine K (1976) Blockgletscher und Blockzungen-Generationen am Nevado de Toluca, Mexico. Die Erde 107:330–352
Heine K (1994) Present and past geocryogenic processes in Mexico. Permafrost Periglacial Processes 5:1–12
Henning I (1975) Die La Sal Mountains, Utah: Ein Beitrag zur Geoökologie der Colorado-Plateau-Provinz und zur vergleichenden Hochgebirgsgeographie. Akad Wiss Literatur Mainz, Abh Math Naturwiss Kl 1975, 2:88
Hermes K (1955) Die Lage der oberen Waldgrenze und ihr Abstand zur Schneegrenze. Kölner Geogr Arb 5:277
Herrmann F (1925) Rock glaciers della Valsavarenche. Natura 16:139–142
Herzog P, Hofer A (1981) Unaxial creep tests on a morainic material from Switzerland. In: Jessberger HL (ed) Ground freezing 1980. Engin Geol 18:79–87
Herzog P, Hofer A (1983) Unaxial creep tests on a morainic material from Switzerland. Mitt Inst Grundbau Bodenmechanik ETH Zürich 116:9
Heuberger H (1966) Gletschergeschichtliche Untersuchungen in den Zentralalpen zwischen Sellrain und Ötztal. Dtsch Alpenverein Wiss Alpenvereinsh Innsbruck 20:125
Heuberger H (1971) Roland Beschel und die Lichenometrie. Z Gletscherkunde Glazialgeol 7:175–184
Heuberger H (1974) Landformen im Kartenbild. Topographisch-geomorphologische Kartenproben 1:25000, Gruppe VII Alpen/Zentralalpen, Kartenproben 3: Gletscher Firnflächen, frische Wallmoränen am Sulztalferner, Ötztal. Topographisch-Geomorphologische Kartenproben VII:3–16 (Westermann)
Higuchi K (1978) Glaciers and climates of Nepal Himalayas. Rep of the glaciological expedition of Nepal, part IV, vol 40. Seppyo, 87 pp
Hinkel KM (1988) Frost mounds formed by degradation at Slope Mountain, Alaska, U.S.A. Arct Alp Res 20:76–85
Hodge SM (1972) The movement and basal sliding of the Nisqually glacier, Mount Rainier. Dep Atmos Sci Univ Washington, Sci Rep, 409 pp
Hodge SM (1974) Variations in the sliding of temperate glaciers. J Glaciol 13:349–369
Hodge SM (1978) USGS monopulse ice radar. US Geol Surv, Project Office Glaciol, internal report, pp 1–9

Högbom B (1914) Über die geologische Bedeutung des Frostes. Bull Geol Inst Univ Upsala 12:257–390

Hoekstra P, McNeill D (1973) Electromagnetic probing of permafrost. In: 2nd Int Conf on Permafrost, N Am Contrib, Natl Acad Sci, Washington DC, pp 517–526

Höllermann P (1964) Rezente Verwitterung, Abtragung und Formenschatz in den Zentralalpen am Beispiel des oberen Suldentales (Ortlergruppe). Z Geomorphol Suppl 4:257

Höllermann P (1968) Die rezenten Gletscher der Pyrenäen. Geogr Helv 23:157–168

Höllermann P (1969) Zur Frage der unteren Strukturbodengrenze in Gebirgen der Trockengebiete. In: Bibus E, Semmel A (eds) Geomorphologie arider Gebiete. Z Geomorphol Suppl 15:156–166

Höllermann P (1980) Naturräumliche Höhengrenzen und die Hochgebirgsstufe in Gebirgen des westlichen Nordamerika. In: Jentsch C, Liedtke H (eds) Höhengrenzen in Hochgebirgen. Arb Geogr Inst Univ Saarland 29:75–117

Höllermann P (1983a) Blockgletscher als Mesoformen der Periglazialstufe. Bonner Geogr Abh 67:73

Höllermann P (1983b) Blockgletscherstudien in europäischen und nord-amerikanischen Gebirgen. In: Poser H, Schunke E (eds) Mesoformen des Reliefs im heutigen Periglazialraum. Abh Akad Wiss Gött Math Phys Kl Folge 3, 35:116–119

Höllermann P (1983c) Probleme der Blockgletscherforschung. Referat des Diskussionsbeitrages. In: Poser H, Schunke E (eds) Abhandlungen der Akademie der Wissenschaften Göttingen, Math Phys Kl Folge 3, 35:151–159

Hölzle M (1992) Permafrost occurrence from BTS measurements and climatic parameters in the eastern Swiss Alps. Permafrost Periglacial Processes 3:143–147

Hölzle M (1994) Permafrost und Gletscher im Oberengadin. Grundlagen und Anwendungsbeispiele für automatisierte Schätzverfahren. Mitt Versuchsanst Wasserbau Hydrol Glaziol ETH Zür 132:121

Höelzle M, Haeberli W, Keller F (1993) Application of BTS-measurements for modelling mountain permafrost distribution. In: Proc 6th Int Conf Permafrost in Beijing, 1:272–277

Hövermann J (1972) Die periglaziale Region des Tibesti und ihr Verhältnis zu angrenzenden Formungsregionen. In: Hövermann J, Oberbeck G (eds). Hans-Poser-Festschrift. Gött Geogr Abh 60:261–283

Hole AD (1912) Glaciation in the Telluride Quadrangle, Colorado. J Geol 20:502–529

Holtmeier K-F (1978) Die bodennahen Winde in den Hochlagen der Indian Peaks Sektion (Colorado Front Range). Münster Geogr Arb 3:1–47

Holtmeier K-F (1979) Luftbild Arapaho-Paß, Colorado Front Range, USA. Die Erde 110:97–106

Hooke RL, LeBaron M, Dahlin BB, Kauper MT (1972) Creep of ice containing dispersed fine sand. J Glaciol 11:327–336

Hopkins DM (1973) Sea level history in Beringia during the past 250,000 years. Quat Res 3:520–540

Howe E (1909) Landslides in the San Juan Mountains, Colorado: including a consideration of their causes and their classification. US Geol Surv Prof Pap 67:58

Hsü KJ (1978) Albert Heim: Observations on landslides and relevance to modern interpretaions. In: Voight B (ed) Rockslides and avalanches 1, Natural phenomena. Dev Geotech Engin Amsterdam 14A:71–93

Huang M, Wang Z, Ren J (1982) On the temperate regime of continental-type glaciers in China. J Glaciol 28:117–128
Huang SL (1985) Temperature and time effects on the closure of a gravel room in permafrost. Bull Assoc Engin Geol 22:53–65
Huder JA (1979) Geotechnische Untersuchungen für das Gefrierverfahren. Mitt Versuchsanst Wasserbau Hydrol Glaziol ETH Zürich 41:77–98
Hughes OL (1966) Logan Mountains YT Measurements on a rock glacier. Ice 20:5
Humlum O (1982) Rock glacier types on Disko, Central West Greenland. Geogr Tidsskr 82:59–66
Humlum O (1984) Altitudinal trends and talus-derived rock glaciers on Disko, Central West Greenland. Geogr Tidsskr 84:35–39
Humlum O (1988a) Natural cairns on rock glaciers as an indication of a solid ice core. Geogr Tidsskr 88:78–82
Humlum O (1988b) Rock glacier appearance level and rock glacier initiation line altitude: a methodological approach to the study of rock glaciers. Arct Alp Res 20:160–178
Humlum O, Hambrey MJ, Swett K (1982) Rock glaciers in northern Spitsbergen: discussion and reply. J Geol 90:214–218
Hunter JA (1973) The application of shallow seismic methods to mapping of frozen surficial materials. In: 2nd Int Conf on Permafrost, N Am Contrib, Natl Acad Sci, Washington DC, pp 527–535
Ichim I (1978) Preliminary observations on the rock glacier phenomenon in the Romanian Carpathians. Rev Romaine Géol Géof Géogr Géogr 23:295–299
Iken A (1981) The effect of the subglacial water pressure on the sliding velocity of a glacier in an idealized numerical model. J Glaciol 27:407–421
Iken A, Roethlisberger H, Flotron A, Haeberli W (1983) The uplift of Unteraargletscher at the beginning of the melt season: a consequence of water storage at the bed? J Glaciol 29:28–47
Ivanovskij LN (1981) Gljacial'naja geomorfologija gor (na primere Sibiri i Dal'nego Vostaka). Akad Nauk SSSR Novosibirsk:102–106
Iveronova MI (1950) Kamennye gletcery severnogo Tjan'-Sanja.-In: Raboty tjan'-sanskoj fiz iko-geograficeskaj Stancii. Trudy instituta geografii. Akad Nauk SSSR 44:69–88
Ives JD (1977) Late- and postglacial glacier fluctuations and sea level changes in Arctic Canada. Geogr Ann 59A:253–256
Ives JD, Fahy BD (1971) Permafrost occurrence in the Front Range, Colorado, Rocky Mountains, USA. J Glaciol 10:105–111
Ives RL (1939) Rock glaciers in the Colorado Front Range. Geol Soc Am Abstr 50:1913–1914
Ives RL (1940) Rock glaciers in the Colorado Front Range. Bull Geol Soc Am 51:1271–1294
Iwata S (1976) Some periglacial morphology in the Sagarmatha (Everest) region, Khumbu Himal. In: Higuchi K, Nakajima C, Kusunoki K (eds) Glaciers and climates of Nepal Himalayas: report of the glaciological expedition to Nepal; Part I. J Jpn Soc Snow Ice 38:115–119
Jackson LE, MacDonald GM (1980) Movement of an ice-cored rock glacier, Tungsten, N.W.T., Canada, 1963–1980. Arctic 33:842–847
Jäckli H (1957) Gegenwartsgeologie des bündnerischen Rheingebietes. Beitr Geol Karte Schweiz Geotech Ser 36:136

Jäckli H (1978) Der Blockstrom in der Val dal Acqua im Schweizerischen Nationalpark. In: Kasser P (ed) Jahrb Schweiz Naturforsch Ges 1978, Wiss Teil, pp 213–221

Jakob M (1992a) Blockgletscher im Khumbu Himalaya, Nepal. Diplomarbeit Geogr Inst Heidelberg 123 pp

Jakob M (1992b) Active rock glaciers and the lower limit of discontinuous alpine permafrost, Khumbu Himalaya, Nepal. Permafrost Periglacial Processes 3:253–256

Jakob M (1994) Comments "Debris-covered glaciers in the Sierra Nevada, and their implications for snowline reconstruction", by DH Clark, MM Clark, and AR Gillespie. Quat Res 42:356–358

Jeanneret F (1975) Blockgletscher in den Südalpen Neuseelands. Z Geomorphol NF 19:83–94

Jegerlehner J (1903) Die Schneegrenze in den Gletschergebieten der Schweiz. Gerlands Beitr Geophys 5:486–566

Jeremic ML, Prudhomme RS (1985) Consolidation of mine fill by freezing. Can Inst Mining Metallurgy Bull 78:35

Jessberger HL (1965) Das Kriechverhalten von gefrorenem Lockergestein in Zusammenhang mit der bautechnischen Anwendung des Gefrierverfahrens. Geol Bavar 55:422–439

Jessberger HL (1980) Ground freezing: mechanical properties processes and design. state-of-the-art-report. Norw Inst Technol 18:1–33

Jochimsen M (1966) Ist die Größe des Flechtenthallus wirklich ein brauchbarer Maß stab zur Datierung von glazialmorphologischen Relikten? Geogr Ann 48A:157–164. See: Arct Alp Res 5:417–424

Johnson JP (1969) Investigation on rock glaciers in Jasper National Park. submitted to the National and Historic Parks Branch, Mimeographed Manuscript, 30 pp

Johnson JP (1972) Rock glacier, one mile north by foot trail from Mile 1061, Alaska Highway. In: Hughes OL, Rampton VN, Rutter NW (eds) Field excursion A-11 guidebook, Quarternary geology and geomorphology, southern and central Yukon (northern Canada). 24th Int Geol Congr, Montreal, pp 17–18

Johnson JP (1973) Some problems in the study of rock glaicers. In: Fahey BD, Thompson RD (eds) Proc 3rd Guelph Symp on Geomorphology, Univ Guelph, Norwich, pp 84–94

Johnson JP, Nickling WG (1979) Englacial temperature and deformation of a rock glacier in the Kluane Range, Yukon Territory, Canada. Can J Earth Sci 16:2275–2283

Johnson PG (1971) Ice cored moraine formation and degradation, Donjek Glacier, Yukon Territory, Canada. Geogr Ann 53A:198–202

Johnson PG (1974) Mass movement of ablation complexes and their relationship to rock glaciers. Geogr Ann 56A:93–101

Johnson PG (1975) Mass movement processes in Metalline Creek, south west Yukon Territory. Arctic 28:130–139

Johnson PG (1978) Rock glacier types and their drainage systems, Grizzly Creek, Yukon Territory. Can Earth Sci 15:1496–1507

Johnson PG (1980a) Rock glaciers: glacial and non-glacial origins. In: World glacier inventory: workshop at Riederalp 1978. Int Assoc Hydrol Sci 126:285–293

Johnson PG (1980b) Glacier-rockglacier transition in the Southwest Yukon Territory, Canada. Arct Alp Res 12:195–204

Johnson PG (1981) The structure of a talus-derived rock glacier deduced from its hydrology. Can J Earth Sci 18:1422–1430

Johnson PG (1983) Rock glaciers, a case for a change in nomenclature. Geogr Ann 65 A:27–34

Johnson PG (1984a) Paraglacial conditions of instability and mass movement: a discussion. Z Geomorphol NF 28:235–250

Johnson PG (1984b) Rock glacier formation by high-magnitude low-frequency slope processes in the southwest Yukon. Ann Assoc Am Geogr 74:408–419

Johnson PG (1987) Rock glacier: glacier debris systems or high-magnitude low-frequency flows. In: Giardino JR, Shroder JF Jr, Vitek JD (eds) Rockglaciers. Allen & Unwin, Boston, pp 175–192

Johnson PG (1992a) Micro-relief on a rock glacier, Dalton Range, Yukon, Canada. Permafrost Periglacial Processes 3:41–47

Johnson PG (1992b) Stagnant glacier ice, St. Elias Mountains, Yukon. Geogr Ann 74 A:13–19

Johnson PG, Lacasse D (1988) Rock glaciers of the Dalton Range, Kluane Ranges, south-west Yukon Territory, Canada. J Glaciol 34:327–332

Johnson RB (1967) Rock streams on Mount Mestas, Sangre de Cristo Mountains, southern Colorado. US Geol Surv Prof Pap 575-D:D217–D220

Jong MGG de, Kwadijk JK (1988) Fossil rock glaciers in central Vorarlberg, Austria. Arct Alp Res 20:86–96

Jorda M (1978) Les formations à blocs des milieux subalpins et alpins des Alpes de Haute Provence: Essai d'interprétation génétique et chronologique. In: Union Géographique Internationale, Comité de Coordination des Recherches Périglaciaires, Colloque sur le Périglaciaire d'Altitude du Domaine Méditerranéen et Abords, Strasbourg 1977, pp 31–34

Jorda M (1980a) Morphogenèse et évolution des paysages dans les Alpes de Haute-Provence depuis le Tardiglaciaire. Bull Assoc Géogr Fr 57:295–304

Jorda M (1980b) Eboulis et glaciers rocheux: Eboulis et Environment géographique, passé et actuel. Actes Coll Paris Univ Paris X:107–119

Jorda M (1983) Eboulis et glaciers rocheux. Bull Assoc Geogr Fr 60:15–24

Jorda M, Gabert P (1984) Recherches francaises récentes sur les phénomènes périglaciaires. In: 25th Congr Géogr Int, Paris, Alpes Occidentales et Sud-Est de la France, pp 65–77

Joyce JRF (1950) Stone runs of the Falkland Islands. Geol Mag 87:105–116

Julian M (1978) Ages et origine des glaciers rocheux des Alpes Maritimes. In: Union Géogr Int, Comité de Coordination des Recherches Périglaciers, Coll sur le Periglaciaire d'Altitude du Domaine Méditerranéen et Abords, Strasbourg 1977, pp 45–52

Julian M (1983) Eboulis et glacier rocheux. In: Actes du Coll tenu le 8 Janvier 1983 à Paris lors de la séance de l'Association des Géographes Francais, pp 153–159

Kaelin M (1971) The active push moraine of the Thompson glacier, Axel Heiberg Island, Canadian Arctic Archipelago, Canada. Dissertation No 4671, McGill Univ Montreal, 68 pp

Kaiser B (1975) Etude géodynamique des versants dans le Massif de la Vanoise: Orientations et résultats récents. Trav Sci Parc Nat Vanoise 6:9–40

Kaiser B (1983) Morphodynamique périglaciaire en Vanoise: Observations et mésures sur deux formes majeures: talus d'éboulis et glaciers rocheux. Trav Sci Parc Nat Vanoise 13:55–80

Kaiser B, LeCompte M (1977) Recherche sur la dynamique des versants de montagne dans le Brianconnais: exemple du versant Sud de la tète Noire du Galibier (1950–2843m). Trav Lab Géogr Phys Univ Paris 7, pp 7–31

Karczewski A, Kostrzewski A, Marks L (1981) Morphogenesis of subslope ridges to the north of Hornsund, Spitsbergen. Polish Polar Res 2:29–38

Karrasch H (1974) Probleme der periglazialen Höhenstufe in den Alpen. In: Hans Graul-Festschrift. Heidelb Geogr Arb 40:15–29

Karte J (1978) Die räumliche Abgrenzung und regionale Differenzierung des Periglaziärs. Bochumer Geogr Abh 35:211

Karte J (1983) Periglacial phenomena and their significance as climatic and edaphic indicators. Geo J 7:329–340

Karte J, Liedtke H (1981) The theoretical and practical definition of the term Periglacial in the geographical and geological meaning. Biul Peryglacjalny 28:123–135

Kasser P (1967) Fluctuations of glaciers 1959–1965. IHAS and UNESCO, Paris, 52 pp

Kasser P (1973) Fluctuations of glaciers 1965–1970. ICSI/IAHS and UNESCO, Paris, 357 pp

Kawasaki K, Osterkamp TE, Jurik RW, Kienle J (1983) Gravity measurements in permafrost terrain containing massive ground ice. Ann Glaciol 4:133–140

Kelletat D (1970) Zum Problem der Verbreitung, des Alters und der Bildungsdauer alter (inaktiver) Periglazialerscheinungen im Schottischen Hochland. Z Geomorphol N F 14:510–519

Keller F (1992) Automated mapping of mountain permafrost using the program Permakart within the Geographical Information System ARC/INFO. Permafrost Periglacial Processes 3:133–138

Keller F, Gubler HU (1993) Interaction between snow cover and high mountain permafrost, Murtèl-Corvatsch, Swiss Alps. Proc 6th Int Conf on Permafrost, Beijing Vol 1 pp 332–337

Kerschner H (1974) Besprechung Hans Graul-Festschrift. Z Gletscherk Glazialgeol 10:277–282

Kerschner H (1976) Das Daun und das Egesen-Stadium in ausgewählten Tälern der Zentralalpen von Nordtirol und Graubünden. Dissertation Univ Innsbruck, 305 pp

Kerschner H (1978) Paleoclimatic inferences from Late Würm rock glaciers, Eastern Central Alps, Western Tyrol, Austria. Arct Alp Res 10:635–644

Kerschner H (1981a) Alpengeographie, Klimaschwankungen, Klimageschichte. Österr Alpenver 3:96–97

Kerschner H (1981b) Alpengeographie, Blockgletscher. Österr Alpenver 4:134–135

Kerschner H (1982) Outlines of the climate during the Egesen advance (Younger Dryas, 11.000–10.000 BP) in the Central Alps of the Western Tyrol, Austria. Z Gletscherk Glazialgeol 16,2:229–240

Kerschner H (1983a) Zeugen der Klimageschichte im Oberen Radurschltal. Alpenvereinsjahrbuch 82/83:23–28

Kerschner H (1983b) Late glacial paleotemperatures and paleoprecipitation as derived from permafrost: glacier relationships in the Tyrolean Alps, Austria. In: Proc 4th Conf on Permafrost Fairbanks, Alaska. National Academic Press, Washington, pp 589–594

Kerschner H (1983c) Überlegungen zum Klima während des Egesenstadiums (Jüngere Dryas, 11.000–10.000 B.P.) in den Zentralen Westalpen (Westtirol, Österreich). In:

Schroder-Lanz H (ed) Late- and postglacial oscillations of glaciers: glacial and periglacial forms (in memoriam H. Kinzl). Balkema, Rotterdam, pp 65–72

Kerschner H (1985) Quantitative paleoclimatic inferences from lateglacial snowline, timberline and rock glacier data, Tyrolean Alps, Austria. Z Gletscherkd Glazialgeol 21:363–369

Kershaw GP (1978) Rock glaciers in the Cirque Lake area of the Yukon-Northwest Territories. Albertan Geogr 14:61–88

Kesseli EJ (1941) Rock Streams in the Sierra Nevada, California. Geogr Rev 31:203–227

Keusen HR, Haeberli W (1983) Site investigation and foundation design aspects of cable car construction in Alpine permafrost at the "Chli Matterhorn", Wallis, Swiss Alps. In: Proc 4th Int Conf on Permafrost, Fairbanks, Alaska, National Academic Press, Washington pp 601–605

King L (1976) Permafrostuntersuchungen in Tarfala (Schwedisch-Lappland) mit Hilfe der Hammerschlagseismik. Z Gletscherkd Glazialgeol 12:187–204

King L (1982) Qualitative und quantitative Erfassung von Permafrost in Tarfala (Schwedisch-Lappland) und Jotunheimen (Norwegen) mit Hilfe geoelektrischer Sondierungen. In: Barsch D (ed) Experimente und Messungen in der Geomorphologie. Z Geomorphol Suppl 43:139–160

King L (1983) High mountain permafrost in Scandinavia. In: Proc 4th Int Conf on Permafrost, Fairbanks, Alaska, National Academic Press, Washington, pp 612–617

King L (1984) Permafrost in Skandinavien: Untersuchungsergebnisse aus Lappland, Jotunheimen und Dovre/Rondane. Heidelb Geogr Arb 76:174

King L (1986) Zonation and ecology of high mountain permafrost in Scandinavia. Geogr Ann 68A:131–139

King L, Hell G (1993) Photogrammetry and geomorphology of high arctic push moraines, examples from Ellesmere Island, Canadian Arctic and Spitzbergen, Svalbard Archipelago. Z Geomorphol N F Suppl 92:21–38

King L, Lehmann R (1973) Beobachtungen zur Ökologie und Morphologie von Rhizocarpon geographicum (L.) DC. und Rhizocarpon alpicola (Hepp.) Rabenh. im Gletschervorfeld des Steingletschers. Ber Schweiz Bot Ges 83:139–147

King L, Fisch W, Haeberli W, Waechter HP (1987) Comparison of resistivity and radio-echo soundings on rockglacier permafrost. Z Gletscherk Glazialgeol 23:77–97

King L, Gorbunov AP, Evin M (1992) Prospecting and mapping of mountain permafrost and associated phenomena. Permafrost Periglacial Processes 3:73–81

Kinzl H (1932) Die größten nacheiszeitlichen Gletschervorstöße in den Schweizer Alpen und der Mont-Blanc-Gruppe. Z Gletscherk Glazialgeol 20:269–397

Klaer W (1962a) Die periglaziale Höhenstufe in den Gebirgen Vorderasiens. Z Geomorphol N F 6:17–32

Klaer W (1962b) Untersuchungen zur klimagenetischen Geomorphologie in den Hochgebirgen Vorderasiens. Heidelb Geogr Arb 11:135

Klaer W (1974) Kritische Anmerkungen zur neueren Literatur über das Blockgletscherproblem. In: Eichler H (ed) Hans Graul-Festschrift. Heidelb Geogr Arb 40:275–291

Klaer W (1983) Die Blockgletscherfrage, ein terminologisches Problem? In: Poser H, Schunke E (eds) Mesoformen des Reliefs im heutigen Periglazialraum. Abh Akad Wiss Math Phys Kl Folge 3, 35:120–132

Knoche W (1932) Verdunstungseis auf dem Anden-Vulkan Oyahue. Z Gletscherk Glazialgeol 20:100–102

Koefoed O (1979) Geosounding principles: resistivity sounding measurements. Elsevier, Amsterdam, 276 pp

Kononova RS (1978) Hydrogeochemical zoning of subsurface water as one of the indicators of paleo-permafrost conditions. In: 2nd Int Conf on Permafrost, USSR Contrib, Natl Acad Sci, Washington DC, 422–424

Kotarba A (1988) Fossil rock glaciers in the Polish Tatra Mountains: origin and age. In: Pécsi M and Starkel L (eds) Paleogeography of Carpathian Regions. Geogr Res Inst Hungarian Acad Sci Budapest, pp 161–169

Kozhevnikov AV (1984) Genetic classification of anthropogene sedimentary formations. Vestnik Moskovskogo Univ Geol 39:24–39

Kramer FL (1962) Rivers of stone. Pac Discovery 15:11–15

Krasnoslobodcev IS (1971) Okamennich gletcerach bol'sogo kavkaza. Vestnik Moskovskogo Univ 1:95–97

Krebs N (1925) Klimatisch bedingte Bodenformen in den Alpen. Geogr Z 31:98–108

Kristiansen KJ, Sollid JL (1985) Börselvfjellet-Lille Porsangen, Nord Norge. Kvartärgeologisk og geomorfologisk kart 1:75000. Geografisk Institut, Universitetet Oslo

Kristjansson FJ (1980) Cirque multiple glaciation, Grizzly Creek, Yukon Territory and palaeoclimatic implications. Master's Thesis, Dep Geol Univ Ottawa, Canada, 55 pp

Kroh H (1991) Zeit-und temperaturabhängige Hangbewegungen am Beispiel von Blockgletschern. Diplomarbeit, Univ Kaiserslautern, 63 pp

Krummenacher B, Budmiger K (1992) Monitoring of periglacial phenomena in the Furggentäli, Swiss Alps. Permafrost Periglacial Processes 3:149–155

Lachenbruch AH, Cladouhos TT, Saltus RW (1988) Permafrost temperature and the changing climate. Proc 5th Int Conf on Permafrost, Trondheim, 3:9–17

Lagasquie JJ (1985) Les formations froides de versant dans la vallée du Lot entre Cajarc et Ambeyrac. In: Notes et comptes rendues du Groupe de Travail "Régionalisation du Périglaciaire", Com Nat Fr Géogr 10:3–17

Lamb B, Haworth LA, Calkin PE, Ellis JM (1983) Holocene landscape alteration in an alpine terrain, Brooks Range, Alaska. Geol Soc Am Abstr Progr 15:144

Lambert JE (1982) Relative age of rock glaciers in Grizzly Creek, St. Elias Mountains, Yukon Territory, Canada. Master's Thesis, Dep Geogr, Univ Guelph, Canada, 207 pp

Lautridou J-P, Francou B, Hall K (1992) Present-day periglacial processes and landforms in mountain areas. Permafrost Periglacial Processes 3:93–101

Lehnerer B (1988) Auswertung und Interpretation von Bewegungsmessungen an einem Blockgletscher. Diplomarbeit, Univ Karlsruhe, 132 pp

Leiva JC (1982) Le glacier "Piloto" du Cajon del Rubio (Andes Centrales Argentines). CNRS, Diss, Univ Grenoble

Lerbekmo JF, Campbell FA (1969) Distribution, composition and source of the White River Ash, Yukon Territory. Earth Sci 6, 12:109–116

Leutelt R (1935) Glazialgeologische Beobachtungen im Lasistanischen Hochgebirge. Z Gletscherk Glazialgeol 23:67–80

Lewis CA (1994) Protalus ramparts and the altitude of the local equilibrium line during the last glacial stage in Bokspruit, East Cape Drakensberg, South Africa. Geogr Ann 76A:37–48

Lewis CA, Hanvey PM (1993) The remains of rock glaciers in Bottleneck, East Cape Drakensberg, South Africa. Trans R Soc Afr 48:265–289
Libby WG (1968) Rock glaciers in the north Cascade Range, Washington. Geol Soc Am Spec Pap Abstr 101:318–319
Lichtenhahn C (1971) Zwei Stollenbauten: Stollen im Eis zur Verhinderung von Ausbrüchen eines Sees im Grubengletschergebiet (Wallis) und Stollen im Felsen zur unterirdischen Entwässerung des Rutschgebietes von Camp Vallemaggia (Tessin). Interpraevent 2:465–475
Lieb GK (1986) Die Blockgletscher der östlichen Schobergruppe (Hohe Tauern, Kärnten). In: Festschrift für Wilhelm Leitner. Arb Inst Geogr Karl-Franzens-Univ Graz 27:123–132
Lieb GK (1987) Zur spätglazialen Gletscher- und Blockgletschergeschichte im Vergleich zwischen den Hohen und Niederen Tauern. Mitt Österr Geogr Ges 129:5–27
Lieb GK (1991) Die horizontale und vertikale Verteilung der Blockgletscher in den Hohen Tauern (Österreich). Geomorphol N F 35:345–365
Lieb GK, Slupetzky H (1993) Der Tauernfleck-Blockgletscher im Hollerbachtal (Venedigergruppe, Salzburg, Österreich). Wiss Mitt Nationalpark Hohe Tauern 1:138–146
Liestöl O (1962) Talus terraces in arctic regions. Nor Polarinst Arbok 1961:102–105
Lindner L, Marks L (1985) Types of debris slope accumulations and rock glaciers in South Spitsbergen. Boreas 14:139–153
Lliboutry L (1953) Internal moraines and rock glaciers. Glaciol 18:255–273
Lliboutry L (1955a) Origine et évolution des glaciers rocheux. Séanc Acad Sci Paris 240:1913–1915
Lliboutry L (1955b) L'incorporation des éboulis dans la glace. C R Séanc Acad Sci
Lliboutry L (1956) Observation d'éboulis à lits de limon en cours de formations anciennes dans les Andes de Santiago. C R Séanc Acad Sci Paris 243:2108–2111
Lliboutry L (1961a) Phénoménes cryoniaux dans les Andes de Santiago (Chile). Biul Peryglacjalny 10:209–224
Lliboutry L (1961b) Les glaciers enterrés et leur rôle morphologique. In: Symp Helsinki, Int Assoc Sci Hydrol Pub 54:272–280
Lliboutry L (1965) Traité de glaciologie 2. Masson, Paris, 612 pp
Lliboutry L (1977) Movement of a covered glacier embeded within a rock glacier. Glaciol 18:255–273
Lliboutry L (1979) Local friction laws for glaciers: a critical review and new openings. Glaciol 23:67–95
Lliboutry L (1986) Discharge of debris by glaciar Hatunraju, Cordillera Blanca, Peru. Glaciol 32:133
Loewenherz DS, Lawrence CJ, Weaver RL (1989) On the development of transverse ridges on rock glaciers. J Glaciol 35:383–391
Loomis SR (1970) Morphology and ablation processes on glacier ice. Arct Inst N Am Res Rep 57:1–65
Lougeay R (1972) Patterns of surface temperature in the alpine/periglacial environment as determined by radiometric measurements. In: Bushnell VC, Ragle RH (eds) Icefield Ranges Research Project. Sci Results Montreal 3:163–176
Lougeay R (1978) Observations of ice-cored moraine with Landsat. Geol Soc Am Abstr Progr 10:447

Lougeay R (1982) Comparative morphology of rock glaciers. Proc Rochestr Acad Sci 15:35

Louis H (1969) Vorwort zum Sonderband "Glazialmorphologie". Geomorphol NF (Suppl) 8:1–7

Luckman BH (1976) Neoglaciation and rock glacier activity in the Jasper area, Alberta. Am Quat Assoc Nat Conf Abstr 4:109

Luckman BH (1981) The geomorphology of the Alberta Rocky Mountains: a review and commentary. In: Slaymaker O (ed) High mountains. Geomorphol Suppl 37:91–119

Luckman BH, Crocket KJ (1978) Distribution and characteristics of rock glaciers in the southern part of Jasper National Park, Alberta. Can J Earth Sci 15:540–550

Luckman BH, Osborn GD (1979) Holocene glacier fluctuations in the middle Canadian Rocky Mountains. Quat Res 11:52–77

Lütschg O (1974) Beitrag zur Kenntnis des Wärmehaushaltes der obersten Bodenschichten der Schweizer Alpen. Denkschr Schweiz Naturforsch Ges 77:83–167

Lundquist G (1949) The orientation of the block material in certain species of earth flow. Geogr Ann 31:335–347

Lutton RJ (1973) Natural rubble slopes and their relevance to crater fallback slopes. Geomech Abstr 4:210

Madole RF (1972) Neoglacial facies in the Colorado Front Range. Arct Alp Res 4:119–130

Madole RF (1976) Glacial geology of the Front Range, Colorado. In: Mahaney WC (ed) Quarternary stratigraphy of North America. New York, pp 297–3178

Madole RF (1980) Time of Pinedale deglaciation in north-central Colorado: Further considerations. Geology 8:118–122

Mahaney WC (1973) Neoglacial chronology in the Fourth of July Cirque, Central Colorado Front Range. Arct Alp Res 4:119–130

Mahaney WC (1980) Late Quaternary rock glaciers, Mount Kenya, Kenya. Glaciol 25:492–497

Maisch M (1981) Glazialmorphologische und gletschergeschichtliche Untersuchungen im Gebiet zwischen Landwasser- und Albultal (Kt. Graubünden, Schweiz). Phys Geogr Zür 3:215

Maisch M, Haeberli W (1982) Interpretation geometrischer Parameter von Spätglazialgletschern im Gebiet Mittelbünden, Schweizer Alpen. In: Furrer G, Keller WA, Gamper M, Suter J (eds) Beiträge zur Quartärforschung in der Schweiz. Phys Geogr Zür 1:111–126

Makshimjak RV, Vyalov SS, Chapaev AA (1983) Methods for determining long-term strength of frozen soils. In: Proc 4th Int Permafrost Conf Fairbanks, Alaska. National Academic Press, Washington, pp 783–786

Maksimov EV (1963) About rock glaciers of Dzhvngarsky Alatau. In: Materials on glaciological investigations. Chronicle 7:158–160

Marangunic C (1976) El glaciar de roca Pedregoso, Rio Colorado, V región. Congr Geol Chil Actas 1:D71–D80

Marnezy A (1977) Glaciers rocheux et phénomènes périglaciaires dans le Vallon de la Vanoise. Rev Géogr Alp 65:147–165

Martin HE, Whalley WB (1987a) Rock glaciers. Part I: Rock glacier morphology: classification and distribution. Progr Phys Geogr 11:260–282

Martin HE, Whalley WB (1987b) A glacier ice-cored rock glacier, Tröllaskagi, Iceland. Jökull 37:47–55

Martinec J, Oeschger H, Schotterer U, Siegthaler U (1982) Snowmelt and groundwater storage in an Alpine basin. In: Hydrological aspects of alpine and high mountain areas. Int Assoc Hydrol Sci Publ 138:169–175
Matthes FE (1900) Glacial sculpture in the Bighorn Mountains, Wyoming. Geol Survey 21st Annual Rep 2:167–190
Matthes FE (1948) Moraines with ice cores in the Sierra Nevada. Sierra Club Bull 33:87–96
Matveev SN (1938) Kamennye potoki. In: Problemy fiziceskoj geografii. Akad Nauk SSSR Inst Geogr 6:91–124
Mayewski PA (1979) Glaciology and glacio-geomorphology in Victoria Land and Queen Maud Mountains. Antarct J US 14:53
Mayewski PA, Hassinger J (1980) Characteristics and significance of rock glaciers in southern Victoria Land, Antarctica. Antarct J US 15:68–69
Mayewski PA, Jeschke PA, Ahmad N (1981) An active rock glacier, Wavbal Pass, Jammu and Kashmir Himalaya, India. J Glaciol 27:201–202
Mayr F, Heuberger H (1968) Type areas of lateglacial and postglacial deposits in Tyrol, Eastern Alps. In: Richmond GM (ed) Glaciations of the Alps. Univ Colorado, Boulder, Colorado pp 143–165
McCall JG (1960) The flow characteristics of a cirque glacier and their effects on glacial structure and cirque formation. In: Lewis, WV (ed) Norwegian cirque glacier. R Geog Soc Ser Lond 4:39–62
McGregor VR (1967) Holocene moraines and rock glaciers in the Central Ben Ohau Range, South Canterbury, New Zealand. J Glaciol 6:737–748
McRoberts EC (1974) Stability of slopes in frozen soil, Mackenzie Valley NWT Can Geotech J 11:554–573
McRoberts E (1975) Some aspects of a simple secondary creep model for deformations in permafrost slopes. Can Geotech J 12:98–105
McSaveney ER (1971) The surficial fabric of rockfall talus. In: Quantitative geomorphology. 2nd Annu Geomorphol Symp Ser, State Univ New York, Binghamton, NY, pp 181–197
Meier K-D (1987) Studien zur periglaziären Landschaftsformung in Finnmark (Nordnorwegen). Jahrb Geogr Ges Hannover Sonderh 13:298
Meissner H (1985) Bearing behavior of frost shells in the construction of tunnels. In: Kinosida S, Fukuda M (eds) 4th Int Symp on Ground Freezing. Sapporo, vol 2 Balkema, Rotterdam, pp 37–45
Mensching H (1953) Morphologische Studien im Hohen Atlas von Marokko. Würzb Geogr Arb 1:104
Mensching H (1955) Das Quartär in den Gebirgen Marokkos. Petermanns Geogr Mitt Ergänzungsh 256:79
Mensching H (1977) Bemerkungen zum Problem einer "periglazialen" Höhenstufe in den Gebirgen der ariden Zone im nördlichen Afrika. In: Poser H (ed) Formen, Formengesellschaften und Untergruppen in den heutigen periglazialen Höhenstufen der Hochgebirge Europas und Afrikas. Abh Akad Wiss Gött Math Phys Kl Folge 3, 31:290–299
Mercanton PL (1923) Les variations périodique des glaciers des Alpes Suisses. Les Variations Périodiques des Glaciers des Alpes Suisses 44:254–278
Mercanton PL (1935) Le cryocinemetre de la Commission Helvetique des glaciers. Z Gletscherk Glazialgeol 22:163–171

Messerli B (1965) Beiträge zur Geomorphologie der Sierra Nevada, Andalusien. Juris Verlag, Zürich, 215 pp

Messerli B (1967) Die eiszeitliche und die gegenwärtige Vergletscherung im Mittelmeerraum. Geogr Helv 22:105–228

Messerli B (1972) Formen und Formungsprozesse in der Hochgebirgsregion des Tibesti. Hochgebirgsforschung 2:23–86

Messerli B (1978) Probleme des Periglazials in den Gebirgen der ariden Zone. In: Raynal R (ed) Colloq sur le Périglaciaire d'Altitude du Domaine Méditerranéen et Abords, Strasbourg-Univ Louis Pasteur, pp 333–345

Messerli B, Zurbuchen M (1968) Blockgletscher im Weissmies und Aletsch und ihre photogrammetrische Kartierung. Die Alpen 3:139–152

Meyer-Harrassowitz HLF (1916) Die Blockfelder im östlichen Vogelsberg. Versammlungsber Niederrhein Geol Ver 1915/16:29–49

Michaud, J (1951) Employ de marques dans l'étude des mouvements du sol au Chambeyron (Basses Alpes). C R Acad Sci 230:314–315

Michaud J, Cailleux A (1950) Vitesse du mouvement du sol au Chambeyron (Basses-Alpes). C R Hebdomadaires Séanc Acad Sci 230:314–315

Miller CD (1973) Chronology of neoglacial deposits in the Northern Sawatch Range, Colorado. Arct Alp Res 5:385–400

Miller GH (1973) Late Quaternary glacial and climatic history of Northern Cumberland Peninsula, Baffin Island, N.W.T., Canada. Quat Res 3:561–583

Miller MN, Andesron JH (1968) Rock glaciers of the Dezadeach District Yukon Territory, Canada. In: 19th Alaska Sci Conf. Whitehorse, Yukon, Canada, 2 pp

Mills HH (1976) Estimated erosion rates on Mt. Rainier, Wahington. Geology 4:401–406

Minetti JL, Corte AE (1984) Zonificación latialtitudinal del clima en la zona andina y su relación el limite inferior del hielo perenne (LIHP) y del límite inferior geocriogénico (LIG). In: Corte AE (ed) Actas 2da reunión Grupo Periglacial Argentino, 1984. An Inst Argent Nivol Glaciol:129–143

Mjagkov SM (1980) Kamennye gletcery transantarkticeskichgor. Antarctica 20:89–92

Mörner NA (1977) Southward displacement of the distribution of glaciation during the three maxima of the last ice age. J Glaciol 18:305–308

Monbaron M, Tenthorey G (1989) Hydrologie en zone périglaciaire au Vallon de Réchy (VS) – essai de coloration sur un glacier rocheux. Bull Soc Neuchâteloise Géogr 32/33:117–128

Moore DW, Friedmann I (1991) Longitudinal section of an alpine rock glacier exposed south of Berthoud Pass, central Colorado Front Range. Geol Soc Am Rocky Mountain and South Central Sections, Abstr Progr 23, 50 (1412)

Morales SE (1965) The Huascaran avalanche in the Santa Valley, Peru. In: Int Assoc Hydrol Sci, Davos Symp. Publ 69:304–315

Morris SE (1979) Holocene facies of the Blanca Massif, southern Colorado. Geol Soc Am Abstr Progr 11:297

Morris SE (1981) Topoclimatic factors and the development of the rockglacier facies, Sangre de Cristo Mountains, southern Colorado. Arct Alp Res 13:329–338

Morris SE (1987) Regional and topoclimatic implications of rock glacier stratigraphy: Blanca Massif, Colorado. In: Giardino JR, Shroder JF, Vitek JD (eds) Rock glaciers. Allen & Unwin, London, pp 107–125

Mougin JP (1974) Stabilité des talus: établissement d'abaques indiquant les modifications du coéfficient de securité en fonction de la position de la nappe. Int Assoc Engin Geol, Int Congr Proc 2, pp V18.1–V18.6
Mountain Agenda (1992): An Appeal for the Mountains. Geogr Inst Univ Bern, 44 pp
Müller F, Calfisch T, Mueller G (1976) Firn und Eis der Schweizer Alpen. Geogr Inst ETH Zür Publ 57:293
Müller H-N, Stötter J, Schubert A, Betzler A (1984) Glacial and periglacial investigations in Skidadalur, Trollaskagi, Northern Iceland. Polarforschung 54:95–109
Muller SW (1947) Permafrost or permanently frozen ground and related engineering problems. JW Edwards, Ann Arbor, 231 pp
Mundry E, Greinwald S, Knödel K, Losecke W, Meiser P, Reitmayer G (1985): Geoelektrik. In: Bender F (ed) Angewandte Geowissenschaften. Enke, Stuttgart, pp 299–434
Nagl H (1971) Zur Erkenntnis quartärer Klimaschwankungen aus geomorphologischen Erscheinungen am Beispiel des Pöllatals (Hafnergruppe, Kärnten). Carinthia 161:9–30
Nagl H (1976) Die Raum-Zeit-Verteilung der Blockgletscher in den Niederen Tauern und die eiszeitliche Vergletscherung der Seckauer Tauern. Mitt Naturwiss Ver Steiermark 106:95–118
Nangeroni G (1929) Grotte e laghi subglaciali, colate e mari di pietre. Natura 20:152–161
Naumann CF (1854) Lehrbuch der Geognosie V. II. Leipzig, 1022 pp
Nemcok A, Mahr T (1974) Kammene ladvoce v Tatrach (Les glaciers de pierre dans la Tatra). Geogr Casopis 26:359–373
Nicholas JW (1994) Fabric analysis of rock glacier debris mantles, La Sal Mountains, Utah. Permafrost Periglacial Processes 5:53–66
Nickling WG, Bennett L (1984) The shear strength characteristics of frozen coarse granular debris. J Glaciol 30:348–357
Nicod J (1983) Essai sur le répartition, la dynamique et l'age des glaciers rocheux du Beaufortin (Savoie). In: Actes du Coll tenu le 8 Janvier 1983 à Paris lors de la séance de l'Assoc Géogr Fr, pp 161–170
Nixon JF (1984) Laterally loaded piles in permafrost. Can Geotech J 21:431–438
Odell NE (1948) Stagnant glacier in British Columbia. J Glaciol 1:191
Olyphant GA (1981) Allometry and cirque evolution. Geol Soc Am Bull 92:679–685
Olyphant GA (1983a) Analysis of the factors controlling cliff burial by talus within Blanca Massif, southern Colorado, U.S.A. Arct Alp Res 15:65–75
Olyphant GA (1983b) Computer simulation of rock glacier development under viscous and pseudoplastic flow. Geol Soc Am Bull 94:499–505
Olyphant GA (1985) Topoclimate and the distribution of neoglacial facies in the Indian Peak section of the Front Range, Colorado, U.S.A. Arct Alp Res 17:69–78
Olyphant GA (1987) Rock glacier response to abrupt changes in talus production. In: Giardino JR, Shroder JF, Vitek JD (eds) Rock glaciers. Allen & Unwin, London, pp 55–64
Ommanney CSL (1976a) Canadian rock glaciers. Can Assoc Geogr Annu Meet Progr Resumes, pp 156–157
Ommanney CSL (1976b) Canadian rock glacier studies. Paper presented at the Can Assoc Geogr Annu Meet, Laval Univ, Quebec City, May 23–27, (unpublished)
Ommanney CSL (1979) Glacier studies. Can Geophys Bull 32:167–174

Ono Y, Watanabe T (1986) A protalus rampart related to alpine debris flows in the Kuranosuke cirque northern Japanese alps. Geogr Ann 68A:213–223

Osborn G (1975) Advancing rock glacier in the Lake Louise Area, Banff National Park, Alberta. Can J Earth Sci 12:1060–1062

Osborn G (1985) Discussion of: "Movement of ice cemented rock glaciers by hydrostatic pressure: an example from Mt. Mestas, Colorado". Z Geomorphol NF 29:373–374

Ostercamp TE (1983) Thermal models for estimating the potential impact of a warmer climate on permafrost in Alaska. Interim rep to the Alaska Counc Sci Technol for grant W-81-03, UAG R-295, 29pp

Ostercamp TE, Payne MW (1981) Estimates of permafrost thickness from well logs in Northern Alaska. Cold Regions Sci Technol 5:13–27

Outcalt SE, Benedict JB (1965) Photo-interpretation of two types of rock glaciers in the Colorado Front Range, USA. J Glaciol 5:849–856

Oxford SP (1985) Protalus ramparts, protalus rock glaciers and soliflucted till in the northwest part of the English Lake District. In: Boardman J (ed) Field guide to the periglacial landforms of northern England. Quat Res Assoc, Cambridge, pp 38–46

Ödegard RS, Sollid JL, Liestöl O (1992) Ground temperature measurements in mountain permafrost, Jotunheimen, South Norway. Permafrost Periglacial Processes 3:231–234

Östrem G (1959) Ice melting under a thin layer of moraine and the existence of ice in moraine ridges. Geogr Ann 41:228–230

Östrem G (1963) Comparative crystallographic studies on ice from ice-cored moraines, snow-banks and glaciers. Geogr Ann 45:210–240

Östrem G (1964a) Ice-cored moraines in Scandinavia. Geogr Ann 46:282–337

Östrem G (1964b) Ice crystals from an ice-cored moraine on Baffin Island, N.W.T. Geogr Bull 22:72–79

Östrem G (1965) Problems of dating ice-cored moraines. Geogr Ann 47A:1–38

Östrem G (1966) Ice-cored moraines in the Kebnekajse area. Biul Peryglacjalny 11:271–278

Östrem G (1967) Laboratory measurements of the resistivity of ice. J Glaciol 6:643–650

Östrem G (1971) Rock glaciers and ice cored moraines, a reply to D. Barsch. Geogr Ann 53A:207–213

Östrem G, Arnold K (1970) Ice-cored moraines in southern British Columbia and Alberta, Canada. Geogr Ann 52A:120–128

Owen LA Benn DI, Derbyshire E, Evans DJA, Mitchell WA, Thompson D, Richardson S, Lloyd M, Holden C (1995): The geomorphology and landscape evolution of the Lahul Himalaya, northern India. Z Geomorphol NF 39:145–174

Papertzian VC (1973) A study of rock glaciers in Banff National Park, Alberta, Canada. BS Thesis, Dept Geol, Queens Univ, Kingston, Ontario, 107 pp

Pariseau WG, Voight B (1978) Rockslides and avalanches: basic principles and perspectives in the realm of civil and mining operations. In: Voight B (ed) Rockslides and avalanches 2, engineering sites. Dev Geotech Engin Amsterdam 14B:1–92

Parson CG (1980) Rock glaciers and site characteristics on the Blanca Massif in South-Central Colorado. PhD Diss, Univ Iowa, Iowa City, 212 pp

Parson CG (1987) Rock glaciers and site characteristics on the Blanca Massif, Colorado, USA In: Giardino JR, Shroder JF, Vitek JD (eds) Rock glaciers. Allen & Unwin, London, pp 127–144

Parsons WH (1939) Glacial geology of the Sunlight Area, Wyoming. J Geol 47:737–747
Partsch J (1882) Die Gletscher der Vorzeit in den Karpathen und den Mittelgebirgen Deutschlands. Moebner, Breslau, 198 pp
Paschinger H (1954) Würmvereisung und Spätglazial in der Sierra Nevada, (Spanien). Z Gletscherkd Glaziageol 3:55–67
Paterson WSB (1981) The physics of glaciers. Pergamon Press, Oxford, 380 pp
Paterson WSB (1994) The physics of glaciers, 3rd edn. Elsevier, Amsterdam, 480 pp
Patton HB (1910) Rock streams of La Veta Peak, Colorado. Bull Geol Soc Am 21:663–676
Patzelt G (1972) Die Spätglazialen Stadien und postglazialen Schwankungen der Ostalpengletscher. Ber Dtsch Bot Ges 85:47–57
Patzelt G (1977a) Statistik der Längenmessungen an den österreichischen Gletschern 1960–1975. Z Gletscherkd Glazialgeol 12:91–94
Patzelt G (1977b) Der zeitliche Ablauf und das Ausmaß postglazialer Klimaschwankungen in den Alpen. In: Frenzel B (ed) Dendrochronologie und postglaziale Kilmaschwankungen in Europa. Steiner, Wiesbaden, pp 248–259
Patzelt G (1983) Die spätglazialen Gletscherstände im Bereich des Mieslkopfes und im Arztal, Tuxer Voralpen, Tirol. Arbeiten zur Quartär- und Klimaforschung Innsbruck, Geogr Stud 8 (Festschrift F, Fliri):35–44
Pavlik HF (1980) A physical framework for describing the genesis of ground ice. Progr Phys Geogr 4,4:531–548
Pekala K (1969) Block fields and recent morphogenetic processes in the Western Bieszczady Mountains. Ann Univ M. Curie-Sklodowska Sec B24:85–87
Peltier LC (1945) Block fields in Pennsylvania. Geol Soc Am Bull 56:1190
Penck A (1894) Morphologie der Erdoberfläche, 2 vols. Engelhorn, Stuttgart, 471, 696 pp
Peulvast JP (1981) Glaciers converts et glaciers rocheux fossiles aux Iles Lofoten (Norvége du Nord). In: Hommages á René Raynal. Rech Géogr Strasbourg 16, 17:147–159
Pewe TL (1983) Alpine permafrost in the contiguous United States: a review. Arct Alp Res 15:145–156
Pierce WG (1961) Permafrost and thaw depressions in a peat deposit in the Beartooth Mountains, Northwestern Wyoming. US Geol Surv Prof Pap 424B:154B–156B
Pillewizer W (1938) Photogrammetrische Gletscheruntersuchungen im Sommer 1938. Z Ges Erdkd Berl 9/1:367–372
Pillewizer W (1957) Untersuchungen an Blockströmen der Ötztaler Alpen. Geomorphol Abh Abh Geogr Inst Freien Univ Berlin 5 (Otto-Maull-Festschrift):37–50
Pissart A (1964) Vitesse des mouvements du sol au Chambeyron. Biul Peryglacjalny 14:303–309
Pissart A (1972) Vitesse des mouvements de pierres dans les sols et sur des versants périglaciaires au Chambeyron (Basses Alpes). Congr Colloq Univ Liége 67:251–268
Pissart A (1975) Glace de segrégation, soulèvement du sol et phénomènes thermokarstiques dans les régions à pergélisol. Bull Soc Géogr Liège 11:89–96
Pissart A, Francou B (1992) Vertical Movements of boulders in a subnival boulder pavement at 2800 m a.s.l. in the Alps (France). Permafrost Periglacial Processes 3:203–208
Polanski J (1964) Description geologica de la Hoja 25a, Volcan San Jose, Provincia de Mendoza. Direcion Nac Geol Mineria Bol 98:94

Polanski J (1965) The maximum glaciation in the Argentine Cordillera. Geol Soc Am Spec Pap 84:453–472

Porter SC (1977) Present and past glaciation threshold in the Cascade Range, Washington, USA: topographic and climatic controls, and paleoclimatic implications. J Glaciol 18:101–115

Porter SC, Denton GH (1967) Chronology of neoglaciation in the North American Cordilliera. Am J Sci 265:177–210

Poser H, Schunke E (1983) Mesoformen des Reliefs im heutigen Periglazialraum. Abh Akad Wiss Gött Math Phys Kl 3, Folge 35:133–150

Potter N (1967) Rock glacier and mass wastage in the Galena Creek area northern Absaroka Mountains, Wyoming. US Army Natick Lab, Tech Rep, Nattick, Mass (mimeographed), 75 pp

Potter N (1968) Galena Creek rock glacier, northern Absaroka Mountains, Wyoming (Abstract). Geol Soc Am Spec Pap 115:438

Potter N (1969) Rock glaciers and mass-wastage in the Galena Creek area, northern Absaroka Mountains. PhD Thesis, Univ Minnesota, 150 pp

Potter N (1972) Ice-cored rock glacier, Galena Creek, northern Absaroka Mountains, Wyoming. Bull Geol Soc Am 83:3025–3058

Potter N (1974) Snow avalanches and rock glaciers, Sunlight Mining Region, northern Absaroka Mountains, Wyoming. In: Voight B, Voight MA (eds) Rock mechanics. Pennsylvania State Univ, University Park, pp 51–55

Potter N, Moss JH (1968) Origin of the Blue Rocks block field and adjacent deposits, Berks County, Pennsylvania. Geol Soc Am Bull 79:255–262

Powell DR (1963) Physical geography of the White Mountains, California-Nevada. Univ California, Berkely, 150 pp unpublished

Pröhl E (1977) Bestimmung von Blockgletscherbewegungen aus Luftbildern. Diplomarbeit, Univ Darmstadt 39 pp

Qiu G, Li S, Ting L (1982) Some new data in high mountain permafrost and periglacial phenomena in the vincinity of Tianshan Station. Annu Rep on the work at Tianshan Glaciological Station I (1980–81), Lanzhou Inst Glaciol Cryopedol: 113–122 (in Chinese)

Rapp A (1960a) Recent development of mountain slopes in Kärkevagge and surroundings, northern Scandinavia. Geogr Ann 42:71–200

Rapp A (1960b) Talus slopes and mountain walls at Tempelfjorden, Spitsbergen. Nor Polarinst Skrifter 119:1–96

Rapp A, Stromquist L (1979) Field experiments on mass movements in the Scandinavian mountains with special reference to Kärkevagge; Swedish Lappland. Stud Geomorphol Carpatho-Balcanica 11:23–40

Rapp A, Nyberg R, Lindh L (1986) Nivation and local glaciation in N and S Sweden: a progress report. Geogr Ann 68A:197–205

Rathjens C (1978) Klimatische Bedingungen der Solifluktionsstufe im sommertrockenen Hochgebirge, am Beispiel des afghanischen Hindukusch. In: Mensching H (ed) Beiträge zur Geomorphologie der ariden Zone. Geomorphol Suppl 30:132–142

Rathjens C (1982) Vergleich und Typenbildung in der Hochgebirgsforschung. In: Meynen E, Plewe E (ed) Beiträge zur Hochgebirgsforschung und zur Allgemeinen Geographie. Festschr Harald Uhlig 2, Erdkdl Wiss Wiesbaden 59:1–8

Raymond CF (1980) Temperate valley glaciers. In: Colbeck SC (ed) Dynamics of snow and ice masses. Academic Press, New York, pp 79–139

Raynal R (1952) La région de la haute Moulouya. In: Aspects de la géomorphologie du Maroc. Notes Mém Serv Géol Maroc 96:53–70

Raynal R (1960) Quelques apercus sur l'éxistance et l'importance de phénomènes périglaciaires wurmiens au Maroc. Congr Coll Univ Liége: "Le Périglaciaire pré-wurmiens", 17:109–122

Renner F (1982) Beiträge zur Gletschergeschichte des Gotthardgebietes und dendroklimatologische Analysen an fossilen Hölzern. Phys Geogr Zür 8:182

Resnati C, Smiraglia C (1990) Determinazione della struttura interna del rock glacier di Val Pisella/(Alta Valtellina) attraverso sondaggi eletrici verticali. Risultati e problemi. Geogr Fis Dinam Quat 13:171–177

Richmond GM (1952) Comparison of rock glaciers and block streams in the La Sal Mountains, Utah. Bull Geol Soc Am 63:1292–1293

Richmond GM (1962) Quaternary stratigraphy of the La Sal Mountains, Utah. US Geol Surv Prof Pap 324:135

Richmond GM (1964) Glaciation of the Little Cottonwood and Bells Canyon, Wasatch Mountains, Utah. US Geol Surv Prof pap 454D:41

Richmond GM (1965) Glaciation of the Rocky Mountains. In: Wright HE, Frey J (eds) The Quaternary of the United States. Univ Press, Princeton, pp 217–230

Richter H, Barthel H, Haase G (1961) Klimamorphologische Höhenstufen des zentralen Changai in der Mongolischen Volksrepublik. Geogr Ber 20:162–168

Rinehart CD, Ross DC (1964) Geological and mineral deposits, Mt. Morrison Quadrangle, California. US Geol Surv Prof Pap 385:106

Röthlisberger H (1967) Electrical resistivity soundings on glaciers: introductory remarks. J Glaciol 15:599–606

Röthlisberger H (1972) Seismic exploration in cold regions. Cold Regions Sci Engin Monogr II-A2a:1–139

Röthlisberger H (1981) Eislawinen und Ausbrüche von Gletscherseen. Jahrb Schweiz Naturforsch Ges 1978, Wiss Teil:170–212

Röthlisberger H, Voegtli K (1967) Recent DC resistivity soundings on Swiss glaciers. J Glaciol 6:607–621

Roggensack WD, Morgenstern N (1978) Direct shear tests on natural fine grained permafrost soils. In: Proc 3rd Int Conf on Permafrost, Edmonton, 1:729–735

Rognon P (1967) Le massif de l'Atokou et ses bordures. Etude géomorphologique. Ed du Centre National de la Recherche Scientifique, Paris 559 pp

Rognon P (1977) Formes périglaciaires dans le Massif de l'Atakor (Hoggar). Assoc Géogr Alsace. Colloq sur le périglaciaire d'altitude du domaine méditerranean et abords, Stasbourg, May 12–14, 1977, pp 347–364

Rohn O (1900) A reconnaissance of the Chitna River and Skolai Mountains, Alaska. US Geol Surv 21st Annu Rep Part 2:399–400

Rolshofen M (1977) Aktualgeomorphologische Höhenstufen – ein Vergleich aus Ost- und Westalpen. Mitt Geogr Ges München 62:103–111

Romanovskii NN, Tyurin AI (1983) Rock stream description. In: Proc 4th Int Conf on Permafrost, Fairbanks, Alaska. National Academic Press, Washington, pp 1078–1082

Roots EF (1954) Geology and mineral deposits of Aiken Lake map-area, British Columbia. Geol Surv Can Mem 274:246

Rother K (1962) Über Blockbildungen im nördlichen Talschwarzwald und Murgtal. Diss, Tübingen 90 pp

Rother K (1965) Ein Beitrag zum Blockmeerproblem. Z Geomorphol NF 9:321–331
Russel IC (1895) The influence of debris on the flow of glaciers. J Geol 3:823–832
Sadovsky AV, Bondarenko GI (1983) Creep of frozen soils on rock slopes. In: Proc 4th Int Conf on Permafrost, Fairbanks, Alaska. National Academic Press, Washington, pp 1101–1104
Sales JK (1983) Heart Mountain blocks in a giant volcanic rock glacier. Wyoming Geol Assoc Guidebook 34:117–165
Salomon W (1916) Die Bedeutung der Solifluktion für die Erklärung deutscher Land schafts- und Bodenformen. Geol Rdsch 7:30–41
Salomon W (1929) Arktische Bodenformen in den Alpen. Sitzungsber Heidelb Akad Wiss Math Naturwiss Kl Berlin 1929, 5:31
Sayles SH (1968) Creep of frozen sand. Cold Regions Res Engin Lab, Tech Rep Hanover, 190 pp
Schmidlin TW (1988) Alpine permafrost in eastern United States; A Review: In: Proc 5th Int Conf on Permafrost. Tapir, Trondheim, 1:241–246
Schneider E (1955) Topographische Karte 1:25.000 Chomolongma-Khumbu Himal. Kartographische Anst Freytag-Berndt und Artaria, Wien
Schoeneich P (1992) Glaciers rocheux fossils dans les Préalpes vaudoises. Bull Soc Vaudoise Sci Nat 82,1:35–55
Schott C (1931) Die Blockmeere in den deutschen Mittelgebirgen. Forsch Dsch Landes Völkerkd 29:1–78
Schröder H (1992) Aktive blockgletscher im zentralen Teil des nördlichen Tienschan. Petermanns Geogr Mitt 136:109–119
Schrott L (1991) Global solar radiation, soil temperature and permafrost in the Central Andes, Argentina: a Progress report. Permafrost Periglacial processes 2:59–66
Schrott L (1993) Die Solarstrahlung als steuernder Faktor im Geosystem der subtropischen semiariden Hochanden (Agua Negra, San Juan, Argentinien). Heidelb Geogr Arb 94:199
Schrott L (1995) The hydrological significance of high mountain permafrost and its relation to solar radiation. A case study in the high Andes of San Juan, Argentina. Bamberger Wirdschafts geogr Arb 13 (in press)
Stone PB (1992) The state of the world's mountains. Zed Books, London, 391 pp
Schunke E (1975) Die Periglazialerscheinungen Islands in Abhängigkeit von Klima und Sustrat. Abh Akad Gött Math Phys Kl 3, Folge 30:273
Schweizer G (1968a) Der Formenschatz des Spät- und Postglazials in den Hohen Seealpen: Aktualgeomorphologische Studien im oberen Tinéetal. Z Geomorphol Suppl 6:167
Schweizer G (1968b) Le tardiglaciaire et le niveau des neiges permanentes dans les hautes montagnes des Alpes Maritimes. L'exemple du bassin supérieur de la Tinée. Mediterranée 9:25–40
Schweizer G (1970) Der Kuh-E-Sabalan (Nordwestiran). In: Blume H, Schröder KH (eds) Beiträge zur Geographie der Tropen und Subtropen. Tübinger Geogr Stud 34, Sonderbd 3:163–178
Scott WJ, Sellmann PV, Hunter JA (1978) Geophysics in the study of permafrost. In: Proc 3rd Int Conf on Permafrost, Fairbanks, Alaska, vol 2. National Academic Press, Washington, pp 93–115

Sellmann PV, Brown J, Lewellen RI, McKim H, Merry C (1975) The classifications and geomorphic implication of thaw lakes on the arctic coastal plain, Alaska. USA Cold Regions Res Engin Lab Res Rep 344, 28 pp

Seret G (1983) The origin and movement of rock glaciers. In: Evenson EB, Schluchter C, Rabassa J (eds) Tills and related deposits. Elsevier, Rotterdam, pp 351–356

Serrat D (1979) Rock glacier morainic deposits in the eastern Pyrenees. In: Schluchter C (ed) Moraines and varves. Elsevier, Rotterdam, pp 93–100

Shakesby RA, Dawson AG, Matthews JA (1987) Rock glaciers, protalus ramparts and related phenomena, Rondane, Norway: a continuum of large-scale talus-derived landforms. Boreas 16:305–317

Shakesby RA, Matthews JA, McCarroll D (1995) Pronival ("ramparts") in the Romsdalalpane, southern Norway: forms, terms, subnival processes, and alternative mechanisms of formation. Arctic Alpine Res 27:271–282

Shields DH, Domaschuk L, Azizi F, Kjartanson B (1989) Primary creep parameters for ice as measured in-situ. Cold Regions Sci Technol 16:281–290

Shroder JF Jr (1972) Rock glaciers on Aquarius Plateau, Utah. In: La Géographie internationale. In: Adams WP, Hellener FM (eds) Papers submitted to the 22nd Int Geogr Congr. Univ Toronto Press, Montreal, pp 63–65

Shroder JF Jr (1973a) Movement of boulder deposits, Table Cliff Plateau, Utah. Geol Soc Am Abstr Progr 5:511–512

Shroder JF Jr (1973b) Tree-ring dating and analysis of movement of boulder deposits, High Plateaus of Utah. In: 9th Congr Int Union Quat Res, Christchurch, New Zealand, pp 328–329

Shroder JF Jr (1975) Dendrogeomorphologic analysis of mass movement. Proc Assoc Am Geogr 7:222–262

Shroder JF Jr (1977) Tree-ring dating of mass-movement periodicity. In: 10th INQUA Congr Abstr: 420. Int Union of Quaternary Research (INQUA), Birmingham

Shroder JF Jr (1978) Dendrogeomorphological analysis of mass movement on Table Cliffs Plateau, Utah. Quat Res 9:168–185

Shroder JF Jr (1987) Rock glaciers and slope failures: High Plateaus and La Sal Mountains, Colorado Plateau, Utah, U.S.A. In: Giardino JR, Shroder JF, Vitek JD (eds) Rock glaciers. Allen & Unwin, London, pp 193–238

Shroder JF Jr, Giardino JR (1978) Progress on rock glacier research. Trans Nebraska Acad Sci 6:51–54

Shroder JF Jr, Giardino JR (1987) Analysis of rock glaciers in Utah and Colorado, U.S.A., using dendrogeomorphological techniques. In: Giardino JR, Shroder JF, Vitek JD (eds) Rock glaciers. Allen & Unwin, pp 151–159

Shroder JF Jr, Sewell RE Jr (1981) Diamicton differentiation: La Sal Mountains and High Plateaus, Utah and Red Rock Pass, Idaho. Geol Soc Am Abstr Progr 13:553

Shroder JF Jr, Sewell RE Jr (1985) Mass movement in the La Sal Mountains, Utah. Utah Geol Miner Surv, Spec Stud 63, Salt Lake City

Shroder JF Jr, Giardino JR, Butler DR (1976) Dendrochronologic analysis of rock glaciers and snow avalanches. American Quaternary Association, Abstr 4th Biennial Meet, Arizona State Univ, Tempe, 115 pp

Shumskii PA (1964) Principles of structural glaciology. Dover Publications, New York, 497 pp

Siebenthal CE (1907) Notes on glaciation in the Sangre de Cristo Range, Colorado. J Geol 15:15–22

Sissons JB (1975) A fossil rock glacier in Western Ross. Scott J Geol 2: 83–86

Sissons JB (1976a) A remarkable protalus rampart complex in Western Ross. Scott Geogr Mag 92:62–90

Sissons JB (1976b) A fossil rock glacier in Western Ross: Reply to WB Whalley. Scott J Geol 12:178–179

Sissons JB (1978) Readvance in the southeast Grampians. In: Quaternary Studies in North-East Scotland. Dept of Geography, Univ of Aberdeen, pp 23–29

Sissons JB (1979) The Loch Lomond advance in the Cairngorm Mountains. Scott Geogr Mag 95:66–82

Sissons JB (1980) The Loch Lomond advance in the Lake District, northern England. Trans R Soc Edinb Earth Sci 71:13–27

Slaymaker HO, McPherson HJ (1977) Mountain geomorphology: geomorphological processes in the Canadian cordillera. BC Geographical Series 14. Tantalus, Vancouver, 274 pp

Smiraglia C (1985) Contributto alla conoscenza dei rock glaciers delle Alpi Italiane. Riv Geogr Ital 92:117–140

Smiraglia C (1989) Misure di velocità superficiale al rock glacier Orientale di Val Pisella (Gruppo del Cevedale (Alta Valtellina)). Geogr Fis Dinam Quat 12:41–44

Smiraglia C, Fabre D, Evin M, Assier A (1992) Les glaciers rocheux de la Manzina (Valfurva, Parc National du Stelvio, Italie). Société Hydrotechnique de la France (Section de Glaciologie, Grenoble, 11–12 Mars 1992), 7pp (unpubl report)

Smith HTU (1951) Rock glaciers in the Norman Wells area, N.W.T., Canada. Geol Soc Am Bull 62:1480

Smith HTU (1962) Periglacial frost features and related phenomena in the United States. Biul Peryglacjalny 11:325–342

Smith HTU (1968) "Piping" in relation to periglacial boulder concentrations Biul Peryglacjalny 17:195–204

Smith HTU (1973) Photogeologic study of geomorphic processes in the Canadian Cordillera. Geogr Ann 55A:69–84

Smith HTU, Smith AP (1945) Periglacial rock streams in the Blue Ridge Area. Geol Soc Am Bull 56:1198

Sollid JL (ed) (1989) Excursion guide northern Finland–Norway–Sweden. 5th Int Conf on Permafrost, Norwegian Commitee on Permafrost. Tapir Publ, Trondheim

Sollid JL, Sörbel L (1992) Rock glaciers in Svalbard and Norway. Permafrost Periglacial Processes 3:215–220

Sommerhoff G (1977) Zur spät- und postglazialen Morphodynamik im oberen Rißachtal, Karwendel. Mitt Geogr Ges München 62:89–102

Sone T (1990) The permafrost environment of the Daisetsu Mountains, Hokkaido, Japan. Permafrost Periglacial Processes 3:235–240

Sone T (1992) Permafrost environment of the Daisetsu Mountains, Hokkaido, Japan. Permafrost Periglacial Processes 3:235–240

Souchez RA (1971) Ice-cored moraines in South-Western Ellesmere Island, N.W.T., Canada J Glaciol 10:245–254

Souchez RA, Lorrain RD (1978) Origin of the basal ice layer from Alpine glaciers indicated by its chemistry. J Glaciol 20:319–328

Souchez RA, Lorrain RD (1991) Ice composition and glacier dynamics. Springer series in physical environment 8. Springer, Berlin Heidelberg New York, 207 pp

Soutadé G (1973/74) Aspects du modèle périglaciaire supraforestier des Pyrenées Orientales. Bull Assoc Etude Fr Quat 10:239–254

Soutadé G (1981) La haute montagne pyrenéenne dans le Massif du Luchonais. In: Hommages à René Raynal. Rech Géogr Strasbourg 16 Spec 17:143–145

Speck ChK (1994) Änderungen des Grundwasserregimes unter dem Einfluß von Gletschern und Permafrost. Mitt Versuchsanst Wasserbau Hydrol Glaziol ETH Zür 134:164

Spencer AC (1900) A peculiar form of talus Science (NS) 11:188

Spönnemann J (1974) Spülstreifen – eine besondere Erscheinung der Flächenspülung: Beobachtungen aus Kenia. In: Geomorphologische Prozesse und Prozeßkombinationen in der Gegenwart unter verschiedenen Klimabedingungen. Abh Akad Wiss Gött Math Phys Kl Folge 3, 29:137–146

Squyres SW (1978) Martian fretted terrain: flow of erosional debris. Icarus 34:600–613

Stäblein G (1977) Periglaziale Höhenstufen zwischen Subarktis und Äquator. Die Erde 108:151–54

Stäblein G (1979) Verbreitung and Probleme des Permafrostes im nördlichen Kanada. In: Pletsch A, Schott C (eds) Kanada Marb Geogr Schr 79:27–43

Stäblein G (1983) Mesoreliefformen des heutigen Periglazialraumes. 114:71–74

Stanton DM (1979) Morphology of rock glaciers in the Three Apostles area, central Sawatch Range, Colorado. Geol Soc Am Abstr Progr 1:303

Stanton DM (1980) Neoglacial cirque-erosion rates in the central Sawatch Range, Colorado. Geol Soc Am Abs Progr 12:528

Steenstrup KJV (1883) Bidrag til Kjendskab til Braeerne og Brae-Isen i Nord-Grönland. Medd Grönl 4:69–112

Steinhauser P (1974) Seismisch bestimmte Eigenschaften des Eises der Dachsteingletscher. Z Gletscherkd Glazialgeol 10:101–109

Stingl H (1969) Ein periglazialmorphologisches Nord-Südprofil durch die Ostalpen. Gött Geogr Abh 49:115

Streiff-Becker R (1958) Toteis und Blockströme. Leben Umwelt 14:259–260

Süsstrunk A (1951) Sondage du glacier par la méthode seismique. La Houille Blanche 6 A:309–319

Sugden DE, Clapperton CM (1980) West Antarctic ice sheet fluctuations in the Antarctic Peninsula area. Nature 286:378–381

Svensson H (1989) The rock glacier, a "new" feature in Scandinavian geomorphology. Geogr Ann 71A:95–97

Swett K, Hambrey MJ, Johnson DB (1978) Lobate rock glaciers and related talus terraces in northern Spitsbergen. Geol Soc Am Abstr Progr 10:502

Swett K, Hambrey M, Johnson D (1980) Rock glaciers in northern Spitsbergen. J Glaciol 88:475–482

Taber S (1929) Frost heaving. J Geol 37:428–461

Tailleffer F (1963) La carte de morphologie glaciaire des Pyrénées au 1/50 000. Rev Géogr Pyrénées Sud-Ouest 34:5–10

Tailleffer F (1968a) La carte de morphologie glaciaire des Pyrenées au 1/50 000. Rev Géogr Pyrenées Sud-Ouest 34:5–10

Tarakanov AG (1982) Kriogennyj rel'ev bassejna verchov'ev r. Sary-Dzaz. Gljaciologicesie i gidrometeorologiceskie issledovania na Tjan'sane Frunze: 82–102

Tatenhove F van, Dikau R (1990) Past and present permafrost distribution in the Turtmanntal, Wallis, Swiss Alps. Arct Alp Res 22:302–316

Tenthorey G (1992) Perennial Névés and the hydrology of rock glaciers. Permafrost Periglacial Processes 3:247–252

Tenthorey G (1993) Paysage géomorphologique du Haut-Val de Réchy (Valais, Suisse) et hydrologie liée aux glaciers rocheux. Univ Fribourg, Fac Sci, Inst Géogr 307 pp

Tenthorey G, Gerber E (1991) Hydrologie du glacier rocheux de Murtel (Grisons), Description et interprétation de tracages d'leau. Ber Forsch Geogr Inst Freiburg/CH 3:119–121

Terzaghi K (152) Permafrost. Boston Soc Civil Engin J 39:1–50

Thompson DE (1981) Mechanics of flow and erosion potential of rock glaciers and glaciers in compression flow. In: Holt HE, Kosters EC (eds) Reports of Planetary Geology Program. NASA Tech Memorandum 82385, NASA Office of Space Sci, Washington, DC pp 384–385

Thompson EG, Sayles FH (1972) In situ creep analysis of room in frozen soil. J Soil Mech Found Div 98:899–916

Thompson HR (1957) The old moraines of Pangnirtung Pass, Baffin Island. J Glaciol 3:42–49

Thomson WF (1962) Preliminary notes on the nature and distribution of rock glaciers relative to true glaciers and other effects of the climate on the ground in North America. In: Coll Obergurgl (10.9–18.9.1962); Assoc Int Hydrol Sci, Comm Neiges et des Glaciers, Publ 58:212–219

Thorn CE, Loewenherz DS (1987) Spatial and temporal trends in alpine periglacial studies implications for paleo reconstruction. In: Boardman J (ed) Periglacial processes and landforms in Britain and Ireland. Cambridge Univ Press, Cambridge, pp 57–65

Tihay J (1983) Les éboulis et leur environnement géomorphologique autre que celui des glaciers rocheux. In: Séance thematique le 8 janvier; les éboulis et leur environnenent géographique passé et actuel. Bull Assoc Géogr Fr 60:5–13

Ting J (1983) Tertiary creep models for frozen sand. J Geotech Eg ASCE 109:932–945

Titkov SN (1988) Rock glaciers and glaciation of the Central Asia Mountains. In: Proc 5th Int Conf on Permafrost in Trondheim, Norway, 1:259–262

Tolgensbakk J, Sollid JL (1988) Kafjord, kvartaergeologi og geomorfologi 1:50 000, 1634 II. Geogr Inst Univ Oslo

Toshio S, Nobuyuki TS, Masami F (1988) Alpine permafrost occurence at Mt. Taisetsu, Central Hokkaido in Northern Japan. In: Proc 5th Int Conf on Permafrost in Trondheim, Norway, 1:253–258

Troll C (1941) Studien zur vergleichenden Geographie der Hochgebirge der Erde. Ber 23, Hauptversammlung der Gesellschaft von Freunden und Förderern der Rheinischen Friedrich-Wilhelms-Universität Bonn 21:49–96

Troll C (1943) Thermische Klimatypen der Erde. Petermanns Geogr Mitt 89:81–89

Troll C (1944) Strukturböden, Solifluktion und Frostklimate der Erde. Geol Rdsch 34:545–694 (English translation, US Army Corps of Engineers, SIPRE, Translation 43 by HE Wright, 1958), 121 pp

Troll C (1954) Über das Wesen der Hochgebirgsnatur. Jahr Dtsch Alpenver 80:142–157

Troll C (1970) Vergleichende Geographie der Hochgebirge der Erde in landschaftsökologischer. Sicht Geogr Rdsch 27:185–198

Troll C (1972) Geoecology and the world wide differentiation of high mountain ecosystems. In: Geoecology of the high mountain regions of Eurasia. International geographical Union Steiner, Wiesbaden, pp 1–16
Troll C (1973) High mountain belts between the polar caps and the equator: their definition and lower limit. Arct Alp Res 5:A19–A27
Troll C, Paffen KH (1964) Jahreszeitenklimate der Erde Erdhunde 18:5–28
Trombotto D (1983) Geocriologia de la Lagunita del Plata. In: Actas lra Reunión del Grupo Periglacial Argention IANIGLIA-CONICET 83:149–156
Trombotto D (1991) Untersuchungen zum periglazialen Formenschatz and zu periglazialen Sedimenten in der Lagunita del Plata, Mendoza, Argentinien. Heidelb Geogr Arb 90:150
Tyrell JB (1910) Rock glaciers or chrystocrenes. J Geol 18:549–533
Tyurin AI (1983) Classification of rock streams. In: proc 4th Conf on Permafrost, Fairbanks, Alaska National Academy Press, pp 1283–1285
Urdea P (1988) Considérations sur les glaciers de pierres des montagnes du Retazat. Analele stiintifice ale Universitatii "Al. I Cuza" din Iasi Tomul 34:65–73
Urdea P (1992) Rock glaciers and periglacial phenomena in the southern Carpathians Permafrost Periglacial Processes 3:267–273
Valades B, Veyret Y (1974) Quelques aspects des modèles d'origine glaciaire, périglaciaire et nivale sur les confins meridionaux de la margeride. Rev Géomorphol Dyn 23:163–177
Van Husen D (1976) Schuttströme als Ausdruck des periglazialen Massenabtrages in den Östlichen Karawanken (Österreich). Z Geomorphol NF 20:97–107
Van Husen D (1981) Geologisch-sedimentologische Aspekte im Quartär von Österreich. Mitt Österr Geol Ges 74:197–230
Vere DM, Matthews JA (1985) Glacier formation from a lateral moraine at Bukkeholsbreen, Jotunheimen, Norway: a sedimentological approach. Z Geomorphol NF 29:397–415
Vernon P, Hughes OL (1966) Surficial geology, Dawson, Larsen Creek, and Nash Creek map areas, Yukon Territory. Geol Sur Can Bull Ottawa 136:25
Veyret Y (1979) Essai sur la terminologie glaciaire. Géogr Phy Quat 33:205–222
Veyret Y (1981) Quelques aspects de l évolution post-wisconsienne du bourelet des Tongats (région de Saglek, Ramah Bay, Québec-Labrador). In: Hommages à René Raynal Rech Géogr Strasburg 16, 17:161–167
Vialov SS (1965) The strength and creep of frozen soils and calculations for ice-soil retaining structures. Cold Regions Res Engin Lab, Translation 76, Hannover
Vialov SS (1978) Kinetic theroy of deformation of frozen soils. In: Proc 2nd Int Conf on Permafrost, Yakutsk, USSR, vol 1 National Academy of Science, Washington, pp 751–755
Vick SG (1981) Morphology and the role of landsliding in formation of some rock glaciers in the Mosquito Range, Colorado. Bull Geol Soc Am 1, 92:75–84
Vick SG (1987) Significance of landsliding in rock glacier formation and movement. In: Giardino JR, Shroder JF, Vitek JD (eds) Rock glaciers. Allen & Unwin, London, pp 239–263
Viers G (1961) La glaciation du massif du Carlit (Py.-Orientales). Rev Géogr Pyrenées Sud-Ouest 32:5–33
Viers G (1962) Les Pyrénées. Que sais-je? No 955, Paris 128 pp

Vietoris L (1972) Über die Blockgletscher des Äußeren Hochebenkars. Z Gletscherk Glazialgeol 8:169–188
Vitek JD, Giardino JR (1987) Rock glaciers: a review of the knowledge base. In: Giardino JR, Shroder JF, Vitek JD (eds) Rock glaciers. Allen & Unwin, London pp 1–26
Vivian R (1975) Les glaciers des Alpes Occidentales. Grenoble, 513 pp
Voight B (1978) Rockslides and avalanches 1, natural phenomena. Dev Geotech Engin Amsterdam 14A:833
Voight B (1979) Rockslides and avalanches 2, engineering sites. Dev Geotech Engin Amsterdam 14B:850
Voight B, Pariseau WG (1978) Rockslides and avalanches: an introduction. In: Voight B (ed) Rockslides and avalanches 1, natural phenomena. Dev Geotech Engin Amsterdam 14A:1–67
Von Engeln OD (1935) Erosion marginal to a plateau glacier. Geol Soc Am Bull 46:985–998
Von Klebelsberg R (1948/49) Handbuch der Gletscherkunde und Glaziologie, vol 2. Springer, Wien Berlin Heidelberg New York, 1028 pp
Von Köppen W (1931) Grundriß der Klimakunde, 2nd edn. DeGruyter, Berlin-Leipzig, 388 pp
Von Lozinski W (1910) Die periglaziale Fazies der mechanischen Verwitterung. C R XI Cong Géol Int Stockholm 2:1039–1053
Von Rudloff H (1980) Die Klima-Entwicklung in den letzten Jahrhunderten im mitteleuropäischen Raume (mit einem Rückblick auf die postglaziale Periode). In: Oeschger H, Messerli B, Svilar M (eds) Das Klima. Springer, Berlin Heidelberg New York, 125–148
Vonder Mühll D (1992) Evidence of intrapermafrost groundwater flow beneath an active rock glacier in the Swiss Apls. Permafrost Periglacial Processes 3:169–173
Vonder Mühll D (1993) Geophysikalische Untersuchungen im Permafrost des Oberengadins. Mitt Versuchsanst Wasserbau Hydrol Glaziol Edigenöss Tech Hochschule Zür 122:222
Vonder Mühll D, Haeberli W (1990) Thermal characteristics of the permafrost within an active rock glacier (Murtèl/Corvatsch, Grisons, Swiss Apls). Glaciol 36:151–158
Vonder Mühll D, Holub P (1992) Borehole logging in Alpine permafrost, Upper Engadin, Swiss Alps. Permafrost Periglacial Processes 3:125–132
Vonder Mühll D, Klingelé E (1994) Gravimetrical investigation of ice rich permafrost within the rock glacier Murtèl-Corvatsch (Upper Engadin, Swiss Alps). Permafrost Periglacial Processes 5:13–24
Vonder Mühll D, Schmid W (1993) Geophysical and photogrammetrical investigations of rock glacier Muragl I, Engadin, Swiss Alps. Proc 6th Int Conf on Permafrost, Beijing
Vorndran E (1969) Untersuchungen über die Schuttentstehung und ablagerungsformen in der Hochregion der Silvretta (Ostalpen). Schr Geog Ins Univ Kiel 29, 3:138
Vorndran G (1977) Hangabtragungsbilanzen. In: Wirthmann A (ed) Hangformen und hangprozesse. Z Geomorphol Suppl 28:124–133
Wanger S (1992) Creep of Alpine permafrost, investigated on the Murtel Rock Glacier. Permafrost Periglacial Processes 3:157–162
Wahrhaftig C (1949) The frost-moved rubbles of Jumbo Dome and their significance in the Pleistocene chronology of Alaska. J Geol 57:216–231

Wahrhaftig C (1954) Observations on rock glaciers in the Alaska Range (abs.). Geol Soc Am Bull 65:1353

Wahrhaftig C (1987) Foreword. In: Giardino JR, Shroder JF, Vitek JD (ed) Rock glaciers. Allen & Unwin, London, pp 7–42

Wahrhaftig C, Cox A (1959) Rock glaciers in the Alaska Range. Geol Soc Am Bull 70:383–436

Waldrop HA, White SE (1965) Arapaho Glacier and Arapaho Rock Glacier, Trip 1. In: 7th Congr Int Assoc Quat Res, Boulde-Denver, Colorado. Univ Nebraska Press, Lincoln, pp 5–10

Wallace RG (1976) Arapaho rock glacier, Colorado Front Range. Trans Ill Acad Sci 69:415–424

Ward P, Hall RD (1982) Rock glaciers of the Tobacco Root Range, southwestern Montana. Geol Soc Am Abstr Progr 14:353

Washburn AL (1973) Periglacial processes and environments. Arnold, London, 320 pp

Washburn AL (1979) Geocryology: a survey of periglacial processes and environments. Arnold, London 406 pp

Washburn AL (1979/1980) Permafrost features as evidence of climatic change. Earth-Sci Rev 15:327–402

Washburn AL (1981) Periglaziale Forschung in revue. Geol Rdsch 70:664–690

Watts RD, England AW (1976) Radio-echo sounding of temperate glaciers: ice properties and sounder design criteria. J Glaciol 17:39–48

Watts RD, Wright DC (1981) Systems for measuring thickness of temperate and polar ice from the ground or from the air. J Glaciol 23:459–469

Wayne WJ (1981a) Ice segregation as an origin for lenses of non-glacial ice in "ice-cemented" rock glaciers. J Glaciol 27:506–510

Wayne WJ (1981b) La evolution de glaciares de escombros y morenas en la cuenca del Rio Blanco, Mendoza. In: 8th Congr Geol Argentino, San Luis, 20–26.9.1981, Actas 4:153–166

Wayne WJ (1982) Fossil patterned ground and rock glaciers in the Ruby Mountains, Nevada. Geol Soc Am Abstr Progr 14:353

Wayne WJ (1983a) Correspondence: ice segregation as an origin for lenses of non-glacial ice in "ice-cemented" rock glaciers. Glaciol 29:524

Wayne WJ (1983b) Paleoclimatic inferences from relict cryogenic features in alpine regions. In: Proc 4th Int Conf on Permafrost, Fairbanks, Alaska, National Academic Press, Washington, pp 1378–1383

Wayne WJ (1984) The Quaternary succession in the Rio Blanco basin, Cordon de la Plata, Mendoza Province, Argentina: an application of multiple relative-dating techniques. In: Mahaney WC (ed) Quaternary dating methods. Elsevier, Amsterdam pp, 389–406

Webber PJ, Andrews JT (1973) Lichenometry: a commentary. Arct Alp Res 5:295–302

Weiher U (1994) Mögliche Ursachen für die Inaktivität von Blockgletschern im Permafrost Alaskas. Diplomarbeit Geographisches Institut der Universität Heidelberg (unpubl)

Weiss EH (1958) Eine eisführende Schutthalde in den Gailtaler Alpen. Carinthia 2, 68:62–63

Wentworth CK, Delo DM (1931) Dinwoody Glaciers, Wind River Mountains, Wyoming: with a brief survey of existing glaciers in the United States. Bull Geol Soc Am 42:605–620

Werenskiold W (1953) The extent of frozen ground under the sea bottom and glacier beds. J Glaciol 2:197–200

Wertman J (1974) The unsolved general glacier sliding problem. J Glaciol 23:97–111

Whalley WB (1974a) Rock glaciers and their formation as a part of glacier debris-transportation system. Geogr Pap, Dep Geogr, Univ Reading 24:60

Whalley WB (1974b) Origin of rock glaciers. J Glaciol 13:323–324

Whalley WB (1974c) The mechanics of high-magnitude, low-frequency rock failure. Geogr Pap Dep Geogr, Univ Reading 27:48

Whalley WB (1976a) A rock glacier and its relation to the mass balance of Corrie glaciers, Strupbreen, Troms Norway. Nor Geogr Tidsskr 30:51–55

Whalley WB (1976b) A fossil rock glacier in Western Ross. Scott J Geol 12:175–179

Whalley WB (1979) The relationship of glacier ice and rock glacier at the Grubengletscher, Kanton Wallis, Switzerland. Geogr Ann 61A:49–61

Whalley WB (1983) Rock Glaciers – permafrost features or glacial relics? In: Proc 4th Int Conf on Permafrost, Fairbanks, Alaska. National Academic Press, Washington, pp 1396–1401

Whalley WB (1992) A rock glacier in south Ellendalen, Lyngen Alps, Troms. Nor Geogr Tidskr 46:29–31

Whalley WB, Azizi F (1994) Rheological models of active rock glaciers: evaluation, critique and a possible test. Permafrost Periglacial Processes 5:37–51

Whalley WB, Martin HE (1992) Rock glaciers: II models and mechanisms. Progr Phys Geogr 16.2:127–186

Whalley WB Douglas GR, Jonsson A (1983) The magnitude and frequency of large rockslides in Iceland in the postglacial. Geogr Ann 65A:99–110

Whalley WB, Martin HE, Gellatly AF (1986) The problem of "hidden" ice in glacier mapping. Ann Glaciol 8:181–183

Whalley WB, Gordon JE, Gellatly AF (1989) Aspects of topographic and climatic controls on 19th and 20th century glacier changes in the Lyngen and Bergsfjord areas, North Norway. In: Oerlemans J (ed) Glacier fluctuations and climatic change. Kluwer, Dordrecht, pp 153–172

White PG (1976) Some observations on the origins of rock glaciers. Geol Soc Am Abstr Progr 8:299

White PG (1979) Rock glacier morphometry, San Juan Mountains, Colorado. Geol Soc Am Bull 90:515–518

White SE (1967) Rockfall, alluvial, and avalanche talus in the Colorado Front Range. Geol Soc Am Spec Pap 115:237

White SE (1971a) Debris falls at the front of Arapaho rock glacier, Colorado Front Range, USA. Geogr Ann 53A:86–91

White SE (1971b) Rock glacier studies in the Colorado Front Range, 1961 to 1968. Arct Alpine Res 3:43–64

White SE (1971) Alpine subnival boulder pavements in Colorado Front Range. Geol Soc Am Bull 83:195–200

White SE (1974a) Comment: relative age dating techniques. Geology 2:326

White Se (1974b) Rock glaciers and block fields. Geol Soc Am, Abstr with Programs 6:1005

White SE (1975) Additional data on Arapaho rock glacier in Colorado Front Range, USA. J Glaciol 14:529–530

White SE (1976) Rock glaciers and block fields, review and new data. Quat Res 6:77–97

White SE (1981) Alpine mass movement forms (non-catastrophic): classification, description and significance. Arct Alpine Res 13:127–137

White SE (1987) Differential movement across transverse ridges on Arapaho rock glaciers, Colorado Front Range, USA In: Giardino JR, Shroder JF, Vitek JD (eds) Rock glaciers. Allen & Unwin, London, pp 145–149

Wilhelmy H (1974a) Zur Genese der Blockmeere, Blockströme und Felsburgen in den deutschen Mittelgebirgen. Ber Dtsch Landeskunde 48:17–41

Wilhelmy H (1974b) Klima-Geomorphologie in Stichworten. Teil IV der Geomorphologie in Stichworten. Hirt Kiel, 375 pp

Willams JR, van Everdingen RO (1973) Groundwater investigations in permafrost regions of North America: a review. In: 2nd Int Conf on Permafrost, North American Contribution, National Academy of Sciences, Washington DC, pp 435–446

Wilson P (1990) Characteristics and significance of protalus ramparts and fossil rock glaciers on Erigal Mountain, Country Donegal. Proc R Irish Acad 90 B:1–21

Worsley P (1987) Permafrost stratigraphy in Britain – a first appromixation. In: Boardman J (ed) Periglacial processes and landforms in Britain and Ireland. Cambridge Univ Press, Cambridge, pp 89–99

Yarnal BM (1982) Surface sediment analysis of a rock glacier. J Int Assoc Math Geol 14:371–386

Yuanlin Z, Carbee DL (1984) Uniaxial compressive strength of frozen silt under constant deformation rates. Cold Regions Sci Technol 9:3–15

Zaretskiy YuK, Shchobolev AG (1983) A mathematical model for the viscoplastic deformation of frozen soils. In: Proc 4th Int Permafrost Conf Fairbanks, Alaska, National Academic Press, Washington, pp 1457–1462

Zielinski GA (1986) Evidence for limited early neoglacial rock glacier formation in the Temple Lake cirque, Wind River Range, Wyoming. Geol Soc Am Abstr Progr 18:425

Zielinski GA (1989) Lacustrine sediment evidence opposing Holocene rock glacier activity in the Temple Lake Valley, Wind River Range, Wyoming, U.S.A. Arc Alp Res 21:22–23

Index of Place Names

Absoraka Mountains 53, 55, 213
Aconagua 211
Admirality Bay 66, 248
Afghanistan 59, 237
Africa 195
Alaska 22, 24, 29, 35, 50, 52, 125, 191
Alaska Range 27, 29, 50, 51, 55, 123, 124, 191, 192, 240
Albana 9, 95, 120, 192, 193
  rockglacier 138
  West 138
Alberta 53, 54, 199, 226, 231, 232, 262
Aletsch 148, 149
  glacier 19, 27
Alma Ata 61
Alp Placer 11
Alps 16, 19, 35, 37, 38, 40, 41, 70, 119, 211, 237
Alps of New Zealand 35, 63
Altai 61
  Ranges 61
American Cordilleras 35
Andes 19, 209, 212, 235, 259
  of Argentina and Chile 35
  of Central and South America 55
  of Mendoza 60, 67, 93, 208
Antarctica 35, 66, 248
Anzere 202
Appalachians 219
Arapahoe 164, 165, 166
  rockglacier 125, 126, 157
Argentina 57, 98
  Andes 187, 189, 191
Arizona 195, 196, 197, 229, 252
Asia 23, 36, 59, 61, 62, 235
Atigun Pass 125
Atlas 35, 195
Australia 219
Austria 38, 41, 43, 90, 233, 235, 246, 251, 261, 266
Austrian Alps 40, 42

Bab Taza 35
Banff 53, 55
  National Park 24, 226
Barney Top 205
Beagle 15
Becs de Bosson 265
Ben Ohau Range 63
Bernina 12
Bettmerhorn 149
Bever 180, 181
Black Forest 219
Blanca Massif 19, 205, 206, 226
Bottleneck 36
Brionçonais 38
Britain 195
British Columbia 54
Bröggerhalvöya 10
Broken River 64
Brooks range 22, 24, 29, 50, 51, 52, 55, 125, 191
Buffalora 181
Bukkehol 72, 226
Burnett Range 64

Calgary 238
Campo 203, 204
Canada 53, 127, 220, 232, 262
Canadian Rocky Mountains 19
Carl Troll glacier 216
Catchment Daberbach 236
Caucasus 61, 237
Cerro Negro 26
Chamberon 38
Changai Mountains 61
Chesney Wold rockglacier 36

Chessi 187
China 61, 237
Chomolongma 212
Cirque Lake 53
Clear Creek rockglacier 123, 124, 125
Clear Grit 208
Colorado 14, 19, 22, 23, 24, 56, 75, 76, 77, 125, 126, 199, 204, 208
Colorado Front Range 53, 55, 128
Combe de Laurichard 167
Combe de Prafleuri 74, 97, 99
Cordillera neovolcanica 255, 256
Corvatsch 98
Cragieburn Range 64

Davis rockglaciers 208
Debanttal 236
Denver 74, 75, 238
Disco 24, 214
Disco fjord 49
Disco Island 23, 49, 51, 227
Dolomites 207
Donjek glacier 211
Drakensberg 36, 255
Dugla 90
Düniberg 199

Edmonton 238
Eggishorn 19, 149
El Salto 25, 97, 98
  rockglacier 68
Ellesmere Island 216
Elm 199, 200, 201
Engadin 7, 9, 11, 19, 29, 37, 84, 85, 90, 93, 95, 120, 138, 139, 142, 143, 218, 259
Escudillo Mountains 195
Esjufjöll 46
Europe 35
  Northern 43

Falkland Islands 15
Falwindes 236
Flüelapass 74, 95, 234, 266
France 38
Frances Lake Area 225
Frank 199
French Alps 93
Front Range 126
  of Colorado 5
Furggentälti 117

Galena Creek 90, 167, 212
  A 167, 168
  B 167, 168
  rockglacier 20, 213, 214
Ghiarce die Schiantala 215
Glockturm
  glacier 217
  mountain 130
  rockglacier 130
Graben 90
Graubünden 7, 9, 11, 12, 25, 37, 38, 39, 74, 78, 84, 93, 98, 100, 106, 110, 112, 116, 123, 131, 138, 153, 180, 181, 193, 221, 229, 233, 234, 242, 266
Greenland 14, 23, 35, 49, 50, 51, 214
Grialetsch 116, 233, 234
Grisons 193
Grizzly Creek 227
Grosses Gufer 19, 27, 127, 148, 149, 158, 168
  rockglacier 148, 149, 160
GRUB 81
Gruben 71, 72, 78, 97, 98, 157, 158, 162, 163, 165, 167, 168, 176, 188, 227, 262, 265, 270
Gruben glacier 126, 160
Gruben rockglacier 100, 103, 110, 111, 126, 142, 144, 145, 151, 156, 160
Gufer 167, 169

Hay River 238
Hess Mountains 53
Hilda 262, 265
Hindukush 35, 59
Hochebenferner 233
Hochebenkar 235, 266
  Äußeres 90, 126, 142, 146, 156, 157, 165, 167, 168, 233, 270
    rockglacier 116, 117, 132, 133, 134, 146, 147, 148
  Inneres 233
Hoggar 35, 195
  Mountains 35
Horcones glacier 211
Hornsund 44
Huerfano Basin 14

Hurrican Basin 176
rockglacier 76, 77

Iceland 46, 237
Imogene Basin 199
India 237

Iowa Gulch 208
Ireland 195
Issy river basin 61
Issyk river (*see* also Prijamosy shcheli) 23
Italy 255
Iztaccihuatl 256, 257

Jaeger rockglacier 146
Japanese Alps 32
Jaspers National Park 53, 55
Jotunheimen 72, 226
Jura mountain 220

Karawanken 207
Kargaul dinka 61
Kärnten 43, 251
Kashmir Himalaya 59
Kaunertal 127
Kaunsertal 127
Kendrick Peak 195, 196, 229, 252
Khumba 90
Khumbu 70
Himalaya 35, 59, 61
King George Island 66, 248
Kintole 176
Kluane Range 79
Krossfjord 47
Kunlun Shan 61

La Sal Mountains 70
La Tscheppa 19, 187, 188
La Veta 206
pass 206
Lac de Tseuzier 202
Laggintal 5, 148
Las Bronces 26
Las Sours 78
Laurichard 128, 129, 130
Lavin 163
Leibnitztal 236
Lesachtal 236
Lintgen 90

Lofoten 44
Lower Engadin 84, 163, 242
Lyngen
Mountains 222
Peninsula 44

Macauley 64
Macun 69, 160, 166, 168, 169, 186, 187, 264
I 25, 26, 84, 85, 86, 87, 118, 158, 163, 167, 176, 179, 242, 246
I rockglacier 139, 141, 156, 159, 161
rockglacier 142, 158
Maiella Mountains 255
Malaspina glacier 20
Marinet 264
Mattertal 188
Mendoza 19, 57, 187, 189, 190, 209, 259, 262
Mexico 35, 48, 55, 255, 256, 257
City 238
Middle East 35
Mischabel 267
Mitrafjellet 47
Mongolia 61, 237
Mont Blanc 38
Moostal 236
Morenas 98
Coloradas 98, 187, 209, 210
Mosquito Range 208
Mount Cook 63
Mount Everest 212
Mount Kenya 35, 36, 195
Mount Mestas 76, 182, 204, 205, 206
Mount Rainier 53
Mount Whitney 55, 59
Mouskorbé 195
Muragl 84, 90, 98, 142, 143
rockglacier 142
Murtegrave 175
Murtèl 98, 169, 258, 259, 265, 270
I 7, 25, 26, 27, 29, 34, 78, 79, 80, 82, 83, 84, 86, 89, 90, 93, 94, 98, 100, 101, 102, 103, 106, 107, 109, 112, 113, 114, 115, 118, 119, 142, 143, 151, 158, 167, 168, 169, 171, 175, 176, 177, 178, 179, 227, 228, 237, 241, 242, 246
I rockglacier 134, 135, 136, 137, 142, 143, 149, 152, 153, 154, 155, 157, 221

Navajo Mountain 195, 197, 224
Nepal 35, 70, 90, 236, 237
Nevada de Toluca 55
New South Wales 219
New Zealand 63
  Alps 237
North America 35, 48, 50, 55, 237
  Arctic 35
North West Teritories 127
Norway 44, 45, 222, 226
Nuptse 70, 90

Obergurgl 266
Odenwald 15
Ogilvie Mountains 241
Ölgruben 270
  rockglacier 127
Olympic Mountain 53
Ötztal 130, 217
  Alpen 133

Pacific northwest 53
Pakistan 237
Pamir 61
Phoenix 238
Pierrier de Saleinaz 204
Pika rockglacier 146
Piz Albana (*see* also Albana) 138, 218
Pontresina-Schafberg 78, 84
Prafleuri 176
Premaju 208
Prijamoy shcheli (*see* also Issyk river) 23, 29, 241
Prins Karls Forland 6
Punt Hennequin 66, 248
Pyrenees 46, 47, 48, 49, 68

Queen Elizabeth Islands 49

Red Mountain 74, 75
RGII 104, 105, 110
Rhine 249
Rif 35, 195, 259
Rocky Mountains 53, 74, 90, 206, 231, 232, 262, 265
Rondane 44, 46
  Mountains 201
Rose Hill rockglacier 36
Rossbodengletscher 100
Rovana 203
  river 204

Sajan 61
Samedan 181
San Francisco Peak 195
San Juan Mountains 14, 22, 23, 24, 53, 55, 76, 77, 199, 228
Sangre de Christo Range 14, 19
Santiago de Chile 57
Scandinavia 35, 44, 69, 216, 220, 237
Scarltal 11
Schiantala 215
  valley 215
Schleinitz 236
Schober 251
  Mountains 41, 42, 43, 236, 237, 241, 261, 266
Scotland 223
Selwyn Mountains 53, 61
Sierra de Aguilar 57
Sierra neovulcanico 55, 257
Sierra Nevada 55, 58, 59
Silverton 14
Simplon 5, 72
Slide Rock Ridge 199
Slim
  river 264
  valley 262
Smoky Mountain 219
South Africa 255
South America 55, 237
South Island 63
South Shetland Islands 66
Southern Wallis 190
Soviet Union 59
Spitzbergen 44, 46, 47, 191
St. Elias Mountains 227, 262, 264
St. Kilda 255
Staniska 236
Steingletscher 211
Stupbreen 44
Stuphallet 10
Sustenpass 68, 211
Suvretta 84, 98
  da S. Murezzan rockglacier 218
  rockglacier 118, 120
Svalbard 10, 44, 46, 47
  Archipelago 6

Swiss Alps 5, 24, 26, 29, 37, 38, 39, 40, 49, 61, 67, 208
Swiss National Park 106, 121, 131
Switzerland 38, 39

Table Cliffs Plateau 204, 205, 206
Tälligrat 189
Talus 79
Tasmania 64, 219
Tauern 43
Temple Lake Valley 229
Tessin 203, 204
Tibesti 35
Tien Schan 61
Tierra del Fuego 35
Tirol 43, 127, 130, 133, 217, 233, 235, 251, 261
Toolong Range 219
Transantarctic Mountains 66
Transbaikalia 61
Troms 44
Tschadinalm 236
Tungsten 167, 262
  rockglacier 127, 219
Two Tumb Range 63, 64

United States 35, 195, 234
Upper Engadin 84, 180, 187, 188, 192, 242
USSR 234, former 237
Utah 70, 197, 204, 205

Val d'Hérémence 74, 99
Val da l'Acqua 37, 123
  rockglacier 156
Val di Campo 12
Val Ferret 204
Val Sassa 37, 95, 142, 193, 258
  rockglacier 106, 120, 121, 123, 124, 131, 132, 151, 153, 157
Valle Maggia 203
Valleys of Montgioie 215
Verkilsdalen 201
Victoria Land 66

Wallis 5, 19, 27, 34, 38, 39, 71, 72, 74, 78, 84, 90, 93, 98, 99, 100, 127, 142, 148, 149, 160, 188, 202, 227, 265, 267
Wasatch Range 53
Weissmies 5, 34, 72, 148, 149, 158, 189, 190
  rockglacier 127, 128, 148, 149, 150, 151
Wernecke 241
West Greenland 227
White Mountain of California 58
White River Ash 227
Wind River Range 229
Wood River area 29
Wyoming 213, 229

Yukon 53, 78, 79, 104, 105, 110, 225, 227, 241, 262, 264

Zailijskiy Alatau 23, 24, 29, 61, 241

Subject Index

Active layer 69, 74, 89, 94, 101, 116, 119–120, 201, 269
Active layer thickness 90, 92, 93, 117, 119, 120
Active rockglacier 35, 36, 38, 93, 184, 269
ADINA 178, 179
Advective heat 110
Age (of rockglaciers) 130, 225–227, 270
Alpine belt 231, 232
Alpine permafrost *see* Mountain permafrost
Altitude (of rockglaciers) 23, 52, 53, 55, 64, 68, 76, 78, 191, 194, 195, 228, 242
Area of permafrost 237
Area (of rockglaciers) *see* Size (of rockglaciers)
Aspect (of rock glaciers) *see* Exposure
Avalanche 185, 186, 222, 224
Avalanche deposits 222
Avalanche snow 74, 82, 132
Average annual temperature 180
Average velocity 128, 130, 171

Balch ventilation 19, 68, 120, 238
Basal shear stress 175, 176
Bergsturz 194, 198, 258
Bergsturz hypothesis 199–202
Block field 15, 69
Block mantle *see* Boulder mantle
Blockstream 15, 16, 219
Borehole deformation 104–108
Borehole drilling 72, 73, 86
Borehole geophysics 102–114
Borehole records 78–83
Borehole temperatures 108–114, 115
Boulder apron 192, 193
Boulder mantle 67, 68, 69, 70, 85, 93, 162, 181, 187, 189, 212, 214, 269
Boulder pavement 69

Caliper log (borehole) 83
Central flow line *see* Stromstrich
Change in activity 157
Change in velocity 131, 132
Chrystocrene 15
Cirque glacier 52
Climate 250–258
Climate (and rockglacier movement) 108, 180
Climatic change 119, 142, 236, 257–258
Climatic inactive rockglacier 8, 9, 93, 132, 189, 193, 227–229
Climatic warming 180, 270
Coarse debris system 240
Complex rockglacier 20
Compression 162, 164
Compressing flow 69, 79, 128, 130, 149, 157, 162, 164, 167
Congelation ice 269
Continuous permafrost 51, 55, 61
Creep 70, 189, 206, 216, 269
Creep law 176, 177, 179
Creep of ice *see* Glen's flow law
Creep of permafrost 69, 179, 182, 184
Creep of rockglacier core 160
Creep rates 176
Crevasses 133, 266
Cross-sectional profile (of velocity) 166
Cross-sectional profiles 148
Cryogenic Belt 267

Dating 270
Daun *see* Older Dryas
Debris fall 186
Debris flow 266
Debris mantle *see* Boulder mantle
Debris rockglacier 11, 32, 33, 34, 51, 54, 148, 186–188, 269
Debris transport 248–250

Debris-covered glacier 210, 211–212
Deformation 172
Deformation of permafrost 169
Deformation rate 104, 108
Density (of rockglacier material) 84, 85, 101, 104, 172
Density of rockglaciers (spatial) 49, 53, 57, 61, 66
Denudation rates 240, 242, 244, 245
Deuterium content 76
Differential movement 127, 158
Discharge (from rockglaciers) 262–265
Discontinuous permafrost 43, 51, 55, 62, 185, 236, 258
Displacement 107
Downwashing of fines 68
Dynamic inactive rockglacier 10, 190, 193

Earth stream 207–208
Egesen *see* Younger Dryas
Environmental warming 138, 140
Equilibrium line altitude (ELA) 39, 41, 215, 235, 250, 252, 254, 255, 267
Exposure (of rockglaciers) 23, 37, 38, 41, 53, 64, 149, 205, 233
Extending flow 7, 19, 153, 157, 162, 163, 168
Extension 164

Fabric 70
Flow law *see* Glen's flow law
Flow lines 158, 159, 162, 163
Forest line *see* Timberline
Formation (of active rockglaciers) 185
Fossil rockglacier *see* Relict rockglacier
Front slope 18, 20, 22, 73, 74, 87, 88, 122, 160, 168, 170, 187, 192, 194, 206
Frost climate 195
Frost heave 68, 69, 82
Frozen core (of rockglaciers) 74, 89, 269
Furrow and ridge topography *see* Surface relief

Genesis (of rockglaciers) 184–195
Geoelectric sounding 73, 93–100, 187, 208, 209
Geometry of movement 158
Geophysical logs 78–83, 103
Geothermal heat 109, 110
Glacial hypothesis 208–219
Glacier 60, 162, 249
Glacieret 269
Glen's flow law 108, 109, 173, 175, 180
Global warming 1, 2, 119, 180, 270
Grain size 68, 71–73, 222
Gravimetry 100–102
Gravity anomaly 101
Ground ice 118

Hazards 265–266
Heat flow 108, 109
High mountain belts 232
High mountain environments 231
High mountain system 2
High viscosity layer 179
Holocene 19, 41, 42, 43, 79, 163, 180, 182, 186, 194, 208, 225, 226, 227, 229, 240, 242, 251, 256
Horizontal movement 121, 123–145, 137, 139, 143, 144, 152, 156, 157, 163, 180, 270
Horizontal velocity 138, 156, 166
Human impact 231
Hydrology 114, 233, 259–265, 270

Ice content 73, 74, 76, 77, 78, 79, 82, 90, 104, 118
Ice core (of rockglaciers) 74, 89, 269
Ice lenses 73, 196
Ice-cored moraine 44, 46, 51, 53, 54, 198, 216, 220
Inactive rockglacier 8, 38, 41, 42, 93, 184, 189–194
Inclination 106
Inclination measurement 169
Internal structure (of rock glaciers) 73–119
Interstitial ice 8, 196

Landslide 194, 198, 202, 258
Last Glacial Maximum 253
Lateral ridges 218
Length (of rockglacier) *see* Size of rockglacier
Lichen cover 18, 76
Lichen growth 180
Lichenometry 142, 160, 181, 225

Limit (of active rockglaciers) 236, 237
Limit (of mountain permafrost) 238
Lithology 67
Little Ice Age 12, 162, 164, 194, 214, 215, 226, 251, 255
Lobate rockglacier 6, 10, 19, 23–24, 55
Loch Lomond Stadial 223, 254, 255
Long-term measurement 131
Longitudinal profile 129
Longitudinal profile (of velocity) 166
Low-viscosity layer 118
Lower limit (of mountain permafrost) 138
Lower limit (of rockglaciers) 39, 52, 55, 149, 256

Mantle (of rockglacier)
see Boulder mantle
Mass balance 160, 212
Mass budget 270
Mass movement 182
Mass movement hypothesis 198–208
Mass transport 248
Material control 185
Maximum velocity 133
Mean annual air temperature (MAAT) 46, 62, 64, 110, 180, 201, 222, 224, 228, 232, 235, 236, 250, 251, 252, 253, 254, 255, 256, 257, 267, 268
Mean annual precipitation 267, 268
Mean summer temperature 181
Mean surface velocity 171, 176
Mean temperature 110
Measurement error 152
Meltwater 73, 77, 119, 147, 189, 258, 263
Meltwater ponds 211
Monthly measurements 142
Moraine 5, 21, 158, 185, 186, 226, 249
Morphometry (of rockglaciers) 20–30
Mountain climate 231
Mountain permafrost 31, 44, 46, 55, 62, 79, 116, 232–240, 269
Movement of rockglaciers
see Horizontal movement and Vertical displacement

Neoglaciation 12
Nival belt 231, 232
Normal stress 175

Older Dryas 41, 43, 229, 246, 253

P-wave velocity 82, 90, 91, 212
Paleoclimate 254
Peak strength (ice-sand mixture) 177
Periglacial hypothesis 219
Permafrost 2, 18, 56, 78, 116
Permafrost creep 216
Permafrost table 74, 103, 110, 112, 263
Permafrost temperature 237
Permafrost, alpine
see Mountain permafrost
Permafrost, continuous
see Continuous permafrost
Permafrost, discontinuous
see Discontinuous permafrost
Photogrammetric measurement 122, 193
Piping 266
Plastic model 172
Precipitation 253
Primary creep 173, 174
Protalus rampart 8, 36, 44, 195, 219, 220–224, 255
Push moraine 198, 216

Radio echo sounding 93, 100
Rate of motion
see Horizontal movement
Reactivation (of rockglaciers) 256
Relative age dating 225, 229
Relict rockglacier 10, 12, 35, 36, 38, 41, 42, 49, 184, 194–195, 197, 229–230
Relief control 185
Remaining problems 270
Residual strength 173
Resistivity 93–100, 104
Resistivity of frozen debris 96
Resistivity of glacier ice 96
Resistivity of unfrozen debris 96
Retreat of rock walls 240, 241, 246
Rheology 27, 67, 121, 140, 171–179
Rhizocarpon 225, 226
Ridge and furrow topography
see Surface relief
Rock stream 14, 15, 31
Rock type 67
Rockfall 70, 222, 248
Rockglacier (definition) 4
Rockglacier cairns 69

Rockglacier core 67, 71, 85, 101, 118, 119
Rockglacier creep 158
Rockglacier distribution 267
Rockglacier mantle *see* Boulder mantle
Rooting zone 94, 119, 128, 149, 160, 186, 187, 208, 215, 216, 227

Scale 2, 22, 184, 225
Seasonal changes 157
Seasonal movement 142, 143, 145
Seasonal temperatures 112, 113
Second layer 91
Secondary creep 164, 173, 174
Sediment concentration 214
Segregation ice 82
Seismic sounding 73, 82–93, 209, 212
Shear stress 109, 172, 175–176, 185
Shearing 162
Size (of rockglaciers) 21, 22–24, 29–30, 176, 195, 225, 242, 246–248
Sliding 144
Snow cover 112, 258, 259
Snow thickness 111
Snowbank 216
Snowline 40, 55, 65
Solar radiation 36, 57, 108, 187, 192, 233
Solifluction 207, 236, 250
Source area 23, 29–30, 194, 241, 242, 245, 246–248
Spatulate rockglacir 20, 55
Special rockglacier 33, 188
Specific rockglacier density (SRD) 61
Specific transport rate 248
Stagnant glacier ice 211
Steady-state creep 142
Storm discharge (water) 263
Strain 270
Strain rates 109, 172, 173, 175–176, 178
Stratigraphy 79, 80, 81, 83, 84, 102, 194
Stress field 162
Stress-strain relationship 172, 178
Stromstrich (central flow line) 21, 86, 89, 157, 163, 166
Structure (internal) 85
Sturzstrom 199, 200, 201
Subnival belt 195, 231, 232
Subsidence 146, 148, 151, 266
Subsurface velocity 168
Surface lowering 144, 153
Surface relief (of rock glaciers) 7, 18, 19, 22, 26–28, 34, 67, 158, 187, 194, 204, 214
Surface temperature 112
Surface velocity 127, 144, 161, 176
Surge 211
Syngenetic ice 269

Talik 114
Talus 5, 186, 220
Talus accumulation 221
Talus apron 169, 171
Talus cone 7, 8
Talus glacier 14, 15
Talus input 147, 185
Talus production 51, 52, 185, 201, 240–246
Talus rockglacier 6, 11, 23, 32, 33, 34, 36, 51, 66, 132, 148, 185–186, 240, 269
Talus slope 8, 21, 34, 47, 74, 158, 185, 217, 224, 241, 249
Talus storage 243, 244, 245
Talus stream (Schuttstrom) 201–207
Talus terrace 44, 220
Tangential stress 175
Taxonomy (of rockglaciers) 32
Temperature 73, 79, 82, 83, 84, 86, 105, 110, 172
Temperature below winter snow (BTS) 114–118, 119
Temperature control 185
Temperature gradient 82, 110, 112, 115
Terminal moraine 7
Tertiary creep 173, 174
Texture 269
Thermal conductivity 110, 112
Thermokarst 31, 187, 210, 211–212
Thickness (of rockglacier material) 20, 21, 22, 23, 24–26, 82, 84, 85, 97, 99, 100, 160, 163, 172, 176, 183
Timberline 53, 65, 204, 229, 231, 232, 252
Tongue-shaped rockglacier 5, 19, 22–23, 55
Tracer experiments 264
Tree-rings 186
Tritium analysis 263

Upper layer 91

Velocity 247
Velocity (of rockglacier motion) 8, 66, 76, 106, 107, 108, 156
Velocity (of water flow) 265
Velocity differences 157
Velocity fluctuations 142
Velocity gradient 168, 169
Velocity profiles 166–171
Velocity variations 270
Vertical displacement 143, 144, 146–156, 270
Viscosity 108, 172

Water circulation 114, 263
Water storage (in permafrost) 260, 261
Weathering pits 225
Weathering rate 247
Weathering rinds 18, 225

Younger Dryas 41, 43, 45, 223, 224, 229, 246, 252, 253, 254

Zero annual amplitude (depth of) 112

Springer-Verlag and the Environment

We at Springer-Verlag firmly believe that an international science publisher has a special obligation to the environment, and our corporate policies consistently reflect this conviction.

We also expect our business partners – paper mills, printers, packaging manufacturers, etc. – to commit themselves to using environmentally friendly materials and production processes.

The paper in this book is made from low- or no-chlorine pulp and is acid free, in conformance with international standards for paper permanency.

GPSR Compliance

The European Union's (EU) General Product Safety Regulation (GPSR) is a set of rules that requires consumer products to be safe and our obligations to ensure this.

If you have any concerns about our products, you can contact us on ProductSafety@springernature.com

In case Publisher is established outside the EU, the EU authorized representative is:

Springer Nature Customer Service Center GmbH
Europaplatz 3
69115 Heidelberg, Germany

Batch number: 10620143

Printed by Printforce, the Netherlands